Climate Management Issues
Economics, Sociology, and Politics

Economics, Sociology, and Politics

Climate Management Issues

Economics, Sociology, and Politics

Julie Kerr Gines

CRC Press
Taylor & Francis Group
Boca Raton London New York

CRC Press is an imprint of the
Taylor & Francis Group, an **informa** business

CRC Press
Taylor & Francis Group
6000 Broken Sound Parkway NW, Suite 300
Boca Raton, FL 33487-2742

First issued in paperback 2019

ISBN-13: 978-1-4398-6106-6 (hbk)
ISBN-13: 978-0-367-38196-7 (pbk)

Library of Congress Cataloging-in-Publication Data

Gines, Julie K.
    Climate management issues : economics, sociology, and politics / by Julie K. Gines.
      p. cm.
    Includes bibliographical references and index.
    ISBN 978-1-4398-6106-6 (hardcover)
    1. Environmental sociology. 2. Climatic changes--Social aspects. 3. Climatic change mitigation--Government policy--International cooperation. 4. Climatic changes--Environmental aspects. 5. Climatic changes--Effect of human beings on. I. Title.

GE195.G54 2012
304.2'5--dc23
                                      2011033737

Visit the Taylor & Francis Web site at
http://www.taylorandfrancis.com

and the CRC Press Web site at
http://www.crcpress.com

Contents

Author

Julie Kerr Gines has been an earth scientist for more than 34 years, promoting a healthier and better-managed environment. She holds a PhD in earth science with an emphasis in remote sensing satellite technology from the University of Utah and has dealt with the hands-on applications of resource and climate change management, focusing on the delicate balance of the multifaceted issues involving the political, economic, and sociological components in both short- and long-term applications with real-world ramifications. She has been involved specifically with projects such as forest management, classification and monitoring of desertification, rangeland change detection and management, vegetation inventory and health assessment, and land-use change.

Dr. Gines approaches the subject with a background uniquely blended with conservation and land-use management expertise with the U.S. Bureau of Land Management, military applications with the U.S. Air Force, and university-based research. She has also had environmental experience serving in the professional community as the president of the Intermountain Chapter of the American Society of Photogrammetry and Remote Sensing and spent years dedicated to the promotion of a healthier environment through participation with various environmental groups. The prolific author has published more than 23 books, many of which are focused on environmental topics such as global warming, climate change management, conservation of natural resources, green living, and the earth sciences.

Introduction

After years of dedicated research, overwhelming scientific evidence, and field observations, it is a well-known fact that climate change is an urgent problem that needs attention from several sectors of society—including scientific, technological, political, environmental, economic, governmental, psychological, and sociological. Although it may be true that future technologies may discover a way to make an easier transition from a fossil fuel-based economy possible—one of the biggest factors in the perpetuation of climate change—there is no time to adopt a *wait and see* attitude. Currently, it remains one of society's most pressing problems. Changes must be made now, and that includes changes in personal perception and attitude and in the way we do business and enact policy as a world community. The fundamental scientific, technical, and industrial knowledge base already exists to enable and empower society to begin combating and correcting the problem now. Each person worldwide has the capability to empower themselves to help meet one of the biggest challenges humankind may ever face, and it begins with personal education, appropriate mindset, economic incentive, effective and intelligent leadership, focused priorities, and the collective working toward a common goal.

It is fundamental to understand that climate change is in actuality a complex collection of issues that extends far beyond the scientific portion of the picture. It not only encompasses the physical and chemical laws of nature but also affects life forms and ecosystems, human lifestyles, cultural values, political systems, economic building blocks, health issues, and human values—in short, it affects fundamental components of all aspects of our lives. Because of that, this book presents the most pertinent issues involving climate change that begin with the scientific basis and then applies that knowledge to the sociological, political, and economic components of the climate change picture, so that you, the reader, can gain new insights into the complex interrelationships that are involved and the decisions that must be made in order to mitigate the climate change problems before it is too late and too much damage has been done.

The book begins by providing an overview of the physical science aspects of climate change, divided into three central categories: the climate system, greenhouse gases, and effects on ecosystems. It is imperative to have a basic, working knowledge of these broad areas so that you can then connect the importance of them to the variability of economic scenarios, the need for global political unification, the diverse sociological ramifications associated with these areas, and what we can expect as a global community if we do not act now to remedy the current situation. From there the book looks next at the inception of climate change management and the role the

United Nations has played: guiding climate talks, conventions, agreements such as the Kyoto Protocol, and scientific review and analysis. It looks at where we have been and where we are headed, as well as what it will take to succeed.

Next, this book explores the sociological connections to climate change and what those ramifications are. It examines the effects of biodiversity, the implications of environmental movements, the concept of climate justice, the ramifications of climate change on society, the reasons for going green, and other important social factors. It then analyzes human psychology and the effect the media has—and can have—on society. It looks at subtle manipulation, truth in reporting, the distinction between journalistic balance and accurate scientific representation, and the ways human perception can be manipulated.

Following that, the book examines the role of international organizations, renewable and energy efficiency partnerships, and other collaborative international endeavors. It also explores the progress that individual countries are making, along with what nations are currently involved in working to achieve sustainability. From there, the book then shifts gears and looks closely at the political arena—first at the current political climate in the United States, President Obama's initiations, and the response of Congress. Then it examines the international playing field and the political actions and contributions of several key countries. Based on that knowledge, it then examines the sociopolitical impacts of climate change, such as the evident and the obscure connections to national security and terrorism that all nations worldwide need to be aware of in order for peace and international cooperation to remain paramount. It also focuses on the problem of inaction and situations that could trigger wars. It defines and gives examples of the nature of conflict and state fragility and how climate change can disrupt even the smallest of countries and pit them against the largest. It then touches on the concept of climate justice and equity: what it is, why it is important, and why some people embrace it while others turn their backs to it.

Then we shift gears and take a good look at an entity that in and of itself deserves a thorough evaluation: the U.S. military. The book focuses on how climate change has affected militaries around the world and how it has forced them to adjust in both peace- and war-time situations. It looks at the new green technologies they have recently embraced, where they have succeeded, how the troops have reacted, how these changes have affected their operations, and what their long-term goals really are. The answers may surprise you.

Next, the economic and socioeconomic implications of climate change are analyzed and considered. A wide-ranging cadre of economic venues is examined, as well as the effects climate change is having, and will likely have, on them. The book focuses on what to expect in both the short and long term, both with and without conscious changes on our parts. Specific fields looked at include the economic impacts of drought and desertification, what will happen to the fishing and forestry industries, changes in monsoon patterns and what that means to millions of people, agriculture and food production, impact on the recreation and tourism industries, impact on healthcare- and labor-related industries, various transportation and energy considerations, and the obstacles and challenges the world's insurance companies are now facing. Following that, it takes a brief look at one of the most well-known economic analyses of climate change, the Stern Review, and also analyzes how humankind can possibly meet the challenges of future climate impacts guaranteed to be coming down the road.

From there, this book presents an analysis of the economics of mitigation options and what can be done to correct the problem, such as the cap-and-trade scheme and what that really entails, the utility and difference between project- and allowance-based markets, as well as the overall economics of cap and trade. Next, it explores solutions to the problem by looking at mitigation options currently available: carbon capture and storage and the impressive results being met in geological formation sequestration, as well as ocean sequestration. Chapter 12 concludes with an explanation of other adaptation strategies.

The focus then turns to the role of climate research, modeling, and data analysis and the exciting advances that are happening in that realm. It discusses the latest discoveries and techniques of mathematical and computer modeling and the special challenges that scientists encounter when trying to build a model of a physical system that is composed of multiple scales, resolutions, components, complexities, time scales, and reactive mechanisms. It analyzes the various types of climate models currently used and outlines the many factors that introduce uncertainty and doubt into the system to create confusion. It also looks at two case studies being currently worked on by the U.S. Bureau of Land Management as they try to come to a better understanding of the natural world and the effect of climate on it.

Finally, the book looks toward the future: who will ultimately win, who will lose. It will give you a clear insight as to what every person must do to help solve the problem and how technology may be able to help out. It will help put in perspective the overall importance of the issue—economically, politically, and socially—and encourage you to lead by example.

1

The Earth's Climate System

Overview

In order to understand the problem and then be able to manage it effectively, it is imperative to have an understanding of the complex theory and science behind climate change. Although much of the existing literature refers to the issue as *global warming*—the term used for the past several years—scientists are transitioning to the term *climate change* because of the multifaceted nature of the phenomenon. This book utilizes the current terminology. As background, the term global warming tends to imply that the only consequence is that the earth's atmosphere is slowly getting warmer, unfortunately causing many to minimize the significance of the issue. When the media reports that scientists predict that the earth's atmosphere is warming, some people expect to hear that the temperature will get tens of degrees warmer. Some may have visions of sitting in a hot, steamy sauna—a vision that for those who live in extremely cold climates, such as Siberia or the Yukon, may be very welcome indeed! Therefore, when climatologists predict a temperature rise of 1.1–6.4°C, the public may be inclined to ask what the urgency about the issue really is. After all, it's just a few degrees, right?

Wrong.

Look at it this way: During the earth's last ice age, the earth was only about 4–6°C *cooler* than it is today. Although that may seem like an insignificant temperature difference, it was enough to blanket huge areas of the earth in thick layers of ice. It had such an enormous impact on ecosystems that it even rendered some species extinct, such as the mammoth and mastodon.

Thus, although a few degrees may seem trivial, the earth's climate is so sensitive that those few degrees can make a significant difference. In addition, although some climate change is natural, scientists have now proven that humans are causing the bulk of the recent changes and rises in temperature, and although those few degrees may not seem like much, it is unfortunately enough to serve as a tipping point—a big enough influence on the climate that, once reached, it will set into motion permanent changes with global ramifications. When climate changes, it affects entire earth systems. Many components of the earth's physical system operate on a global scale, such as the hydrologic, carbon, nitrogen, and phosphorus cycles. The earth's various biogeochemical cycles are always changing and represent the continual interactions between the biosphere (life), lithosphere (land), hydrosphere (water), and atmosphere (air). Various substances move endlessly throughout these four spheres. Of the four spheres, the atmosphere transports elements the most rapidly, and climate change will negatively affect these spheres if left unchecked.

This chapter is the first of three parts that lay the foundation needed to understand how the climate system works as a whole and what climate change is capable of and why. It looks at climate as a global system and the fact that harmful environmental practices at one location can negatively affect other locations. It then examines the effect that human impacts have on the earth's natural carbon cycle—commonly referred to as the human-enhanced carbon cycle. From there, it focuses on global circulation patterns of the atmosphere and oceans and their role in climate change, specifically the consequences of the disruption of the Great Ocean Conveyor Belt, a major circulation pattern responsible for western Europe's mild climate and possibly for abrupt climate change. Finally, this chapter presents current research information about the effects of sea-level rise and the consequences of a warming world.

Introduction

Climate change is one of today's most urgent topics. Despite all the controversy and hype the entire climate change topic has generated, there now exists an overwhelming body of scientific evidence that the problem is real, that its effects are being felt right now on a global scale, that some geographic areas are more vulnerable than others, and that it will take the concerted effort of every person working toward the same goals to put a halt to the rise in both temperature and atmospheric greenhouse gases. The fact is that even if all necessary steps are taken right now to stop the increase in greenhouse gases entering the earth's atmosphere, the effects of climate change will still be present for centuries to come. The complicated process has already been set into irreversible motion. The critical key to understanding the effects of climate change is to be knowledgeable about how to control and mitigate the causes of climate change now in order to minimize future damage to the environment and the life living in it. An unhealthy environment affects all forms of life on earth and through understanding first the causes and effects—the science behind climate change—it will be possible to empower the world's population to solve the problem through political leadership, realistic sociological mindsets, and proper economic measures. The scientific consensus is that people's behavior is largely responsible for today's climate change problem. Another factor that makes this a volatile and controversial issue is that it is not just confined to the realms of the scientific community, nor does it have just one simple, predefined solution—it has multifaceted dimensions involving economic, sociological, political, psychological, and personal issues, making this a topic affecting the future outcome of the life of every person on earth now and in the future. To make the problem even more sensitive, it is projected that those who will suffer the worst effects are not even those who have caused the bulk of the problem—the undeveloped nations will take the brunt of what the developed nations have largely caused.

The scientific community has many theories about climate change. The topic is extremely challenging in nature because there are so many factors involved in it, making it difficult to pinpoint its exact causes and solutions across the board. The earth's climate is extremely complicated, and climatologists are conducting daily research in order to improve their understanding of all the interrelated components.

Each year, about 6.4 metric tons of carbon is released into the atmosphere. Studies show that concentrations of carbon dioxide (CO_2) have increased by about one-third since 1900. During this same time period, experts say the earth has warmed rapidly. Because of this, a connection has been made that humans are contributing significantly to climate change. Even scientists who are skeptical about the climate change issue recognize that there is much more CO_2 in the atmosphere than ever before.

Although it is certainly true that the atmosphere is warming up, that is only one part of what is going on. As the earth's atmosphere continues to warm, it is setting off an avalanche of other mechanisms that will do even greater harm to the earth's natural ecosystems. Glaciers and ice caps are melting, sea levels are rising, and ocean circulation patterns are changing, which then changes the traditional heat distributions around the globe. Seasons are shifting, and storms are becoming more intense, leading to severe weather events. Droughts are causing desertification, crops are dying, and disease is spreading. Some ecosystems are shifting where they still can; others are beginning to fail. In short, humans are changing the earth's climate—and not for the better.

Development

In order to understand climate change management issues, it is necessary to have a good understanding and working knowledge of the basic scientific theory behind climate change: how the earth's atmosphere works, how it interacts with the earth's surface, what causes climate change, how it affects the various earth systems, and why it is an issue that must be addressed today. In practical land management, principally overseen by the world's government agencies and appropriated entities, much of the knowledge relied on in order to manage the land and make appropriate policy decisions to properly provide for present and future management capabilities and expertise is current research, largely provided by the world's principal research institutions—chiefly governmental, academic, and private, such as National Aeronautics and Space Administration (NASA), numerous universities worldwide, and esteemed organizations such as the Pew Center on Global Climate Change, respectively. Therefore, the beginning chapters of this book, while merely touching on some of the most prominent key issues, provide an overview of the recent scientific research and work that has been done on the climate change issue, laying the basic groundwork to promote an understanding of the complex, interwoven issues and key concepts of the climate change problem, thereby creating a firm foundation for why this issue is critical to both society and the environment's future, as well as why politics, sociology, and economics play such a critical role in its management. Consulting the references listed at the conclusions of each chapter will also provide a greater depth of detail in each subissue presented.

This research has been crucial toward enabling managers to gain an understanding of all the individual facets of the problem in order to be able to put proper and effective policy in place to provide efficient land-stewardship practices for both the present and the future (which will be presented and discussed in detail in subsequent chapters). It is crucial that you, the reader, understand the scientific concepts behind the complex climate change issue in order to gain an appreciation for the immense challenge that land managers face today in planning for future climate change situations based on a plethora of interactive scenarios that each have the ability to change the long-term outcome, making effective management one of the most challenging tasks of this century—a task that will affect many future generations. These first three chapters provide a brief but comprehensive introduction to climate change science and present the key scientific concepts behind it. Only through understanding the depth of the scientific principles and processes involved is it possible to grasp the interwoven complexity of the issue and gain an appreciation for why informed climate management practices are critical for the future of both humanity and the environment when dealing with sociological, political, and economic issues.

What Is Climate Change?

Climate change is a term used to refer to the increase of the earth's average surface temperature, largely caused by a buildup of greenhouse gases in the earth's atmosphere, primarily from the burning of fossil fuels and the destruction of the world's rain forests. The term was coined by Svante Arrhenius, a Swedish scientist and Nobel Laureate, in 1913. The term is often used to convey the concept that there is actually more going on than just rising temperatures. There is a general misunderstanding that the issue involves just the atmosphere and temperature rising a few degrees; hence the name change from global warming to climate change in order to more accurately convey the multifaceted nature of the phenomenon. Climate change encompasses long-term changes in climate, which include temperature, precipitation amounts, and types of precipitation, humidity, and other factors.

Today, climate change has become one of the most controversial issues in the public eye, appearing frequently in print and televised news reports, documentaries, scientific and political debates, and other venues and economic issues, and the messages can be contradictory and confusing. One goal of this book is to set the record straight by presenting various points of view and clarifying them with the facts. Climate change also receives a lot of attention because it is more than just a scientific issue—it also affects economics, sociology, and people's personal lifestyles and standards of living. It is one of the hottest current political issues, not only in the United States but worldwide. Political positions on climate change have become major platforms and debating issues as public demands toward a solution have intensified in recent years.

More than 2,500 of the world's most renowned climatologists, represented by the Intergovernmental Panel on Climate Change (IPCC), support the concept of climate change and agree that there is absolutely no scientific doubt that the atmosphere is warming. They also believe that human activities—especially burning fossil fuels (oil, gas, and coal), deforestation, and environmentally unfriendly farming practices—are playing significant roles in the problem. The IPCC consolidates their most recent scientific findings every 5–7 years into a single report, which is then presented to the world's political leaders. The World Meteorological Organization (WMO) and the United Nations Environment Programme established the IPCC in 1988 to specifically address the issue of climate change. As a result of their comprehensive analysis, the IPCC determined that this steady warming has had a significant impact on at least 420 animal and plant species in addition to natural processes. Furthermore, this has not just occurred in one geographical location but worldwide.

Unfortunately, the science of climate change does not come with a crystal ball. Scientists do not know exactly what will happen, such as what the specific impacts to specific areas will be, nor can they say with certainty when or where the impacts will hit the hardest, making it all the more difficult for land-resource managers, planners, politicians, and economists to do the best job possible in long-range planning. However, experts are certain that the effects will be serious and globally far-reaching. According to the National Oceanic and Atmospheric Administration (NOAA)/NASA/ Environmental Protection Agency Climate Change Partnership, potential impacts include increased human mortality, extinction of plants and animal species, increased severe weather, drought, and dangerous rises in sea levels. Although climatologists still argue about how quickly the earth is warming and how much it will ultimately warm, they do agree that climate change is currently happening and that the earth will continue to keep warming if something is not done soon to stop it.

DID YOU KNOW?

There are several interesting facts concerning climate change. For instance, did you know that:

- According to the WMO, this past decade has been the warmest on record. In addition, the top 11 warmest years ever recorded have all occurred in the last 13 years (WMO, 2009).
- Scientists attribute the atmosphere's rapid warming principally to human activities, mainly the burning of fossil fuels.
- Every year the CO_2 concentration level in the atmosphere steadily climbs higher, and the earth will reach a "tipping point" of no return if humans do not reverse the trend. In order to avoid a rise of 6°C by the end of this century, which is the critical level identified by climate change research scientists, total global emissions need to have peaked by 2015 and reduced by at least 80 percent by 2050 to prevent nonreversible damage.
- Although every person plays a role in the problem and has a responsibility for its solution, the lion's share of the problem can be attributed to twenty-three of the world's most wealthy nations, which constitute only 14 percent of the world's population. These "developed" countries are guilty of producing 60 percent of the world's carbon emissions since 1850 when industrialization became an integral part of their lifestyles. Even though many of these nations pledged to reduce CO_2 below 1990 levels by 2012 through the ratification of the Kyoto Protocol, their emissions have risen instead.
- Up until just recently, the United States emitted the highest levels of greenhouse gases. China is now the largest emitter of CO_2, surpassing the United States in 2007 by 8 percent. As a comparison, according to the CIA World Factbook, a U.S. citizen emits seven times as much in a year as an Ethiopian does in a lifetime.
- Climate change is a problem that affects every nation sociologically, politically, and economically. The World Health Organization estimates that climate change causes 150,000 deaths per year worldwide from malaria, malnutrition, diarrhea, and flooding. These deaths affect the economic and political stability and structure of a country, weakening its resources and ability to cope with the challenges of climate change, further exacerbating the problem (Kumaresan and Sathiakumar, 2010).
- Climate change is linked to the natural disasters such as hurricanes, flooding, and drought that are increasing in frequency and severity. Between 1996 and 2005, disasters caused $667 billion in direct losses to people worldwide. According to a World Bank Independent Evaluation Group, losses were twenty times greater in developing countries (Parker, 2007).
- In 2005, Hurricane Katrina hit the U.S. Gulf Coast, causing 1,836 deaths.
- In October 2005, Hurricane Stan hit Guatemala, Mexico, El Salvador, Nicaragua, and Costa Rica, causing more than 1,500 deaths.
- According to the Anuradha Mittal, founder of the international policy think tank, *The Oakland Institute*, rising world food prices caused in part by climate change and expanded biofuel production led to food riots and protests in more than fifty countries between January 2007 and July 2008 (Mittal, 2009).
- The thawing of the Arctic permafrost is affecting the traditional way of life of its indigenous people, making hunting and traveling difficult and dangerous
- The low-lying island of Tuvalu in the Pacific has already evacuated three thousand of its inhabitants to New Zealand
- By 2020, up to 250 million people in Africa and 77 million in South America will be under increased water stress because supplies will no longer meet demand
- By 2025, tens of millions more will go hungry because of low crop yields and rising global food prices.
- A rise in sea level of just 1 meter would displace ten million people in Vietnam and eight to ten million in Egypt
- The number of people in Africa at risk of coastal flooding will rise from one million in 1990 to seventy million by 2080 according to the United Nations Fact Sheet on Climate Change (UNFCCC, 2006).

The Global System Concept

Although different types of climate exist in different parts of the world, the climate does work as a complete unified global system—conditions and actions in one area can impact conditions and actions in others. For example, because the earth shares one atmosphere, what goes on in China will affect the United States. Similarly, what goes on in South America eventually affects Africa, and so on. NASA has reported that soot (black carbon) originating from industrial practices is being deposited on Arctic snow and ice, causing incoming *electromagnetic radiation* to heat up the now-darker surfaces that the soot is covering, causing more snow and ice to melt, creating a cycle of increased melting.

Because of these interactions, climate change is also a global issue—it is affecting the earth in different ways, such as destabilizing major ice sheets, melting the world's glaciers, raising sea levels, contributing to extreme weather, and shifting biological species northward and higher in elevation, ensuring that no area of the earth is immune from its effects. In fact, NASA scientists have determined that the atmospheric concentration of carbon dioxide, the principal greenhouse gas, is now higher than it has been for the past 650,000 years.

A remark often heard is that the earth's temperature and CO_2 levels have risen in the past, so these cyclic variations are normal. This is true, as can be seen in Figure 1.1. As paleoclimatologists have studied ice age cycles, they have determined that the atmospheric CO_2 closely follows the earth's surface

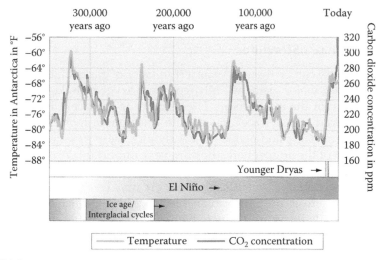

FIGURE 1.1
(See color insert.) Fluctuations in temperature (orange line) and in the amount of carbon dioxide concentrations in the atmosphere (blue line) over the past 350,000 years. The temperature and carbon dioxide concentrations at the South Pole run roughly parallel to each other, showing the strong correlation between the two. (From Casper, J. K., *Global Warming Trends: Ecological Footprints*, Facts on File, New York, 2009.)

temperature. It is a known fact that as biological activity rises (with increases in temperature), CO_2 levels also rise because more CO_2 is produced and is entered into the atmosphere. This increases warming and encourages more CO_2 production. It is critical to note, however, that under natural processes (nonanthropogenic) CO_2 concentrations never exceeded 300 ppm (parts per million) until the Industrial Revolution occurred in the late eighteenth century. It was at that point that CO_2 levels began to continually rise. In 1958, when CO_2 concentrations began to be recorded, the atmosphere was at 315 ppm. Concentrations in March 2011 are at 392.4 ppm and still steadily rising (CO_2 Now, 2011). In addition, NASA's scientists from the Goddard Institute for Space Studies in New York released a report in 2005 outlining that their scientific studies provided clear evidence that an enhanced greenhouse effect is ongoing. Their study involved satellites, data from buoys, and computer models to study the earth's oceans. Scientists were able to conclude that more energy from the sun is being absorbed by the atmosphere than is emitted back to space, which is causing the earth's energy to become unbalanced and is now significantly warming the earth (NASA Goddard Institute for Space Studies, 2011). Scientists claim that the world is entering "largely uncharted territory as atmospheric levels of greenhouse gases continue to rise" (Pew Center on Global Climate Change, 2011). Today, scientists have no doubt that the recent spike in the graph showing the rise of CO_2 concentration is due to anthropogenic activity. In summary, because the earth is a global system, climate change must be approached as a global issue with far-reaching cause-and-effect issues, and it will take all nations working toward a common solution to effectively mitigate the problem (ScienceDaily, 2010).

The Atmosphere's Structure

The atmosphere can be thought of as a thin layer of gases that surrounds the earth. The two major elements, nitrogen and oxygen, make up 99 percent of the volume of the atmosphere. The remaining 1 percent is composed of what is referred to as "trace" gases. The trace gases include water vapor, methane, argon, CO_2, and ozone (O_3). Although these gases only make up a small portion of the atmosphere, they are very important. Water vapor in the atmosphere is variable; arid regions may have less than 1 percent, the tropics may have 3 percent, and over the ocean, there may be 4 percent.

The atmosphere is not uniform from the earth's surface to the top; it is divided vertically into four distinct layers: the troposphere, the stratosphere, the mesosphere, and the thermosphere, designated by height and temperature. The lowest level, closest to the surface of the earth, is the troposphere. It extends upward to an average height of 12 kilometers. This level is critical to humans because all of the earth's weather occurs in this layer. In this level, the temperature gets cooler with increasing height. In fact, temperature at the surface of the earth averages 15°C, and at the tropopause (the top of the troposphere), the temperature is just –57°C. In addition, moisture content decreases with altitude in this layer; above the troposphere there is not enough oxygen in the atmosphere to sustain life, and winds increase with height. To illustrate this, one of the most pronounced wind systems, the Jet Stream, is located at the top of the troposphere.

The climate of an area is the result of both natural and *anthropogenic* factors. Natural elements come from the following four principal inputs:

- Atmosphere
- Lithosphere
- Biosphere
- Hydrosphere

The human factor influences climate when it alters land and resource uses, a principal concern for land, wildlife, and natural-resource managers worldwide. For example, when a natural forested area is converted to a city, it has a direct effect on climate—one of the most notable is the *urban heat island effect*.

Even though climate does change naturally, those types of changes occur so slowly (such as over millennia) that they are not readily detectable by humans. Climate changes caused by humans, however, are occurring much more quickly (such as in centuries or even decades) and are becoming noticeable within a few generations. Often, a change in one part of the climate will produce changes in other areas of climate as well. Since the Industrial Revolution began in the late eighteenth century—and especially since the introduction and use of fossil fuels involved in the rapid modernization of the twentieth and twenty-first centuries in developed countries—the global average temperature and atmospheric CO_2 concentrations have increased notably.

Because CO_2 levels are higher now than they have been in the past 650,000 years and because surface temperatures on earth have risen significantly during that same time, this has led scientists to the conclusion that humans are responsible for some of the unusual warmth that exists today. In support of this notion, climatologists at both NOAA and NASA have run two types of computer models: one of climate systems with natural climate processes alone and another with natural climatic processes combined with human activities. The models that include the human activities more accurately resemble the actual climate measurements of the twentieth century, giving scientists further conviction that human activity does play a significant role in the climate change issue that is apparent today.

Since the 1970s, instruments on satellites have also monitored the earth's climate. In addition, measurements of the atmospheric CO_2 have been obtained since 1958, when the world's first monitoring station was built on top of Mauna Loa, the highest mountain in Hawaii, to enable scientists to monitor the atmospheric CO_2 levels in order to document any increases. They were able to determine a documentable increase—one that is still occurring today—known worldwide as the famous "Keeling Curve." Currently, the CO_2 monitoring network has expanded to more than 100 stations globally in order to track concentrations of carbon dioxide, methane, and other greenhouse gases.

Even though scientists know natural climate variability and cycles will continue to occur, they also expect CO_2 levels to rise and climate change to increase because of human interactions with climate. How much variability actually occurs will ultimately depend on the choices humans make in future population growth, energy choices, technological developments, and global policy decisions.

Dr. Rajendra Pachauri, the chairman of the IPCC (which will be discussed in detail in Chapter 4), said at an international conference attended by 114 governments in Mauritius in January 2005 that the world has "already reached the level of dangerous concentrations of carbon dioxide in the atmosphere." He recommended immediate and "very deep" cuts in the pollution levels if humanity is to survive. He also said: "Climate change is for real. We have just a small window of opportunity and it is closing rather rapidly. There is not a moment to lose" (Lean, 2005).

In a news report in *The Guardian UK* in February 2006, Dave Stainforth, a climate modeler at Oxford University, said: "This is something of a hot topic but it comes down to what you think is a small chance—even if there's just a half percent chance of destruction of society, I would class that as a very big risk" (Adam, 2006).

Chris Rapley, head of the British Antarctic Survey, commented that the huge west Antarctic ice sheet may be starting to disintegrate, an event that would raise sea levels around the world by 16 feet (5 m). He said, "The IPCC report characterized Antarctica as a slumbering giant in terms of climate change. I would say it is now an awakened giant. There is real concern" (CICERO, 2006). According to the American Geophysical Union, "Natural influences cannot explain the rapid increase in the global near-surface temperatures observed during the second half of the 20th century" (American Institute of Physics, 2010).

The Carbon Cycle: Natural versus Human Amplification

The carbon cycle is an extremely important earth cycle and plays a critical role in climate change. Carbon dioxide enters the atmosphere during the carbon cycle. Because it is so plentiful, it originates from several sources. Vast amounts of carbon are stored in the earth's soil, oceans, and sediments at the bottoms of oceans. Carbon is stored in the earth's rocks and is released when the rocks are eroded. Carbon exists in all living matter. Every time animals and plants breathe, they exhale CO_2.

When examining the earth's natural carbon cycle, it is important to understand that the earth maintains a natural carbon balance. Throughout geologic time, when concentrations of CO_2 have been disturbed, the system has always gradually returned to its natural (balanced) state. This natural readjustment operates very slowly.

Through a process called diffusion, various gasses that contain carbon move between the ocean's surface and the atmosphere. Because of this, plants in the ocean use CO_2 from the water for photosynthesis, which means that ocean plants store carbon, just as land plants do. When ocean animals consume these plants, they then store the carbon. Then when they die, they sink to the bottom and decompose, and their remains become incorporated in the sediments on the bottom of the ocean.

Once in the ocean, the carbon can go through various processes: it can form rocks, then weather, and it can also be used in the formation of shells. Carbon can move to and from different depths of the ocean and also exchange with the atmosphere. As carbon moves through the system, different components can move at different speeds. Their reaction times can be broken down into two categories: short-term cycles and long-term cycles.

In short-term cycles, carbon is exchanged rapidly. One example of this is the gas exchange between the oceans and atmosphere (evaporation). Long-term cycles can take years to millions of years to complete. Examples of this would be carbon stored for years in trees; or carbon weathered from a rock being carried to an ocean, being buried, incorporated into plate tectonic systems, and then later being released into the atmosphere through a volcanic eruption. Throughout geologic time, the earth has been able to maintain a balanced carbon cycle. Unfortunately, this natural balance has been upset by recent human activity. Over the past 200 years, fossil fuel emissions, land-use change, and other human activities have increased atmospheric carbon dioxide by 30 percent (and methane, another greenhouse gas, by 150 percent) to concentrations not seen in the past 420,000 years (which is the time span of the longest fully documented ice core record).

Humans are adding CO_2 to the atmosphere much faster than the earth's natural system can remove it. Figure 1.2 illustrates the greenhouse effect. Prior to the Industrial Revolution, atmospheric carbon levels remained constant at around 280 ppm. This meant that the natural carbon sinks were balanced between what was being emitted and what was being stored. After the Industrial Revolution began and carbon dioxide levels began to increase—from 315 ppm in 1958 to 391 ppm in 2010—the "balancing act" became unbalanced, and the natural sinks could no longer store as much carbon as was being introduced into the atmosphere by human activities, such as transportation and industrial processes.

In addition, according to Dr. Pep Canadell of the National Academy of Sciences, 50 years ago, for every ton of CO_2 emitted, natural sinks removed 600 kg (Global Carbon Project, 2007). In 2006, only 550 kg were removed per ton, and the amount continues to fall today, which indicates the natural sinks are losing their carbon storage efficiency. This means that although the world's oceans and land plants are absorbing great amounts of carbon, they cannot keep up with what humans are adding. The natural processes work much more slowly than the human ones do. The earth's natural cycling usually takes millions of years to move large amounts from one system to

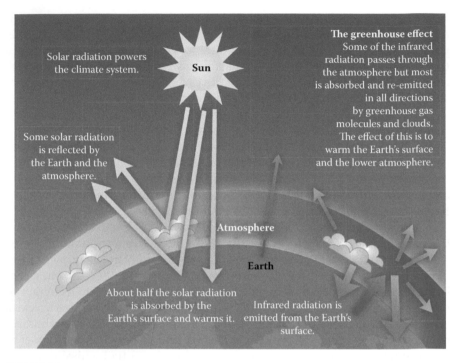

FIGURE 1.2
(See color insert.) This diagram illustrates the earth's natural greenhouse effect. The natural greenhouse effect is what makes earth a habitable planet—if the greenhouse effect did not exist, the earth would be too cold, and life could not exist. As more greenhouse gases are added to the atmosphere through the burning of fossil fuels, deforestation, and other measures, however, less heat is able to escape to space, warming the atmosphere unnaturally—causing a situation called the enhanced greenhouse effect. This is what is contributing to climate change today.

another. The problem with human interference is that the impacts are happening in only centuries or decades, and the earth cannot keep up with the fast pace. The result is that each year the measured CO_2 concentration of the atmosphere gets higher, making the earth's atmosphere warmer.

Another way that humans are contributing to climate change is through deforestation. By burning or cutting down the rain forests, two things happen: the forest can no longer store carbon, and burning the trees releases the carbon that had been stored long term in the wood back to the atmosphere, further exacerbating the problem.

The Hydrologic Cycle and the Relationships between the Land, Ocean, and Atmosphere

The hydrologic cycle is also important because it plays a direct role in the function of healthy ecosystems. The hydrologic cycle describes the movement of all the earth's water. It has no starting point and involves the existence and movement of water on, in, and above the earth. The earth's water is always moving and changing states—from liquid to vapor to ice and back again. This cycle, illustrated in Figure 1.3, has been in operation for billions of years, and all life on earth depends on its continued existence. What has scientists concerned is that climate change is affecting major components of the water cycle, and it is having a negative impact.

When the cycle is in equilibrium, water is stored as a liquid in oceans, lakes, and rivers; in the soil; and in underground aquifers in the rocks. Water is stored frozen in glaciers, ice caps, and snow and also in the atmosphere as water vapor, droplets, and ice crystals in clouds. Water can change states and move to different locations. For instance, water can move from the ocean to the atmosphere when it evaporates and turns from a liquid to a gas (vapor). Plants release water as a gas through transpiration. Water in the atmosphere condenses to form clouds, which can then form rain, snow, or hail and return to the earth's surface. Water that comes back to earth can be stored where it lands (in an ocean), flow above ground (river), or infiltrate and move underground (as groundwater).

That is a recount of how the earth's natural water cycle works. When climate change becomes an issue, however, it enhances the water cycle, making it more "extreme." Because climate change causes the earth's atmosphere to be warmer, the atmosphere develops a higher saturation point, enabling it to evaporate and hold more water from the earth's surface. This may have a two-fold effect. First, in areas where more water vapor exists, more clouds will form, causing more rain and snow. In other areas, especially those farther away from water sources, more evaporation and transpiration (together referred to as evapotranspiration) could dry out the soil and vegetation. The result would be fewer clouds and less precipitation, which could cause drought problems for farmers, ranchers, cities, and wildlife habitat. All of the ecosystems in these areas on an international scale could be negatively impacted.

Areas that are receiving increasing amounts of precipitation, such as portions of Japan, Russia, China, and Indonesia, will also experience more water infiltrating the ground and over the surface. This could increase the levels of lakes and rivers, causing serious flooding, and maybe even the formation of new lakes. Wetter conditions will also affect the plants and animals in those areas.

The drier areas, such as Africa and the southwestern United States, will also experience serious impacts. As the ground dries out from evapotranspiration, the atmosphere loses an important source of moisture. This, in turn, creates fewer clouds, which means there is less rain, making the area more arid. As less water is available to infiltrate the ground, less will be able to live off, on, or in the soil. Rivers and lakes would dry up, vegetation would die off, and the land would no longer be able to support humans, animals, and other life.

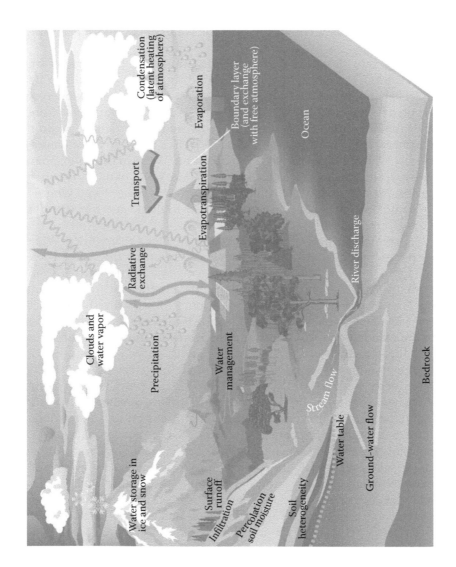

FIGURE 1.3

(See color insert.) The earth's hydrologic cycle. (Courtesy of NASA, NASA energy and water cycle study, http://news.cisc.gmu.edu/NEWS%2005%20 Discovery%20and%20Product%20projects.htm, 2011.)

There are many cycles operating on earth, with elements continually passing in and out of them. All natural cycles on earth must work together in a state of dynamic equilibrium. This means that as substances move and change at different times and places, they do it in a way that does not negatively impact the entire working system—all of the components in the system work together and complement each other. This dynamic equilibrium changes as the seasons change, because different needs must be met at different times. In the spring and summer when plants grow, they need carbon so they take it from the atmosphere and the soil. Then when the growing season is over in the fall and winter, plants release carbon back to the soil and atmosphere. This plays a significant role, especially in the earth's northern hemisphere because most of the landmasses are located there, creating a global seasonal change of carbon dioxide in the atmosphere. The oceans and atmosphere also interact extensively. Oceans are more than a moisture source for the atmosphere. They also act as a heat source and a heat sink (storage), as well as a carbon sink.

Global Energy Balance

Interactions between energy from the sun, the earth, and the atmosphere all have an effect on the earth. This is called the global energy balance, an energy balance that also plays a role in climate. If the energy balance changed and the atmosphere began retaining more heat, this could trigger climate change. It is this scenario that climatologists are concerned about now and in the future.

The global energy balance regulates the state of the earth's climate. Modifications to it—a concept called *forcings*—can be either natural sources or human-made sources and could cause the global climate to change. Natural forcings might include variations in the sun's intensity, a shift in the earth's orbit around the sun, a shift in the earth's tilt, or an increase in volcanic activity. Human-caused forcings could include burning fossil fuels, changing land-use patterns, or deforestation.

Greenhouse gases in the atmosphere do have an effect on the global energy balance. Without the natural amount of greenhouse gases in the atmosphere, the earth would not be an inhabitable planet because it would be too cold. The natural amounts of carbon dioxide, water vapor, and other greenhouse gases make life possible on earth because they keep the atmosphere warm.

Several things can have an effect on the energy balance, such as clouds and atmospheric aerosols. Clouds can interact in several ways with energy. They can block much of the incoming sunlight and reflect it back to space. In this way, they have a cooling effect. Clouds also act like greenhouse gases, and they block the emission of heat to space and keep the earth from releasing its absorbed solar energy. The altitude of the cloud in the atmosphere can affect the energy budget. High clouds are colder and can absorb more surface-emitted heat in the atmosphere, yet they do not emit much heat to space because they are so cold. Clouds can cool or warm the earth, depending on how many clouds there are, how thick they are, and how high they are. Because of this, it is not yet fully understood what effect clouds will have on surface temperatures if climate change continues into the next century and beyond.

Climatologists have proposed different opinions. Some think that clouds may help to decrease the effects of climate change by increasing cloud cover, increasing thickness, or by decreasing in altitude. Others think that clouds could act to increase the warming on earth if the opposite conditions occurred. According to Anthony Del Genio of NASA's Goddard Institute for Space Studies, when air temperatures are higher, clouds are thinner and less capable of reflecting sunlight, which increases temperatures on earth, exacerbating climate change. Del Genio, in an interview with CNN, said that "in the larger context of the climate change debate I'd say we should not look for clouds to get us out of this mess. This is just one aspect of clouds, but this is the part people assumed would make climate change less severe" (Environmental News Network, 2000).

One way that scientists try to predict future climate change and the effects of climate change is through the analysis and interpretation of mathematical climate models. In these computer models, climatologists attempt to account for all items that affect climate. Cloud cover is one of those variables. Today, this is still one of the most difficult variables to control and interpret. The climate is so sensitive as to how clouds might change, that even the most complicated, precise models developed today often vary in their climate change prediction under all the different methods available for cloud modeling.

The main reason clouds are so difficult to model is because they are so unpredictable. They can form rapidly and complete their life cycle in a matter of hours. Other climate variables work on a much slower time scale. Clouds also occur in a relatively small geographic area. Other climate variables operate on a much larger—regional or global—scale. According to climate research scientists at NASA, the world's fastest supercomputers can only track a single column of the surface and the atmosphere every 80–322 kilometers. In comparison, a massive thunderstorm system might cover only 32 kilometers. Features that are small, fast, and short lived are hard to predict. This is one of the reasons why predicting specific individual weather events is more difficult than predicting long-term climate changes over broad areas.

Clouds are just one thing that can change the global energy balance; snow and ice can as well. If the earth becomes cold enough, allowing large amounts of snow and ice to form, then more of the sun's energy will be directly reflected back to space because snow and ice have a high albedo. Over a period of time, this will change the global energy balance and the global temperature. Conversely, if the earth warms, the snow and ice will melt. This lowers the surface albedo, allowing more sunlight to be absorbed, which will warm the earth more.

Deforestation can also upset the global energy balance; if the forested area is removed and land is left bare, the ground can then reflect more sunlight back to space, causing a net cooling effect. On the other hand, if the forest material is burned, then the CO_2 stored in the trees is released into the atmosphere, contributing to climate change. Also, forests are good reservoirs of existing CO_2. The plants store and hold the CO_2, keeping it out of the air. If the forest is burned, not only does the already stored CO_2 now enter the atmosphere, but any future storage potential of CO_2 in that forest is now destroyed, creating two negative conditions of CO_2 contributing to climate change.

Atmospheric aerosols (tiny smoke particles) can be added to the atmosphere by sources such as fossil fuels, biomass burning, and industrial pollution. Aerosols can either cool or warm the atmospheric temperature depending on how much solar radiation they absorb versus how much they scatter back to space. Fossil fuel aerosols can also pollute clouds. Scientists need to do a lot more research on aerosols before they fully understand the full impact of aerosols on climate change. The composition of the aerosol, its absorptive properties, the size of the aerosol particles, the number of particles, and how high in the atmosphere are all variables that have an effect on whether aerosols cool or warm the atmospheric temperature and by how much.

Another effect aerosols have on clouds is that as they increase, the water in the clouds gets spread over more particles, and smaller particles fall more slowly, resulting in a decrease in the amount of rainfall. Scientists believe aerosols have the potential to change the frequency of cloud occurrences, cloud thickness, and amount of rainfall in a region. Like clouds, aerosols are also a challenge to accommodate in climate models.

Rates of Change

Climate change "drivers" (causes) often trigger additional changes (feedbacks) within the climate system that can amplify or subdue the climate's initial response to them. For instance, if changes in the earth's orbit trigger an interglacial (warm) period, increasing CO_2 may amplify the warming by enhancing the greenhouse effect. When temperatures get cooler, CO_2 enters the oceans, and the atmosphere becomes cooler. Sometimes the earth's climate seems to be quite stable; other times it seems to have periods of rapid change. According to the U.S. Environmental Protection Agency, interglacial climates (such as the climate today) tend to be more stable than cooler, glacial climates. Abrupt or rapid climate changes often occur between glacial and interglacial periods.

There are many components in a climate system, such as the atmosphere, the earth's surface, the ocean surface, vegetation, sea ice, mountain glaciers, deep ocean, and ice sheets. All of these components affect, and are affected by, the climate. They all have different response times, however, as shown in Table 1.1.

TABLE 1.1

Climate-System Components and Response Times

Climate-System Component	Response Time	Example
Fast responses		
Land surface	Hours to months	Heating of the earth's surface
Ocean surface	Days to months	Afternoon heating of the water's surface
Atmosphere	Hours to weeks	Daily heating; winter inversions
Sea ice	Weeks to years	Early summer breakup
Vegetation	Hours to centuries	Growth of trees in a rain forest
Slow responses		
Ice sheets	100–10,000 years	Advances of ice sheets over Greenland
Mountain glaciers	10–100 years	Loss of glaciers in Glacier National Park
Deep ocean	100–1,500 years	Deep-water replacement

The amount of change applied, as well as the component's ability to naturally respond determines largely what the climate actually ends up doing. For instance, if there is a slow climate change but the system component reacts quickly (in a short period of time), then the response will be visible. If the climate change is rapid but then reverts back to its previous condition and the component's response time is naturally slow, then there will be no response. If the climate change alternates from one extreme to another at a rate that the components can keep up with, these changes will be seen as visible adaptations. It is these types of rates of change that are most enlightening for climatologists because it allows them to more efficiently model all the subtle components of the climate system.

Atmospheric Circulation and Climate Change

When pressure differences alone are responsible for moving air, the air—or wind—will be pushed in a straight path. Winds do follow curved paths across the earth, however, as illustrated in Figure 1.4. Named after the scientist who discovered this effect, Gustave-Gaspard Coriolis, this phenomenon is called the Coriolis force and is an apparent drifting sideways (a property called deflection) of a freely moving object as seen by an observer on earth.

Given in nonvector terms, the property can be described as follows: at a given rate of rotation of the observer, the magnitude of the Coriolis acceleration of the object is proportional to the velocity of the object and also to the sine of the angle between the direction of movement of the object and the axis of rotation. In mathematical terms, the vector formula for the magnitude and direction of the Coriolis acceleration is,

$$a_c = -2\Omega \times v$$

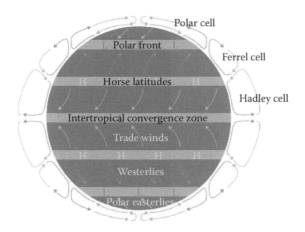

FIGURE 1.4
(See color insert.) This diagram represents the earth's major atmospheric circulation patterns. Major wind systems, such as the trade winds and westerlies lie between permanent bands of high or low pressure located at specific latitudes. (Courtesy of NASA, The water planet: Meteorological, oceanographic and hydrographic applications of remote sensing, Section 14 of *Remote Sensing Tutorial*, http://rst.gsfc.nasa.gov/Front/tofc.html, 2011.)

where a_c is the acceleration of the particle in the rotating system; v is the velocity of the particle in the rotating system; Ω is the angular velocity vector that has a magnitude equal to the rotation rate w and is directed along the axis of rotation of the rotating reference frame; and $\times$ is the multiplication operator. The equation may be multiplied by the mass of the object in question to produce the Coriolis force, as follows.

$$F_c = -2m\,\Omega \times v$$

The *Coriolis effect* is the behavior added by the *Coriolis acceleration*. The formula implies that the Coriolis acceleration is perpendicular to both the direction of the velocity of the moving mass and to the frame's rotation axis. Therefore, the following conditions apply:

- If the velocity is parallel to the rotation axis, the Coriolis acceleration is *zero*.
- If the velocity is straight inward to the axis, the acceleration is *in the direction* of local rotation.
- If the velocity is straight outward from the axis, the acceleration is *against the direction* of local rotation.
- If the velocity is in the direction of local rotation, the acceleration is *outward from the axis*.
- If the velocity is against the direction of local rotation, the acceleration is *inward to the axis*.

Therefore, the Coriolis force is the tendency for any moving body on or above the earth's surface, such as an ocean current, an air mass, or a ballistic missile, to drift sideways from its course because of the earth's rotation underneath. In other words, a moving object appears to veer from its original path. In the northern hemisphere, the deflection is to the right of the motion; in the southern hemisphere, it is to the left.

The Coriolis deflection of a body moving toward the north or south results from the fact that the earth's surface is rotating eastward at greater speed near the equator than near the poles, because a point on the equator traces out a larger circle per day than a point on another latitude nearer either the North or South Pole (the equator is the largest circle, and other latitudes are smaller). A body traveling toward the equator with the slower rotational speed of higher latitudes tends to fall behind or veer to the west relative to the more rapidly rotating earth below it at lower latitudes. Similarly, a body traveling toward either the north or south pole veers eastward because it retains the greater eastward rotational speed of the lower latitudes as it passes over the more slowly rotating earth closer to the pole.

The practical applications of the Coriolis force are important when calculating terrestrial wind systems and ocean currents. Scientists studying the weather, ocean dynamics, and other related earth phenomena must take the Coriolis force into account. The Coriolis force also affects regional and global weather patterns because it interacts with jet streams.

According to a study conducted in 2006 by Gabriel Vecchi of the University Corporation for Atmospheric Research, the trade winds in the Pacific Ocean are weakening as a result of climate change (Vecchi, 2006). This conclusion is based on the findings of a study that showed that the biology in the area may be changing, which could be harmful to marine life and have a long-term effect of disrupting the marine food chain. Researchers predict that it could also reduce the biological productivity of the Pacific Ocean, which would impact not only the natural ecosystem and balance but also the food supply for millions of people.

The study used climate data consisting of sea-level atmospheric pressure over the past 150 years and combined that with computer modeling to conclude that the wind has weakened by about 3.5 percent since the mid-1800s. The researchers predict that another 10 percent decrease is possible by the end of the twenty-first century.

Some of the computer-modeling simulations included variables such as the effects of human greenhouse gas emissions, whereas other simulations included only natural factors that affect climate such as volcanic eruptions and solar variations. Vecchi concluded that the only way to account for the observed weakening of the trade winds was through the model that included human activity—specifically from greenhouse gases and the burning of fossil fuels. According to an interview in LiveScience, MSNBC, Vecchi believes "this is evidence supporting global warming and also evidence of our ability to make reasonable predictions of at least the large scale changes that we should expect from global warming" (Vecchi, 2006).

Vecchi believes climate change is to blame because in order for the ocean and atmosphere to maintain an energy balance, the rate at which the atmosphere absorbs water from the ocean must equal the amount that it loses to rainfall. As climate change increases the air temperature, more water evaporates from the ocean into the air. The atmosphere cannot convert it to rainfall and return it back fast enough. Because the air is gaining water faster than it can release it, it gets overloaded, and the natural system compensates by slowing the trade wind down, decreasing the amount of water being drawn up into the atmosphere in order to maintain the energy balance.

The drop in winds reduces the strength for both the surface and subsurface ocean currents and interferes with the cold-water upwelling at the equator that is responsible for supplying ocean ecosystems with valuable nutrients, which are the lifeblood of the fishing industry. This is one example of how climate change could affect atmospheric circulation. Climate change can affect other atmospheric circulation patterns as well, such as the seasonal monsoon systems. Much of the world's population relies on the monsoon rains. If that balance were upset, it could cause disastrous effects for food and water supplies, ecosystems, health issues, and more. The sometimes obvious, and other times subtle, effects of climate change are serious and can be far reaching.

Extreme Weather

A serious concern about climate change is the potential damage that can be done to humans, property, and the environment as a result of extreme weather events, such as severe drought and storms. Today, scientists, such as Dr. Christopher Landsea at NOAA, are trying to understand just how much impact climate change may have on the occurrence and frequency of drought, hurricanes, and tornadoes (Landsea, 2000). It still remains difficult to assess because climate change will have different impacts on different areas of the earth. Although experts cannot predict exactly where hurricanes or other severe storms will occur in the future, they are fairly certain that as the atmosphere continues to warm under the influence of climate change, it will cause an increase in heat waves. Because warmer air can hold more moisture, it will change the hydrological cycle, which will alter flooding and drought patterns. There is a great concern that increasing ocean temperatures will also increase the likelihood of tropical cyclones or hurricanes.

The WMO has warned that extreme weather events, such as drought, hurricanes, and heavy rainfall may very likely increase because of climate change. The WMO is an organization of meteorologists from 189 countries. It is a specialized agency of the United Nations and serves as its voice on the state and behavior of the earth's atmosphere, its interaction with the oceans, the resulting climate as a product from the interaction, and the resulting distribution of water resources. There are not many forces in nature that can compare to the destructive capability of a hurricane. These storms can have winds blow for long periods of time at 249 kilometers per hour or higher. Not only is the wind destructive, but also the rainfall and storm surges can cause significant damage and loss of life (Climateaction, 2010).

Hurricane Katrina, which formed August 23, 2005, and dissipated August 30, 2005, affecting the Bahamas, South Florida, Cuba, Louisiana, Mississippi, Alabama, and the Florida Panhandle was the deadliest hurricane in the history of the United States, killing more than 1,800 people and destroying more than 200,000 homes. There were more than 900,000 evacuees; many relocated to states in the western United States. It was also the costliest hurricane in United States history with more than $75 billion in estimated damages. Today, only about 40 percent of the New Orleans pre-Katrina residents have returned to the city.

A hurricane—or tropical cyclone—forms over tropical waters, between latitudes of 8° and 20° in areas of high humidity, light winds, and where the sea surface is warm. Typically, temperatures must be 26.5°C or warmer to start a hurricane, which is why climate change and the heating of the ocean is such a concern.

Protecting life and the environment from severe weather triggered by climate change currently has many research scientists at the National Hurricane Center at NOAA and elsewhere engaged in theoretical studies, computer modeling, and collection and analysis of field data in an effort to gain a better understanding of the mechanics of climate change and its interaction with the environment to improve forecasting, response, and safety.

In an article published in Switzerland by the Environmental News Service (2010), extreme weather events, such as wildfire outbreaks across Russia, record monsoonal flooding in Pakistan, rain-induced landslides in China, and the calving of a huge chunk of ice off the Petermann Glacier in Greenland, fall in line with what WMO scientists are warning of when they project "more frequent and intense extreme weather events due to global warming" (WMO/UNFCCC, 2011).

These identified unprecedented events that are occurring concurrently around the world are causing a loss of human life and property. The WMO has brought to public attention the fact that the similar timing of all these incidences closely coincides with the IPCC's Fourth Assessment Report published in 2007, suggesting a scientific connection in support of climate change (IPCC, 2007). In the IPCC's Summary for Policy Makers it states: "the frequency and intensity of extreme events are expected to change as earth's climate changes, and these changes could occur even with relatively small mean climate changes" (IPCC, 2001).

The summary points out that some of these changes have already occurred, specifically the examples previously mentioned. In Moscow, they experienced 30 daily record temperature highs in June 2010, leading to massive forest and peat fires. In Pakistan, the monsoonal flooding was so intense that they received 300 millimeters over a 36-hour period, and the Indus River in the northern portion of the country reached its highest water levels in 110 years. Southern and central areas in the country were also flooded, killing over 1,600 people. More than six million people have been driven from their homes as a result. The Pakistan government reports more than 40 million of their citizens have been adversely affected by flooding thus far. China is also being negatively impacted by the worst flooding in decades. A mudslide in the Gansu province on August 7, 2010, killed more than 700 and left more than 1,000 others unaccounted for. In China, their government has reported that 12 million people have lost their homes to unexpected flooding.

NASA's Aqua satellite, via the MODIS sensor, discovered the calving of a major iceberg—the size of those typically found in Antarctica—on August 5, 2010. It is the largest chunk of ice to calve from the Greenland ice sheet in the last 50 years, dwarfing the tens of thousands of much smaller icebergs that normally calve from the glaciers of Greenland.

What is significant about these extreme weather events, according to the WMO, is that all these events compare with, or exceed in intensity, duration or geographical extent, the previous largest historical events, leaving human interference as the critical link (WMO, 2011). The photo in Figure 1.5 shows

FIGURE 1.5
(See color insert.) On August 5, 2010, an enormous chunk of ice, approximately 251 square kilometers in size, broke off the Petermann Glacier along the northwestern coast of Greenland. According to climate experts at the University of Delaware, the Petermann Glacier lost about one-fourth of its 70-kilometer-long floating ice shelf. The recently calved iceberg is the largest to form in the Arctic in 50 years. (Courtesy of NASA, Ice island caves off Petermann Glacier, http://earthobservatory.nasa.gov/NaturalHazards/view.php?id=45207, 2010.)

the calving of the Petermann Glacier in Greenland on August 5, 2010. It is a chunk of ice roughly 251 square kilometers in size (Environment News Service, 2010).

The Role of Ocean Circulation in Climate Change

There are two factors that make water more dense (which causes it to sink) or less dense (which keeps it on the surface): salt content and temperature. Atmospheric flow and ocean currents are the mechanisms that carry heat from the equator to the poles. There are many processes that can alter the circulation patterns, and when this happens it can change the weather of an area. If the ocean did not distribute heat throughout the world, the equator would be much warmer, and the poles would be much colder.

The role of the oceans is equally as important in transporting heat from the equator to the polar regions as the atmosphere as shown in Figure 1.6. In terms of how much heat and water the oceans can hold, their capacity is much greater than the atmosphere's. In fact, the world's oceans can store approximately 1,100 times more heat than the atmosphere. The oceans also contain 90,000 times more water than the atmosphere does.

As more knowledge is gained about climate change, also gained is a better appreciation of the role the oceans play in shaping the earth's climate. Because of this, much more research has been done on the oceans in the past 15 years, leading to the discovery that the oceans' depths have warmed considerably since 1950. According to scientists at the Woods Hole Oceanographic Institution, up until recently, scientific models predicting climate change could not account for where the projected warmth had gone; it was unaccounted for in the atmosphere. This discrepancy in the model had caused much confusion until researchers finally figured out that the world's oceans were storing the "missing" greenhouse warming. Water has a tremendous capacity to hold heat. The warming had occurred, but up until then, no one had thought to look toward the oceans for the answer (Schmitt, 2010).

Now that scientists understand this relationship, those who study climate change agree that including the ocean system in climate change studies is critical. Not only do the oceans have an enormous thermal capacity, but the constant movement of slow and fast water can affect the weather for months at a time. It is important for climatologists to understand these interactions in the ocean in order to be able to predict regional trends in climate. It is also important to understand the deep-water processes as well as processes that occur near the surface in order to understand what mechanisms drive climate. The oceans also play a critical role in balancing the CO_2 levels. The CO_2 levels in both the atmosphere and dissolved in the ocean reach a joint equilibrium. If something happens to upset this balance—such as changes in chemistry—then sudden shifts in the CO_2 levels can affect the climate.

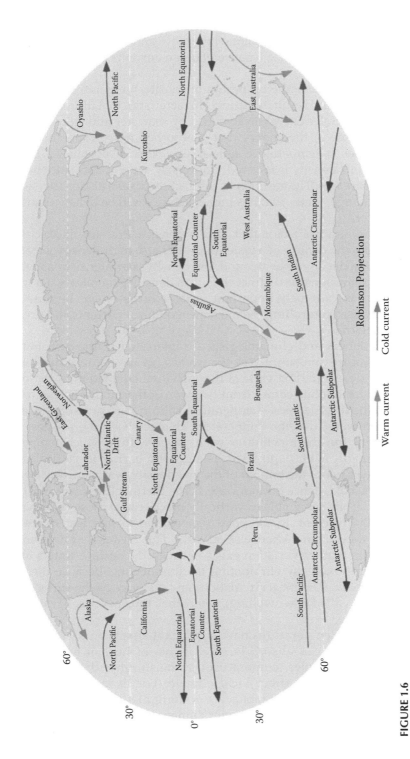

FIGURE 1.6

(See color insert.) The earth's major ocean currents are responsible for the global transport of heat. Without them, many areas would be much cooler than they currently are.

This is one of the concerns about the steadily increasing levels of greenhouse gases. If the oceans reach the point where they cannot continue to absorb any more CO_2, it could upset the balance of ocean currents and climate patterns on a global scale.

In the world's oceans, the properties of density, temperature, and salinity all work together and result in distinct characteristics that ultimately relate to climate change. Solar energy is absorbed by seawater and stored as heat in the oceans. Some of the energy that is absorbed may evaporate seawater, which increases its temperature and salinity. When a substance is heated, it expands, and its density is lowered. Conversely, when a substance is cooled, its density increases. The addition or subtraction of salts also causes seawater density to change. Water that has higher salinity will be denser than lower-salinity water.

Pressure is another factor that affects density. Pressure increases with depth, as does the density of a water mass. Because high-density water sinks below average density seawater and low-density seawater rises above average density seawater, this distinct change in density generates water motion. This concept is extremely important in the world's oceans because it is a chief mechanism controlling the movement of major currents and ocean circulation patterns.

Oceanographers and climatologists are interested in the distribution of both temperature and salinity in the world's oceans because they are two factors that determine the vertical thermohaline circulation in the oceans. The term thermohaline comes from two words: thermo for heat and haline for salt. Of the three factors—temperature, salinity, and pressure—that have an effect on water density, temperature changes have the greatest effect. In the ocean, the thermocline (a water layer within which temperature decreases rapidly with depth) acts as a density barrier to vertical circulation. This layer lies at the bottom of the low-density, warm surface layer and the top of the cold, dense bottom waters. The thermocline keeps most of the ocean water from being able to vertically mix because these two layers are so drastically different. In the polar regions, the surface waters are much colder than they are anywhere else on earth. This means they are denser, so that little temperature variation exists between the surface waters and the deeper waters—basically eliminating the thermocline. Because there is no thermocline barrier, vertical circulation can take place as the surface waters sink (a process called downwelling), where they replenish deep waters in the major oceans.

Water-surface temperatures have significant effects on coastal climates. Because seawater can absorb large amounts of heat, it enables coastal locations to have cooler temperatures in the summer than inland areas. Coastal currents also affect local climate. For example, Los Angeles, California, and Phoenix, Arizona, are at similar latitudes, yet Los Angeles has a much more moderate summer climate because of the effect of the ocean.

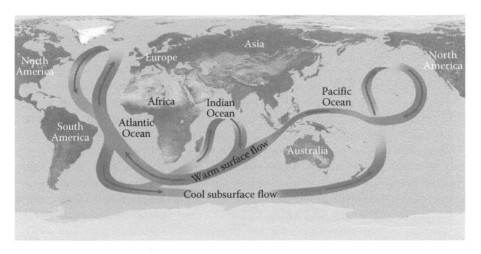

FIGURE 1.7
(See color insert.) The Great Ocean Conveyor Belt is the major transport mechanism of heat in the ocean. If its flow were disrupted, it could trigger an abrupt climate change, such as an ice age in western Europe. (Courtesy of NOAA, Ocean facts, http://oceanservice.noaa.gov/facts/coldocean.html, 2011.)

The Great Ocean Conveyor Belt and Consequences of Destabilization: Abrupt Climate Change

One of the most significant features in the ocean is the thermohaline circulation—more commonly referred to as the Great Ocean Conveyor Belt. This massive, continuous loop of flow, shown in Figure 1.7, plays a critical role in determining the world's climate. The two mechanisms that make this conveyor belt work are heat and salt content.

The Great Ocean Conveyor Belt plays a major part in distributing the sun's thermal energy around the globe after the ocean has absorbed it. In fact, if it were not for this flow, the equator would be much hotter, the poles would be much colder, and western Europe would not enjoy as warm a climate as it currently does.

The Great Ocean Conveyor Belt does not move fast, but it is enormous. It carries as much water within it as one hundred Amazon Rivers could hold. The mechanism that drives it is the differences in density that range throughout it. The ratios of salt and temperature determine the density. The colder and saltier water is, the denser it is, and therefore it tends to sink. When it is warm and fresh, it is less dense and rises to the surface.

The Conveyor Belt literally travels the world. In general, in a continuous loop, it transfers warm water from the Pacific Ocean to the Atlantic as a shallow current, then returns cold water from the Atlantic to the Pacific as a deep current that flows farther south. Specifically, as it travels past the north of Australia, it is a warm current. It travels around the southern tip of Africa, then moves up into the Atlantic Ocean. At this point, it turns into the Gulf

Stream, which is a very warm, north-flowing current critical for providing warmth to western Europe and the northeast coast of the United States. After it passes western Europe and heads to the Arctic, the surface water evaporates, and the water cools down, releasing its heat into the atmosphere. It is this released heat that western Europe enjoys as part of its moderate climate.

At this point, the water becomes very cold and increases in salinity, becoming very dense. As its density increases, it begins to sink. The cold, dense water descends hundreds of meters below the surface of the ocean. It now travels slowly southward through the deep ocean abyss in the Atlantic and flows to the southern hemisphere south of Australia and heads north again, until it eventually mixes upward to the surface of the ocean and starts the process again. This entire conveyor belt cycle takes about 1,000 years to complete.

Scientists at Argonne National Laboratory (ANL) have been actively researching how long it takes water to move through this conveyor system by tracking levels of the radioactive argon-39 isotope in the earth's ocean circulation. They are interested in tracking changes in ocean circulation because scientists have long believed that this system plays an important role in climate moderation. According to Ernst Rehm, a physicist at ANL, "We have some idea that if the 'conveyor belt' stops, then the warm water that is brought to Europe will stop. We have some idea that this may cause an ice age in Europe" (Frontiers Research Highlights, 2003). In addition, scientists believe that climate change could modify the ocean's circulation if water temperatures were to rise (the heat component of thermohaline) and glaciers were to melt and release more of their locked-up fresh water into the oceans, lowering the salinity levels (the salt component, or haline). If climate change caused this to happen—and some scientists claim it already is happening—it could slow, or stop, the thermohaline circulation (Argonne National Laboratory, 2003).

The Great Ocean Conveyor Belt plays an extremely important role in shaping the earth's climate; a slight disruption in it could destabilize the current and trigger an abrupt climate change. Climatologists at NOAA and NASA believe that as the earth's atmosphere continues to heat from the effects of climate change, there could be an increase in precipitation as well as an influx of fresh water added to the polar oceans as a result of the rapid melting of glaciers and ice sheets in the Arctic Ocean. They believe that large amounts of fresh water could dilute the Atlantic Gulf Stream to the point where it would no longer be saline—and, hence, dense enough—to sink to the ocean depths to begin its return from the polar latitudes back to the equator.

Measurements taken over the past 40 years have shown that salinity levels within the North Atlantic region are slowly decreasing. What makes this situation so serious is that if cold water stopped sinking—which means the Gulf Stream would slow and stop—there would be nothing left to push the deep, cold current at the bottom of the Atlantic along, which is what ultimately drives the worldwide ocean current system today. If this were to happen, the results would be dramatic. Western Europe and the eastern part of

North America would cool off. Temperatures could plummet up to 5°C. This is about the same temperature difference as between the average global temperatures during the last Ice Age and today.

Effects of Sea-Level Rise

There are several impacts associated with rising sea levels, making the world's coastlines vulnerable. As climate change continues and sea levels rise, storm surges will increase in intensity, destroying land farther inland from the coastal regions. Flooding will become one of the major problems, and associated with flooding are several other negative impacts. As ocean waters move inland, freshwater areas will become contaminated with saltwater. As saline water intrudes rivers, bays, estuaries, and coastal aquifers, they will become contaminated and unusable. Wildlife that depends on freshwater will have their habitat negatively impacted, and drinking water will become unusable. Erosion will increase along coastlines, causing disaster for many of the world's population that currently reside along the coasts. It will leave many people homeless and be economically devastating, especially in undeveloped countries.

As wetlands, mangroves, and estuaries are impacted, fragile habitats will be lost worldwide. Species will become threatened, endangered, and extinct. Other marine ecosystems will also be harmed, such as coral reefs. Reef habitats are extremely fragile, and significant physical changes in their environment can quickly destroy them. The most vulnerable areas are the low-lying countries of the world with extremely large coastal populations, such as Bangladesh, the Maldives, Vietnam, China, Indonesia, Senegal, Tuvalu, Mozambique, Egypt, the Marshall Islands, Pakistan, and Thailand. Affected developing countries do not have the economic resources to implement adaptation measures, such as building sea walls to hold back rising waters. If sea levels rise, the inhabitants of the coastal areas will have no other choice but to move inland to higher ground, if possible, losing what they have at lower levels. If mass migrations result, this could lead to a host of other negative issues, such as hunger, disease, and civil unrest.

Island states are particularly vulnerable. One of the nations most at risk is the Maldives. This nation lies in the Indian Ocean and is comprised of nearly 1,200 individual islands. Their elevation above sea level is only two meters. With a population of more than 200,000 people, if sea levels were to rise significantly, the entire country could become uninhabitable, leaving the entire country's population homeless. The Marshall Islands and Tuvalu, in the Pacific, face a similar situation. Rising sea levels there would first contaminate drinking water supplies and then drown the landmasses, leaving the population homeless. Other vulnerable locations include London, Amsterdam, Shanghai, New York City, many of the Caribbean islands, and Jakarta.

In a study conducted by Sugata Hazra, an oceanographer at Jadavpur University in Calcutta, India, it was discovered that over the past 30 years, 80 square kilometers of the Sundarbans have disappeared because of rising sea level, displacing more than six hundred families. Another area, Ghoramara, has had all but 5 square kilometers of its land submerged and is now half the size it was in 1969. The Sundarbans represent one of the world's biggest collection of river-delta islands that lie between India and Bangladesh. Sea-level rise has contaminated their drinking water and destroyed forested areas in the ecosystem. It has also threatened the existence of the wildlife, including the Bengal tiger. More than four million people live on the tiny island state, and hundreds of families have already been pushed out of their homes and forced to move to refugee camps on neighboring islands. This is just one example of how rising sea levels are presently impacting developing countries (George, 2010).

The impacts are not just limited to other countries; they will also be felt in the United States. Both the Atlantic and Gulf Coasts face serious impacts in the face of encroaching ocean levels and saline waters. Washington, DC, is one of the more vulnerable areas. Higher sea levels would flood the Potomac and encroach on many famous historical landmarks. Baltimore and Annapolis are in a similar situation.

In the Mississippi delta, the loss of wetlands is a serious issue. Changes in sea level can cause wetlands to migrate landward. The Atlantic coast is one of the more sensitive areas to wetland vulnerability. Not only is this a problem for natural habitats, but historically, these areas have been one of the most rich commercial fisheries in the world. If wetlands are endangered or destroyed, it would also have significant economic ramifications. These issues make the monitoring and control of rising sea levels a critical concern. Areas particularly in danger include Florida, Mississippi, Louisiana, North Carolina, South Carolina, Alabama, Georgia, and Texas.

Future Projections

One of the key issues is that of future sea-level rise. Because the ocean's thermal inertia is so great, it will take decades for the oceans to adjust their levels to the heat absorbed. In fact, for the heating caused by greenhouse gas emissions already released into the atmosphere, sea levels are still trying to find a point of equilibrium. Therefore, even if all greenhouse emissions stopped today, there would still be a lag time for the oceans to stop rising. During this lag time, the oceans will likely rise another 13 to 30 centimeters by 2100. In the 2007 Fourth Assessment Report of the IPCC, a sea-level rise of 18–58 centimeters by 2100 was projected.

Based on the extensive work that has been done to date, scientists have a clear idea of where conditions are going in the future. According to the United States Geological Survey (USGS), based on information obtained from both tidal gauges and satellite measurements worldwide, scientists can say with confidence that sea-level rise has increased during the twentieth century. Based on data acquired from Australia's Commonwealth Scientific and Industrial Research Organization (CSIRO), data gathered from January 1993 to April 2008 shows that sea level has risen on average 3.3 mm/year (Figure 1.8). Increased scientific knowledge has also clarified some issues that were not well understood previously, such as that the large polar ice sheets are far more sensitive to surface warming than initially thought, with significant changes currently being observed on the Greenland and west Antarctic ice sheets. Scientists now realize that these melting ice sheets can add water mass much more quickly to the oceans than previously assumed and play a significant part in overall global sea-level rise.

Today also marks a notable consensus among specialists in climate change at USGS. It is largely recognized and accepted that there could be a rapid collapse of the polar ice sheets, and scientists have keyed in on the fact that anthropogenic actions, such as burning fossil fuels, could result in triggering an abrupt sea-level rise before the end of this century. They stress public education and political policy be brought to the forefront in order to deal most effectively with a situation that affects every person living on earth now and in the future. Figure 1.8(a) shows the recent change in old sea ice in the Arctic, and Figure 1.8(b) shows the drastic decrease in ice coverage in the Arctic over the past 30 years. Figure 1.8(c) shows a global land/ocean temperature index. From 1880 to the present (industrialization), it can be seen that both the earth's land surface and oceans have steadily warmed over time— a significant finding, because it points to the anthropogenic component of the problem. Based on data provided from the National Snow and Ice Data Center, from 1979 to 2009, the extent of Arctic sea ice has decreased roughly 2.5 million square kilometers. Because of this, it makes it all the more important that the scientific basis of climate change be well understood when dealing with the economic, political, and sociological ramifications caused by climate change.

Conclusions

The information presented in this chapter was geared to provide insight into the physical processes of climate change as well as the anthropogenic component and its overall significance to the problem. Based on the nature of the earth's natural processes, there is already substantial evidence that climate change is well under way and that society as a whole can no longer ignore the issue and sit on the sidelines. In addition, the days of passing the buck are over—it is no longer "somebody else's problem." We all have a vested interest in its outcome. In order to become empowered and contribute meaningfully

(a)

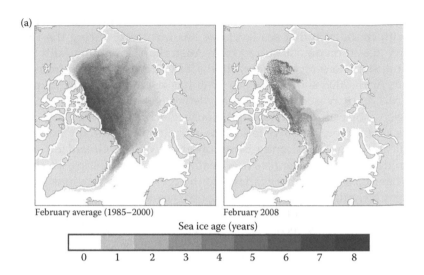

February average (1985–2000)　　　　February 2008

Sea ice age (years)

0　1　2　3　4　5　6　7　8

(b)　　　　　　　1979　　　　　　　　　　　　　　　2007

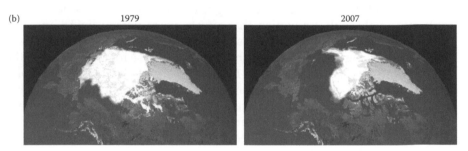

FIGURE 1.8

(See color insert.) In the Arctic, sea-ice extent fluctuates with the seasons. It reaches its peak extent in March, near the end of Northern Hemisphere winter, and its minimum extent in September, at the end of the summer thaw. In September 2007, Arctic sea-ice extent was the smallest area on record since satellites began collecting measurements about 30 years ago. (a) Although a cold winter allowed sea ice to cover much of the Arctic in the following months, this pair of images shows the drastic change in conditions. On the right (February 2008), the ice pack contained much more young ice than the long-term average (left). In the past, more ice survived the summer melt season and had an opportunity to thicken over the following winter. The area and thickness of sea ice that survives the summer has been declining over the past decade. (Courtesy of NASA, NASA and the International Polar Year, http://www.nasa.gov/mission_pages/IPY/multimedia/ipyimg_20080326.html, 2008.) (b) This shows the comparison between the September annual minimum of sea ice in 1979 and the September image from 2007, illustrating the drastic decline in ice. (Courtesy of NASA, "Remarkable" drop in Arctic Sea ice raises questions, http://www.nasa.gov/vision/earth/environment/arctic_minimum.html, 2008.) (c) Since 1880 and the rise of industrialization, the land and ocean temperatures have steadily climbed upward, emphasizing the anthropogenic effect of climate change. (Courtesy of NASA/GISS, Global annual mean surface air temperature change, http://data.giss.nasa.gov/gistemp/graphs/, 2006.) (d) The average monthly Arctic sea-ice extent has steadily decreased by 2.5 million square kilometers since 1979. (Courtesy of National Snow and Ice Data Center, Arctic sea ice extent remains low; 2009 Sees third-lowest mark, http://nsidc.org/news/press/20091005_minimumpr.html, 2009.)

(c)

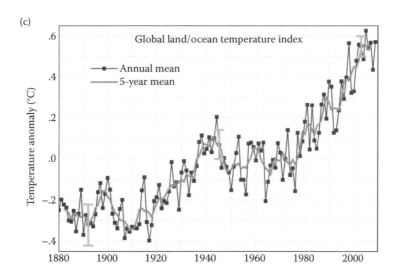

(d)

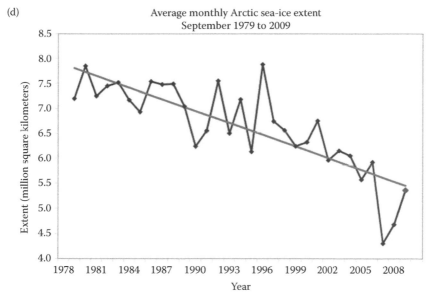

FIGURE 1.8   (CONTINUED)

to the solution, however, it is necessary to understand the pertinent multifaceted issues—the scientific, political, economic, social, and technological components. Although each is different in nature, they are all important pieces to the complete issue—a phenomenon so immense it touches nearly every fabric of society from broad, international scales to personal ones. Similar to just reading a language out loud without truly understanding the words and meanings, going without a working knowledge of the scientific foundations of climate change makes it impossible to comprehend the problem's vastness and relevance to other critical components of our lives—namely those that encompass the economic, political, and social aspects. For example, without a working understanding of the Great Ocean Conveyor Belt and the possibility of abrupt climate change, it is not possible to understand the significance and urgency of economic issues and the loss of homes, food, natural resources, incomes, and basic security; the social aspects of personal loss, fear, riot, unrest, and migration; or the political aspects of security, defense, order, peace, and leadership. Understanding first the scientific basis is critical to finding workable, meaningful, long-term solutions that will find an effective fit in all cultures and societies.

Not often have problems had such far-reaching consequences. Being the first world society to be faced with this dilemma, there is, unfortunately, no prior historical experience to fall back on. This is a modern problem. Our decisions and actions will not only affect lifestyles and opportunities today but will also affect those of generations long into the future. For that, we owe it to ourselves and future generations to obtain a good grasp of the basic scientific theory behind climate change. Without this crucial background, it is not possible to take a convincing stand backed by facts, examples, and viable suggestions for solutions. Knowledge is power, and personal and community education is critical to successful solutions. The correct scientific concepts need to be taught and understood so that the misinformation also being distributed by procrastinators and those in denial is put to rest in order to stop muddling the issue and delaying action. Perhaps one of the best ways to look at and approach the issue is from the standpoint of a global community where it is necessary to reach across borders, boundaries, personal differences, and comfort levels to solve this immense problem together—for this is a war unlike any fought before, and we are all on the front lines.

Discussion

Consider the following issues and their importance in the understanding of the complex phenomenon scientists refer to as climate change:

1. The issue started out being labeled as *global warming*. Recently, experts have begun referring to it as *climate change*. Do you think this is a more appropriate term for the issue? Why or why not?

2. Why is it necessary to study the earth as a global system—the atmosphere, hydrosphere, biosphere, and geosphere—rather than just the atmosphere when dealing with climate change? What effect does society have on each of these systems?

3. If scientists can benefit from studying the earth's climatic past to understand the earth's climate today and make predictions about the future, can they do the same about society? If so, what lessons and insights can be gained?

4. If the ocean's thermohaline circulation were disrupted, how many environmental, economic, and human endangerment consequences can you think of that would occur as a result? If you were in a leadership position (political leader, land manager, business leader, economic advisor, stock broker, academic advisor, etc.), how would this possibility influence your future decisions?

5. Why is personal research just as important as following research and news reports being currently released about climate change? Why may it be even more important?

6. Some groups dismiss climate change as a myth. If you found yourself in a discussion with an individual with that point of view, how would you address these issues?

   a. What factors do you believe cause these various myths to start?

   b. What social groups propagate the myths? What are their value systems based on?

   c. What prevents the majority of the population from researching the facts themselves? (Compare this to political campaign candidate research, etc.)

   d. What personal responsibilities do you think individuals should take upon themselves to research relevant issues on their own and come to their "earned" conclusions? Why?

7. As a head of state of a wealthy nation, how would you provide aid to other countries falling victim to climate change related sea-level rise? How would you approach providing mitigation measures and financial aid to countries?

References

Adam, D. 2006. Climate scientists issue dire warning. *The Guardian*, February 28. http://www.guardian.co.uk/environment/2006/feb/28/science.frontpage news (accessed December 3, 2010).

American Institute of Physics. 2010. Statement on human impacts on climate change. American Institute of Physics Policy Statements. http://www.aip.org/gov/policy12.html (accessed October 3, 2010).

Argonne National Laboratory. 2003. Physicists track great ocean conveyor belt. *Frontiers* (Argonne National Laboratory). http://www.anl.gov/Media_Center/Frontiers/2003/d8ee.html (accessed November 2, 2010).

CICERO. 2006. UK report warns of global catastrophes. *KLIMA Climate Magazine*, January 31. http://www.cicero.uio.no/webnews/index_e.aspx?id=10601 (accessed November 2, 2010).

Climateaction. 2010. Extreme weather events evidence of global warming, research suggests. *Climateaction/UNEP*, August 19. http://www.climateaction programme.org/news/extreme_weather_events_evidence_of_global_warming_research_suggests/ (accessed September 8, 2010).

CO_2 Now. 2011. Earth's CO_2 home page. http://co2now.org/ (accessed June 27, 2011).

Environmental News Network. 2000. Clouds' role in global warming studied. *CNN.com*, October 9. http://archives.cnn.com/2000/NATURE/10/09/clouds.warming.enn/ (accessed October 19, 2010).

Environment News Service. 2010. Extreme weather events signal global warming to world's meteorologists. *Environmental News Service*, August 17. http://www.ens-newsire.com/ens/aug2010/2010-08-17-01.html (accessed December 26, 2010).

Frontiers Research Highlights. 2003. Physicists track great ocean conveyor belt. http://www.anl.gov/Media_Center/Frontiers/2003/d8ee.html (accessed September 1, 2011).

George, N. 2010. Disputed isle in Bay of Bengal disappears into sea. *U.S. News*, March 24. http://www.usnews.com/science/articles/2010/03/24/disputed-isle-in-bay-of-bengal-disappears-into-sea (accessed September 28, 2010).

Global Carbon Project. 2007. Contributions to accelerating atmospheric CO_2 growth from economic activity, carbon intensity, and efficiency of natural sinks. *CSIRO*, October 22. http://www.globalcarbonproject.org/global/doc/Press_GCP_Canadelletal2007.doc (accessed November 8, 2010).

Intergovernmental Panel on Climate Change. 2001. Climate change 2001: Synthesis report: Summary for policymakers. http://www.ipcc.ch/pub/un/syreng/spm.pdf (accessed December 9, 2010).

Intergovernmental Panel on Climate Change. 2007. *WG1: The Physical Science Basis*. Cambridge, UK: Cambridge University Press.

Kumaresan, J., and Sathiakumar, S. 2010. Climate change and its potential impact on health: A call for integrated action. *Bulletin of the World Health Organization*. 88: 163–163. http://www.who.int/bulletin/volumes/88/3/10-076034/en/index.html.

Landsea, C. W. 2000. Climate variability of tropical cyclones: Past, present, and future. In *Storms*, edited by R. A. Pielke, Sr., and R. A. Pielke, Jr. New York: Routledge, 220–241. http://www.aoml.noaa.gov/hrd/Landsea/climvari/ (accessed November 12, 2010).

Lean, G. 2005. Global warming approaching point of no return, warns leading climate expert. *Independent/UK*, January 23. http://www.commondreams.org/headlines05/0123-01.htm (accessed October 15, 2010).

Mittal, A. 2009. The 2008 food price crisis: Rethinking food security policies. *United Nations Conference on Trade and Development*. 56: June. http://www.unctad.org/en/docs/gdsmdpg2420093_en.pdf (accessed August 27, 2011).

NASA Goddard Institute for Space Studies. 2011. GISS surface temperature analysis. http://data.giss.nasa.gov/gistemp/ (accessed September 1, 2011).

Parker, R. 2007. *Development Actions and the Rising Incidence of Disasters: Evaluation Brief 4*. Washington, DC: World Bank Independent Evaluation Group.

Pew Center on Global Climate Change. 2001. The basics: Long-term trends in carbon dioxide and surface temperature. http://www.pewclimate.org/global-warming-basics/facts_and_figures/temp_ghg_trends/longco2temp.cfm (accessed September 1, 2011).

Schmitt, R. 2010. The ocean's role in climate. Woods Hole Oceanographic Institution, November 19. http://www.whoi.edu/page.do?pid=12455&tid=282&cid=10146 (accessed December 1, 2010).

ScienceDaily. 2010. Carbon dioxide controls earth's temperature, new modeling study shows. *Science Daily*, October 15. http://www.sciencedaily.com/releases/2010/10/101014171146.htm.

United Nations Framework Convention on Climate Change. 2006. United Nations fact sheet on climate change: Climate change and adaptation. http://unfccc .int/files/press/backgrounders/application/pdf/factsheet_adaptation.pdf (accessed August 27, 2011).

Vecchi, G. 2006. Trade winds weaken with global warming. University Corporation for Atmospheric Research. http://www.vsp.ucar.edu/about/stories/gVecchi. html (accessed December 7, 2010).

World Meteorological Organization. 2009. 2000–2009: The warmest decade. Press release 869, August 26. http://www.wmo.int/pages/mediacentre/press _releases/pr_869_en.html (accessed August 26, 2011).

World Meteorological Organization. 2011. Weather extremes in a changing climate: Hindsight on foresight. WMO-No.1075. http://www.wmo.int/pages/ mediacentre/news/documents/1075_en.pdf (accessed August 28, 2011).

World Meteorological Organization/United Nations Framework Convention on Climate Change. 2011. Fact sheet: Climate change science: The status of climate change science today. http://unfccc.int/press/fact_sheets/items/4987.php (accessed August 29, 2011).

Suggested Reading

Alley, R. B. 2004. Abrupt climate change. *Scientific American*. 291(5): 62–69.

Amos, J. 2007. Arctic summers ice-free by 2013. *BBC News*, December 12. http://news .bbc.co.uk/2/hi/7139797.stm (accessed December 1, 2010).

Boswell, R. 2007. Northwest Passage in unprecedented ice melt, experts report. *CanWest News Service*, August 28. http://www.canada.com/nationalpost/ news/story.html?id=57868baf-87b0-4f29-9a4f-b6251b48582d&k=92663 (accessed November 3, 2010).

Boyd, R. S. 2002. Glaciers melting worldwide, study finds. *National Geographic News*, August 21. http://news.nationalgeographic.com/news/2002/08/0821_020821_ wireglaciers.html (accessed October 18, 2010).

Britt, R. R. 2005. Scientists put melting mystery on ice. *LiveScience*, June 30. http:// www.msnbc.msn.com/id/8421342/40041707 (accessed December 2, 2010).

Christianson, G. 1999. *Greenhouse: The 200-Year Story of Global Warming*. New York: Walker.

Culotta, E. 1995. Will plants profit from high CO_2? *Science*. 268(5211): 654–656.

D'Agnese, J. 2000. Why has our weather gone wild? *Discover.* June: 72–78.

Eilperin, J. 2005. Severe hurricanes increasing, study finds. *Washington Post*, September 16. http://www.washingtonpost.com/wp-dyn/content/article/2005/09/15/AR2005091502234.html (accessed November 11, 2010).

Gordon, C. 2007. Tracking glacial activity in Norway with photogrammetry software. *Imaging Notes.* 22(1): 24–29.

Hoffman, P. F., and D. P. Schrag. January 2000. Snowball Earth. *Scientific American.* 282(1): 68–75.

Hogan, J. 2005. Antarctic Ice Sheet is an "awakened giant." *New Scientist*, February 2. http://www.newscientist.com/article/dn6962-antarctic-ice-sheet-is-an-awakened-giant.html (accessed November 28, 2010).

Houghton, J. 2004. *Global Warming: The Complete Briefing.* New York: Cambridge University Press.

Karl, T., and K. Trenberth. 1999. The human impact on climate. *Scientific American.* 281(6): 100–105.

Revkin, A. C. 2008. In Greenland, ice and instability. *The New York Times*, January 8. http://www.nytimes.com/2008/01/08/science/earth/08gree.html (accessed September 19, 2010).

Revkin, A. C. 2007. Arctic melt unnerves the experts. *The New York Times*, October 2. http://www.nytimes.com/2007/10/02/science/earth/02arct.html (accessed December 3, 2010).

Weart, S. R. 2004. *The Discovery of Global Warming* (New Histories of Science, Technology, and Medicine). Cambridge: Harvard University Press.

2

The Role of Greenhouse Gases

Overview

Climate change is undoubtedly one of the biggest challenges humankind has ever faced. The bad news is this: A large share of it is our own doing. The good news is this: There is still a window of opportunity left to fix it if we act now. This is a significant, global problem, but with a transition of habits and some dedicated work, there is still a window of hope. However, the clock is ticking, and we are quickly running out of time.

This chapter discusses the greenhouse gases: the chief contributor and reason for the rise in temperature and trigger for the negative global impacts the earth is now experiencing and will experience for thousands of years to come. The chapter begins by presenting the basics of solar radiation and the processes the sun's energy takes to get to the earth and continues by explaining the various pathways the energy takes once it reaches the planet and the complications that then occur because of the greenhouse gases with which humans have overloaded the atmosphere. This introduces the important concept of radiative forcing and how that has now permanently impacted the earth's energy balance.

Next, the chapter introduces the various greenhouse gases, their properties, and why their collection and buildup in the atmosphere is so detrimental. The concept of the climate change potential of each of the greenhouse gases is presented as an illustration of why this whole process has such long and far-reaching effects—why some of the greenhouse gases can remain in the atmosphere and cause damage for thousands of years. The chapter then presents the concept of carbon sequestration and why this process is so important, as well as various carbon sinks and sources and why these must be managed in order to help mitigate the rate of greenhouse gas influx into the atmosphere. Next, it examines the impacts of deforestation, the burning of fossil fuels, and coal emissions and lays the foundation as to why these activities are so environmentally detrimental. In conclusion, the chapter discusses various health concerns and the key impacts to the earth's ecosystems as a result of the increasing concentration of greenhouse gases in the atmosphere and why it is imperative that the world's nations come together now in order to effectively mitigate the problem before it is too late.

Introduction

According to Dr. James E. Hansen of National Aeronautics and Space Administration (NASA), one of the world's foremost climate experts, there is still a small window of time left where actions can be taken to slow down the climate change processes that have been put in motion in order to keep temperatures from climbing over the 2°C mark. However, everyone has to play a part and make a commitment—large or small. As Canadian philosopher Marshall McLuhan said: "There are no passengers on Spaceship Earth. We are all crew."

One of the areas where humans can have the biggest impact in mitigating the rise in atmospheric temperature is through the control of greenhouse gas emissions. The emission of greenhouse gases is the most significant factor in climate change and the area where attention is most needed. Most greenhouse gases are emitted through the burning of fossil fuels—coal, oil, and gas. Fossil fuels are burned not only for public and private transportation, but also in the manufacture of most commodities, the generation of most electricity, most manufacturing and industrial processes, and commerce and commercial transportation. Developed countries are pumping greenhouse gases into the atmosphere at alarming rates. In order to halt the destruction being caused by greenhouse gases, action needs to be taken immediately. To make this happen, however, public education is critical. This chapter lays the foundation as to why greenhouse gases are so detrimental and why their mitigation is critical.

Development

As we saw in Chapter 1, the earth needs the natural greenhouse effect in order for life to survive. It is the process in which the emission of infrared radiation by the atmosphere warms the planet's surface. The atmosphere naturally acts as an insulating blanket, which is able to trap enough solar energy to keep the global average temperature in a comfortable range in which to support life. This insulating blanket is actually a collection of several atmospheric gases, some of them in such small amounts that they are referred to as trace gases.

The framework in which this system works is often referred to as the greenhouse effect because this global system of insulation is similar to that which occurs in a greenhouse nursery for plants. The gases are relatively transparent to incoming visible light from the sun yet opaque to the energy radiated from the earth.

These gases are the reason why the earth's atmosphere does not scorch during the day and freeze at night. Instead, the atmosphere contains molecules that absorb the heat and reradiate it in all directions, which reduces the heat lost back to space. The greenhouse gas molecules keep the earth's temperature ranges within comfortable limits. Without the natural greenhouse effect, life would not be possible on earth. In fact, without the greenhouse effect to regulate the atmospheric temperature, the sun's heat would escape, and the average temperature would drop from 14°C to –18°C, a temperature much too cold to support the diversity of life that exists today on the planet.

Radiation Transmission

Understanding radiation transmission is important to understanding how and why specific gases in the atmosphere are affected during the greenhouse gas process. The earth reflects 30 percent of the incoming solar radiation back out into space. The other 70 percent is absorbed and warms the atmosphere, land, and oceans. In order to maintain an energy balance, keeping the earth from getting too hot or cold, the amount of incoming energy must roughly equal the amount of outgoing energy (Figure 2.1).

The majority of insolation is in the short and medium wavelength region of the spectrum. The highest energy, short wavelengths, such as gamma rays, x-rays, and ultraviolet (UV) light, are absorbed by the mid to high levels of the atmosphere. This is desirable because if these wavelengths traveled to the surface, they could cause harm to life on earth. For example, exposure to UV light can lead to cancer. The medium wavelengths—referred to as visible light—travel to the earth's surface. These waves can be absorbed or reflected at the earth's surface as well as by the CO_2 and water vapor in the atmosphere.

When wavelengths from the electromagnetic spectrum reach the earth, several things can happen. They can get reflected; the earth's surface, the ocean, the clouds, or the atmosphere can absorb them; and they can be scattered. Of that which does enter the earth's atmosphere, nearly one-quarter is reflected by clouds and particulates (tiny suspended particles) in the atmosphere. Of the remaining visible light that does reach the earth's surface, approximately one-tenth is reflected upwards by snow and ice, because of their extremely high albedo (high reflectivity). As the infrared radiation (heat) is emitted upward, it does not simply escape to space. Water vapor and other gases in the atmosphere absorb it and trap it in the atmosphere. This trapped heat then radiates in all directions: upward, down toward the earth's

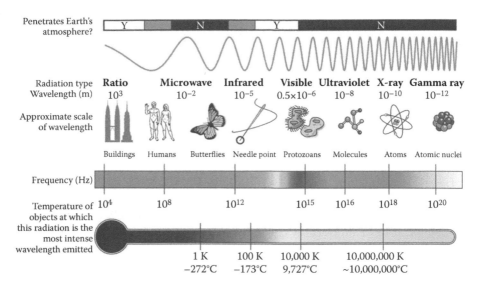

FIGURE 2.1
(See color insert) The sun's electromagnetic spectrum ranges from short wavelengths, such as x-rays, to long wavelengths, such as radio waves. The majority of the sun's energy is concentrated in the visible and nearly visible portion of the spectrum—the wavelengths located between 400 and 700 nm.

surface again, and sideways. This trapped heat energy is what constitutes the greenhouse effect.

The Natural and Enhanced Greenhouse Effect

In this "greenhouse environment," it is the combination of water vapor and trace gases that is responsible for trapping the heat radiated from the sun. As stated, this natural amount keeps the earth habitable instead of at –18°C, in which case the earth would look much different—there would most likely be very little, if any, liquid water available. The entire earth would have an ecosystem similar to that of the harshest areas in Antarctica.

The earth's natural greenhouse effect is critical for the survival and diversity of life. Since the start of the Industrial Revolution in the mid-nineteenth century, the natural greenhouse effect has been augmented by human interference. Carbon dioxide (CO_2), one of the atmosphere's principal greenhouse gases, has been altered to such an extent by human activity that the natural greenhouse effect is no longer in balance. The earth's energy balance now must contend with what is referred to as the enhanced greenhouse effect or anthropogenic greenhouse warming. CO_2 is being added in voluminous amounts as a result of human activity, such as deforestation, agricultural practices, and the burning of fossil fuels for transportation, urban development, the heating and cooling of homes, and industrial processes. In fact, CO_2 in the atmosphere has increased 31 percent since 1895. Concentration of other greenhouse gases, such as methane and nitrous oxides (also related to human activity) have increased 151 and 17 percent, respectively.

At the beginning of the Industrial Revolution, the CO_2 content in the atmosphere was 280 parts per million (ppm). By 1958, it had increased to 315 ppm; by 2004, 378 ppm; by 2005, 379 ppm; by 2007, 383 ppm; and in March 2011, the level was at 392.4 ppm. According to Dr. James E. Hansen, a world-renowned expert on climate change at NASA's Goddard Institute for Space Studies (GISS) in New York, "Climate is nearing dangerous tipping points. We have already gone too far. We must draw down atmospheric CO_2 to preserve the planet we know. A level of no more than 350 ppm is still feasible, with the help of reforestation and improved agricultural practices, but just barely—time is running out" (Hansen, 2008).

Although not the only greenhouse gas, CO_2 does seem to be the one most focused on because it is one of the most important and prevalent. It makes up about 25 percent of the natural greenhouse effect because there are several natural processes that put it in the atmosphere, such as the following:

- Forest fires: When trees are burned, the CO_2 stored within them is released into the atmosphere.
- Volcanic eruptions: CO_2 is one of the gases released in abundance by volcanoes.
- Ocean equilibrium: Oceans both absorb and release enormous amounts of CO_2. Historically, they have served as a major storage facility of CO_2 put into the atmosphere from human sources. Recently, however, scientists at NASA and National Oceanic and Atmospheric Administration (NOAA) have determined that the oceans have become nearly saturated and are approaching their limits as a carbon store.
- Vegetation storage: Trees, plants, grasses, and other vegetation serve as significant stores of carbon. When they die and decompose, half of their stored carbon is released into the atmosphere in the form of CO_2.
- Soil processes: When vegetation dies, the other half of their stored carbon is absorbed by the soil. Over time, some of this carbon is released into the atmosphere as CO_2.
- Biological factors: Any life form that consumes plants that contain carbon also emit CO_2 into the atmosphere through breathing. This includes animals, insects, and even humans.

With this amount of CO_2 entering the atmosphere, fortunately there are some processes that help keep it in check. Plants and trees remove CO_2 from the atmosphere through photosynthesis; the oceans absorb large amounts of CO_2; and phytoplankton take in CO_2 through photosynthesis. Ideally, keeping CO_2 in balance is desirable, but throughout the earth's history, this has not always been possible. When CO_2 levels have dropped, the earth has consistently experienced an ice age. Since the last ice age, CO_2 levels remained fairly constant, however, until the Industrial Revolution, when billions of tons of extra CO_2 began to be added to the earth's atmosphere.

The turning point was the use of fossil fuels—coal, oil, and gas—made from the carbon of plants and animals that decomposed millions of years ago and were buried deep under the earth's surface, subjected to enormous amounts of heat and pressure. When this carbon is converted into usable energy forms and is burned as fuel, the carbon combines with oxygen and releases enormous amounts of CO_2 into the atmosphere.

According to NASA/GISS, CO_2 levels have not been as high as they are today over the last 400,000 years (Hansen, 2007). The enhanced greenhouse effect has not come as a surprise to climate scientists, other experts, and decision makers, however. Climate scientist Charles David Keeling of the Scripps Institution of Oceanography made enormous strides in establishing the rising levels of CO_2 in the earth's atmosphere and the subsequent issue of the enhanced greenhouse effect and climate change. Keeling set up a CO_2 monitoring station at Mauna Loa, Hawaii, in 1957. Beginning in 1958, Keeling made continuous measurements of CO_2 that still continue today. He chose the remote Mauna Loa site so that the proximity of large cities and industrial areas would not compromise the atmospheric readings he was collecting. Air samples at Mauna Loa are collected continuously from air intakes at the top of four 7-meter towers and one 27-meter tower. Four air samples are collected each hour to determine the CO_2 level.

The measurements he gathered show a steady increase in mean atmospheric concentration from 315 parts per million by volume (ppmv) in 1958 to more than 392 ppmv by 2011. The increase is considered to be largely due to the enhanced greenhouse effect and the burning of fossil fuels and deforestation. This same upward trend also matches the paleoclimatic data shown in ice cores obtained from Antarctica and Greenland. The ice cores also show that the CO_2 concentration remained at approximately 280 ppmv for several thousand years but began to rise sharply at the beginning of the 1800s (Figure 2.2).

The Keeling Curve also shows a cyclic variation of about 5 ppmv each year. This is what gives it the sawtooth appearance. It represents the seasonal uptake of CO_2 by the earth's vegetation (principally in the Northern Hemisphere because that is where most of the earth's vegetation is located). The CO_2 lowers in the spring because new plant growth takes CO_2 out of the atmosphere through photosynthesis, and the level of CO_2 rises in the autumn as plants and leaves die off and decay, releasing CO_2 gas back into the atmosphere.

Mauna Loa is considered the premier long-term atmospheric monitoring facility on earth. It is also one of the world's most favorable locations for measuring undisturbed air because vegetation and human influences are minimal, and any influences from volcanic vents have been calculated and excluded from the measurement. The methods and equipment used to create

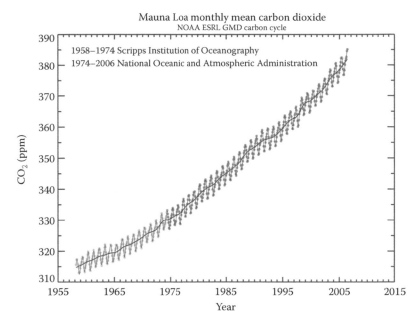

FIGURE 2.2
Collected since 1958, the data in the Keeling Curve have been instrumental in providing convincing evidence that CO_2 levels are rising worldwide. The jagged line depicts the readings collected from the Mauna Loa observation station, and the smoother line depicts data collected from the South Pole. Without these continuous data, it would be much more difficult to determine the existence of climate change.

the continuous monitoring record have been the same for the past 50 years, rendering the monitoring program highly controlled and consistent.

Charles Keeling was a true pioneer in establishing the existence of climate change. His work was instrumental in showing scientists worldwide that the CO_2 level in the atmosphere was indeed continually rising. This was the first strongly compelling evidence that climate change not only existed but was accelerating. Because of the worthwhile merits and significance of Keeling's work, NOAA began monitoring CO_2 levels worldwide in the 1970s. There are currently about 100 sites continuously monitored today because of what was started on Mauna Loa.

Ralph Keeling, Charles Keeling's son, says, "The Keeling Curve has become an icon of the human imprint on the planet and a continuing resource for the study of the changing global carbon cycle. Mauna Loa provides a valuable lesson on the importance of continuous earth observations in a time of accelerating global change" (Keeling, 2008).

The Intergovernmental Panel on Climate Change (IPCC) has projected that with continued climate change, temperatures will rise an average of 1.4–5.8°C in the next 100 years. The polar regions, however, are expected to rise more than the average. Currently, with much of the polar areas being covered by snow and ice, they have a high albedo, because so much of the insolation is reflected back to space. As temperatures warm, however, the snow and ice will melt, exposing darker surfaces (land and water), which will naturally absorb more heat, which will melt more ice and uncover more dark surface area. This could continue in a cycle until all of the snow and ice are melted, greatly changing the heat balance in the polar regions, resulting in a rapid warming (IPCC, 2007).

Radiative Forcing

When discussing the enhanced greenhouse effect and climate change, climate scientists often refer to a concept called "radiative forcing." This is a measure of the influence that an independent factor (ice albedo, aerosols, land use, carbon dioxide) has in altering the balance of incoming and outgoing energy in the earth-atmosphere system. It is also used as an index of the influence a factor has as a potential climate change mechanism. Radiative forcings can be positive or negative. A positive forcing (corresponding to more incoming energy) warms the climate system. A negative forcing (corresponding to more outgoing energy) cools the climate system.

As shown in Figure 2.3, examples of positive forcings (those which warm the climate) include greenhouse gases, tropospheric ozone, water vapor,

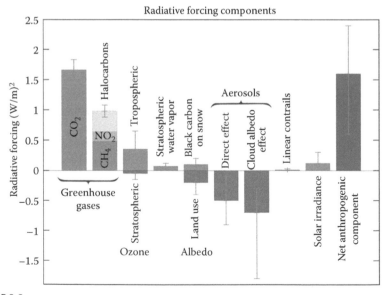

FIGURE 2.3
(See color insert.) This graphic illustrates the concept of radiative forcing. In order to understand climate change, it is necessary to understand the components in the atmosphere that control the overall warming and cooling on short- and long-term bases. (Adapted from NASA/GISS, Forcings in GISS climate model, http://data.giss.nasa.gov/modelforce/.)

solar irradiance, and anthropogenic effects. Those that have a negative rating (they cool the climate) include stratospheric ozone and certain types of land use. Climatologists often use radiative forcing data to compare various cause-and-effect scenarios in climate systems where some type of change (warmer or cooler) has taken place. Radiative forcings and their effects on climate are also built into climate models in order to enable climatologists to determine the effects from multiple input factors on climate change. The greenhouse gases that remain in the atmosphere the longest will have the greatest impact on climate change, making it imperative that climatologists not only be able to identify them but also understand them and how they react with the atmosphere.

The Earth's Energy Balance

Even if greenhouse gas emissions were stabilized at today's rates, climate change would not stop. According to scientists at NASA, CO_2 levels are rising because there is more CO_2 being emitted into the atmosphere than natural processes such as photosynthesis and absorption into the oceans can presently remove. There is no longer a carbon balance; there is a net gain that continues to increase each year. As the CO_2 levels increase, the temperatures continue to rise.

Because the earth's carbon system is already so out of balance, in order to stop climate change, emissions must actually be reduced over the coming years, not just stabilized. Even with the best of intentions, the earth's temperature would not react immediately—there is a delayed reaction. This occurs because there is already an abundance of excess energy stored in the world's oceans. This lag time responding to the reaction is referred to as thermal inertia. Because of thermal inertia, scientists at NASA have determined that the 0.6–0.9°C of climate change that has already occurred this past century is not the full amount of warming the environment will eventually reach from the greenhouse gases that have already been emitted into the atmosphere. Even in the extreme case that all greenhouse emissions stopped immediately, the earth's average surface temperature would continue to climb another 0.6°C or more over the next several decades before temperatures leveled off.

The half-life of CO_2 is about 100 years. Therefore, most of the CO_2 being released today will still be in the atmosphere in 2111. Although there are natural factors that control the CO_2 levels and climate, NASA supports the notion that human influence is changing the climate on earth—many agree that today's documented climate change trend is at least partly anthropogenic in origin. The IPCC says: "The balance of evidence suggests that there is a discernible human influence on global climate" (IPCC, 2007).

According to Dr. James E. Hansen, this lag time is a key reason why it is risky to hold off any longer trying to control greenhouse gas emissions (Hansen, Lacis, Rind, and Russell, 2011). The longer society waits to take positive action to stop climate change, the more severe and long-lasting the consequences will become. Hansen stresses that the time to act is now, not sometime in the future.

If climate change is allowed to continue, there will be many negative effects, such as the disappearance of ecosystems, leading to the extinction of many species. There will be a greater number of heat-related deaths; there will be a greater spread of infectious diseases, such as encephalitis, malaria, and dengue fever through the proliferation of disease-carrying mosquitoes. Malnutrition and starvation will increase as a result of droughts. There will be loss of coastal areas because of sea-level rise from melting ice caps and thermal expansion of the oceans. Agricultural production could become unpredictable, threatening food supplies, and water could be contaminated. Impacts will be health related, social, political, ecological, environmental, and economic.

Greenhouse Gases

Trace gases in the atmosphere act like the glass in a greenhouse. The trace gases serve to trap the heat energy from the sun close to earth. Most greenhouse gases occur naturally and are cycled through the global biogeochemical system. It is the greenhouse gases added by human activity that are trapping too much heat today and causing the atmosphere to overheat. There are several different types of greenhouse gases; some exist in greater quantities than others. They include water vapor, CO_2, methane, nitrous oxide, and halocarbons. They capture 70–85 percent of the energy in rising thermal radiation emitted from the earth's surface.

Water vapor—the most common greenhouse gas—accounts for roughly 65 percent of the natural greenhouse effect. When water heats up, it evaporates into vapor and rises from the earth's surface into the atmosphere, where it forms clouds and acts as an insulating blanket to help keep the earth warm. Water vapor can also reflect and scatter incoming sunlight. This is why cloudy nights are warmer than clear nights. As water vapor condenses and cools, it then comes back to the earth as snow or rain and continues on its way through the water cycle.

The second most prevalent greenhouse gas, CO_2, comprises one-fourth of the natural greenhouse effect. Humans and animals exhale CO_2, vegetation releases CO_2 when it dies and decomposes, and burning trees in a forest fire or burning during deforestation release CO_2. Burning fossil fuels (such as exhaust from cars and industrial processes) are another common source of CO_2.

Methane is a colorless, odorless, flammable gas, formed when plants decay in an environment of restricted air. It is the third most common greenhouse gas and is created when organic matter decomposes without the presence of oxygen—a process called anaerobic decomposition. One of the most common sources is from "ruminants"—grazing animals that have multiple stomachs in which to digest their food. These include cattle, sheep, goats, camels, bison, and musk ox. In their digestive system, their large fore-stomach hosts tiny microbes that break down their food. This process creates methane gas, which is released as flatulence. Livestock also emit methane when they belch. In fact, in one day a single cow can emit one-half pound of methane into the air. Each day 1.3 billion cattle burp methane several times per minute.

Methane is also a by-product of natural gas and decomposing organic matter, such as food and vegetation. Also present in wetlands, it is commonly referred to as "swamp gas." Since 1750, methane has doubled its concentration in the atmosphere, and it is projected to double again by 2050. According to Nick Hopwood and Jordan Cohen at the University of Michigan, every year 350–500 million tons of methane are added to the atmosphere through various activities, such as the raising of livestock, coal mining, drilling for oil and natural gas, garbage sitting in landfills, and rice cultivation (Hopwood and Cohen, 2008). Because rice is grown in waterlogged soils, like swamps, they release methane as a by-product.

Nitrous oxide (N_2O), another greenhouse gas, is released from manure and chemical fertilizers that are nitrogen-based. As the fertilizer breaks down, N_2O is released into the atmosphere. Nitrous oxide is also contained in soil by bacteria. When farmers plow the soil and disturb the surface layer, N_2O is released into the atmosphere. It is also released from catalytic converters in cars and from the ocean. According to Hopwood and Cohen, nitrous oxide has risen more than 15 percent since 1750. Each year 6–12 million metric tons is added to the atmosphere principally through the use of nitrogen-based fertilizers, the disposal of human and animal waste in sewage treatment plants, automobile exhaust, and other sources that have not yet been identified.

Halocarbons include the fluorocarbons, methylhalides, carbon tetrachloride, carbon tetrafluoride, and halons. They are all considered to be powerful greenhouse gases because they strongly absorb terrestrial infrared radiation and stay in the atmosphere for many decades.

Fluorocarbons are a group of synthetic organic compounds that contain fluorine and carbon. A common compound is chlorofluorocarbon (CFC). This class contains chlorine atoms and has been used in industry as refrigerants, cleaning solvents, and propellants in spray cans. These fluorocarbons are harmful to the atmosphere, however, because they deplete the ozone layer, and their use has been banned in most areas of the world, including the United States.

Hydrofluorocarbons (HFCs) contain fluorine and do not damage the ozone layer. Fluorocarbon polymers are chemically inert and electrically insulating. They are used in place of CFCs because they do not harm or break down ozone molecules, but they do trap heat in the atmosphere. HFCs are used in air conditioning and refrigerators. The best way to keep them out of the atmosphere is to recycle the coolant from the equipment in which they are used.

Fluorocarbons have several practical uses. They are used in anesthetics in surgery, as coolants in refrigerators, as industrial solvents, and as lubricants, water repellants, stain repellants, and chemical reagents. They are used to manufacture fishing line and are contained in products like Gore-Tex and Teflon.

Climate Change Potential

Although there are several types of greenhouse gases, they are not the same. They all have different properties associated with them, and the differences are critical, especially their half-lives—the duration of time they remain active in the atmosphere. They also differ significantly in the amount of heat they can trap.

Because many greenhouse gases are extremely potent and can remain in the atmosphere for thousands of years the U.S. Environmental Protection Agency (EPA, 2011) says they can can be 140–23,900 times more potent that CO_2 in terms of their ability to trap and hold heat—it is important to note that their effects can remain very long term. It is important to note that these gases and their effects will continue to increase in the atmosphere as long as they continue to be emitted and remain there. Even though these gases represent a very small proportion of the atmosphere—less than 2 percent of the total—because of their enormous heat-holding potential, they are a significant component to the atmosphere and represent a serious problem for climate change.

The EPA has identified three major groups of high global warming potential (GWP) gases: (1) HFCs, (2) perfluorocarbons (PFCs), and (3) sulfur hexafluoride (SF_6). They represent the most potent greenhouse gases, and the PFCs and SF_6 also have extremely long atmospheric lifetimes—up to 23,900 years. Because their lifetimes are so incredibly long, for practical management purposes, once they are emitted into the atmosphere, they are considered to be there permanently. According to the EPA, once present in the atmosphere, the result is "an essentially irreversible accumulation" (EPA, 2011).

HFCs are man-made chemicals; most of them were developed as replacements for the prior used ozone-depleting substances that were common in industrial, commercial, and consumer products. The GWP index for HFCs ranges from 140 to 11,700, depending on which one is used. Their lifetime in the atmosphere ranges from 1 to 260 years. Most of the commercially used HFCs have a lifetime of less than 15 years, such as HFC-134a, which is used in automobile air conditioning and refrigeration.

Perfluorocarbons generally originate from the production of aluminum and semiconductors. PFCs have very stable molecular structures and usually do not get broken down in the lower atmosphere. When they reach the mesosphere 60 kilometers above the earth's surface, high-energy ultraviolet electromagnetic energy destroys them, but it is a very slow process that enables them to accumulate in the atmosphere for up to 50,000 years.

Sulfur hexafluoride has a GWP of 23,900, making it the most potent greenhouse gas. It is used in insulation, in electric-power transmission equipment, in the magnesium industry, in semiconductor manufacturing to create circuitry patterns on silicon wafers, and also as a tracer gas for leak detection. Its accumulation in the atmosphere shows the global average concentration has increased by 7 percent per year during the 1980s and 1990s—according to the IPCC—from less than one part per trillion (ppt) in 1980 to almost 4 ppt in the late 1990s (IPCC, 2007).

In order to understand the potential impact from specific greenhouse gases, they are rated as to their GWP. The GWP of a greenhouse gas is the ratio of climate change—or radiative forcing—from one unit mass of a greenhouse gas to that of one unit mass of CO_2 over a period of time, making the GWP a measure of the *potential for climate change per unit mass relative to CO_2*. In other words, greenhouse gases are rated on how potent they are compared to CO_2.

GWPs take into account the absorption strength of a molecule and its atmospheric lifetime. Therefore, if methane has a GWP of 23 and carbon has a GWP of 1 (the standard), this means that methane is 23 times more powerful than CO_2 as a greenhouse gas. The IPCC has published reference values for GWPs of several greenhouse gases. Reference standards are also issued and supported by the United Nations Framework Convention on Climate Change, as shown in Table 2.1. The higher the GWP value, the larger the infrared absorption and the longer the atmospheric lifetime. As shown in this table, even small amounts of SF_6 and HFC-23 can contribute a significant amount to climate change.

In response to climate change, the EPA is working to reduce the emission of high GWP gases because of their extreme potency and long atmospheric lifetimes. High GWP gases are emitted from several different sources. Major emission sources of these today are from industries such as electric-power generation, magnesium production, semiconductor manufacturing, and aluminum production.

In electric-power generation, SF_6 is used in circuit breakers, gas-insulated substations, and switchgear. During magnesium metal production and casting, SF_6 serves as a protective cover gas during the processing. It improves safety and metal quality by preventing the oxidation and potential burning of molten magnesium in the presence of air. It replaced sulfur dioxide (SO_2), which was more environmentally toxic. The semiconductor industry uses many high GWP gases in plasma etching and in cleaning chemical vapor-deposition tool chambers. They are used to create circuitry patterns. During primary aluminum production, GWP gases are emitted as by-products of the smelting process.

The best solution found to date to solve the negative impact to the environment and combat climate change is by the EPA working with private industry in business partnerships that involve developing and implementing new processes that are environmentally friendly. In addition, the EPA

TABLE 2.1

Climate Change Potential of Greenhouse Gases

Greenhouse Gas	Lifetime in the Atmosphere	GWP over 100 Years (Compared with CO_2)
Carbon dioxide	50–200 years	1
Methane	12+/–3 years	21
Nitrous oxide	120 years	296
CFC 115	550 years	7,000
HFC-23	264 years	11,700
HFC-32	5.6 years	650
HFC-41	3.7 years	150
HFC-43-10mee	17.1 years	1,300
HFC-125	32.6 years	2,800
HFC-134	10.6 years	1,000
HFC-134a	14.6 years	1,300
HFC-152a	1.5 years	140
HFC-143	3.8 years	300
HFC-143a	48.3 years	3,800
HFC-227ea	36.5 years	2,900
HFC-236fa	209 years	6,300
HFC-245ca	6.6 years	560
Sulfur hexafluoride	3,200 years	23,900
Perfluoromethane	50,000 years	6,500
Perfluoroethane	10,000 years	9,200
Perfluoropropane	2,600 years	7,000
Perfluorobutane	2,600 years	7,000
Perfluorocyclobutane	3,200 years	8,700
Perfluoropentane	4,100 years	7,500
Perfluorohexane	3,200 years	7,400

Source: Houghton, J.T. et al. (eds.) *Climate Change 1995: The Science of Climate Change.* Cambridge, U.K.: Cambridge University Press.

is also working to limit high GWP gases through mandatory recycling programs and restrictions. If a greenhouse gas can remain in the atmosphere for several hundred years, even though it may be in a small amount, it can do a substantial amount of damage. Some of the greenhouse effect today is due to greenhouse gases put in the atmosphere decades ago. Even trace amounts can add up significantly.

Carbon Sequestration

Because CO_2 is the most prevalent greenhouse gas in the atmosphere, it is important to understand carbon: what it is used for, where it comes from, where it goes, and how it interacts with other elements. As it relates to climate change, the concept of carbon sequestration—or carbon storage—is especially important. If carbon can be stored, it is removed from the environment and therefore removed from being a potential source of greenhouse gas. It becomes a sink—a storage place for carbon. For effective management issues, it is important to understand which places store carbon (carbon sink) and which places release carbon (carbon source). Carbon sinks play a direct role in climate change—the more that can be sequestered, the less there is available to contribute to climate change.

Sinks and Sources

In terms of climate change, storing CO_2 in reservoirs—or sinks—is desirable in order to keep the CO_2 out of the atmosphere so that it does not contribute to ever-increasing atmospheric temperatures. There are several mechanisms whereby carbon can be put into storage.

During the process of photosynthesis, plants convert CO_2 from the atmosphere into carbohydrates and release oxygen during the process. Trees in forests over a period of years are able to store large amounts of carbon, which is why promoting the regrowth of old forests and encouraging the growth of new forests are positive methods of combating climate change. Vegetation sequesters 544 billion metric tons of carbon each year. In fact, the regrowth of forests in the Northern Hemisphere is the most significant anthropogenic sink of atmospheric CO_2. When forests are destroyed during the process of deforestation and the land is cleared for other uses such as grazing, it removes a valuable carbon sink from the ecosystem. In areas where the forests are cleared through burning, this process serves to release the CO_2 back into the atmosphere.

In the oceans, as living organisms with shells die, pieces of the shells break apart and fall toward the bottom of the ocean, slowly accumulating as a sediment layer on the bottom. Small amounts of plankton settle through the deep water and come to rest of the floors of the ocean basins. Even though this source does not represent a huge source of carbon, over thousands or millions of years it does add up, becoming another useful long-term carbon store. The world's oceans are currently holding huge amounts of CO_2, helping to minimize the effects of climate change. As the oceans become saturated, however, they will not be able to store as much CO_2 as they have previously. Currently about half of all the anthropogenic carbon emissions are absorbed by the ocean and the land.

A sink can also become a source if the situation changes. Carbon can enter the cycle from several different sources. One of the most common is through the respiration process of plants and animals. Once a plant or animal dies, carbon enters the system. Through the decaying process, the bacteria and fungi break down the carbon compounds. If oxygen is present, the carbon is converted into CO_2. If oxygen is not present, then methane is produced instead. Either way, both are greenhouse gases. A major source into the carbon cycle is through the combustion of biomass. This oxidizes the carbon, which produces CO_2. The most prevalent source of this is through the use of fossil fuels—coal, petroleum products, and natural gas (Figure 2.4). Fossil fuels are large deposits of biomass that have commonly been preserved inside geologic rock formations in the earth's crust for millions of years. Although they are preserved in rock formations, there is no release of carbon to the atmosphere, but as soon as they are processed into fuel products and used for transportation and industrial processes, millions of tons are

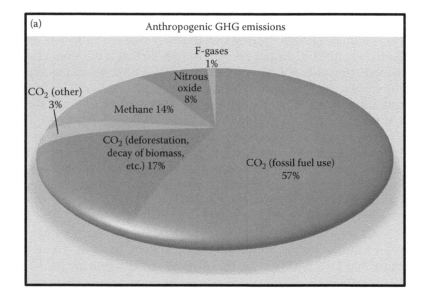

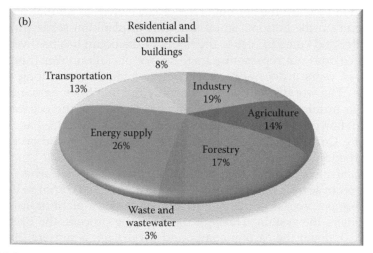

FIGURE 2.4

(a) Anthropogenic greenhouse gas emissions by type. (b) Global greenhouse-gas concentrations by sector.

released into the atmosphere. According to the EPA, the burning of fossil fuels and deforestation releases nearly eight billion metric tons of carbon into the atmosphere each year (Hotinski, 2009).

Another major source of CO_2 to the atmosphere is from forest fires. When trees burn, the stored carbon is released to the atmosphere (Figure 2.5). Other gases, such as carbon monoxide and methane, are also released. After a fire has occurred and deadfall is left behind, it will further decompose and release additional carbon to the atmosphere. One of the predictions with climate change is that as temperatures rise, certain areas will

FIGURE 2.5
Increasing temperatures will cause mountain environments to become drier, increasing the highly flammable tinder in forest areas. Both natural lightning strikes and human carelessness will increase the risk of forest fires. The drier conditions will make the fires spread more quickly and be harder to extinguish, putting both the natural environment and bordering urban areas at extreme risk and causing millions of dollars in damage. (Courtesy of Kelly Rigby, Bureau of Land Management, Utah State Office, Salt Lake City, Utah.)

experience increased drought conditions. When this occurs, vegetated areas will become extremely dry; then, when lightning strikes, wildfires can easily start, destroying vegetation, increasing the concentration of CO_2 gases in the atmosphere, and warming it further.

Another source of carbon occurs near urban settings. When cement is produced, one of its principal components—lime (produced from the heating of limestone)—produces CO_2 as a by-product. Therefore, in urban areas this is another factor to consider; areas of increasing urbanization are areas of increasing CO_2 concentrations.

Volcanic eruptions can be a significant source of CO_2 during a major event. Enormous amounts of gases, including greenhouse gases such as water vapor, CO_2, and sulfur dioxide, can be released into the atmosphere. CO_2 can also be released from the ocean's surface into the atmosphere from bursting bubbles. Phytoplankton, little organisms floating in the world's oceans, also use some of the CO_2 to make their food through photosynthesis.

The earth's zones of permafrost are another potential source of CO_2. According to the Oak Ridge National Laboratory in Tennessee, if climate temperatures continue to rise and cause the permafrost soils in the boreal (northern) forests to thaw, the carbon released from them could dramatically increase atmospheric CO_2 levels and ultimately affect the global carbon balance.

According to Mike Goulden, a scientist working on a NASA project to determine the net exchange of carbon in a Canadian forest, "The soils of boreal forests store an enormous amount of carbon" (Cheshire, 1998).

Goulden says the amount of carbon stored in the soil is greater than both the carbon stored in the moss layer and wood of the boreal forests. Boreal forests cover about 10 percent of the earth's land surface and contain 15–30 percent of the carbon stored in the terrestrial biosphere. Most of the carbon is located in deep layers 40–80 centimeters below the surface, where it remains frozen—the zone called permafrost.

The reason that the permafrost layer is so high in carbon is that the decomposing plant material accumulates on the forest floor and becomes buried before it decomposes. With the steady increase in atmospheric temperatures over recent years, some of the permafrost has been experiencing deep-soil warming and been melting in mid-summer. This process releases CO_2 through increased respiration.

According to Goulden, "Assessing the carbon balance of forest soil is tricky, but if you've got several pieces of evidence all showing the same thing, then you can really start to believe that the site is losing soil carbon. Our findings on the importance of soil thaw and the general importance of the soil carbon balance is already influencing the development of models for carbon exchange." What Goulden is hoping to accomplish in the future is to find a way to use satellite data to improve carbon exchange models (Cheshire, 1998).

There are several human-related sources of CO_2. By far, the largest human source is from the combustion of fossil fuels, such as coal, oil, and gas. This occurs in electricity generation, in industrial facilities, from automobiles in the transportation sector, and for heating homes and commercial buildings. According to the EPA, in 2006, petroleum supplied the largest share of domestic energy demands in the United States. This equates to 47 percent of the total fossil-fuel-based energy consumption during that year. Coal and natural gas followed, at 27 and 26 percent, respectively. These are all heavy producers of CO_2. Electricity generation is the single largest source of CO_2 emissions in the United States, according to the EPA. This is one reason why the EPA began their Energy Star® program, geared toward the production and use of energy efficient appliances.

Industrial processes that add significant amounts of CO_2 to the atmosphere include manufacturing, construction, and mining activities. The EPA has identified six industries in particular that use the majority of energy sources, thereby contributing the largest share of CO_2: (1) chemical production, (2) petroleum refining, (3) primary metal production, (4) mineral production, (5) paper production, and (6) the food processing industry.

In the residential and commercial sector, the main source of direct CO_2 emissions is the burning of natural gas and oil for the heating and cooling of buildings. The transportation sector is the second largest source of CO_2 emissions, primarily because of the use of petroleum products such as gasoline, diesel, and jet fuel. Of this, 65 percent of the CO_2 emissions are from personal automobiles and light-duty trucks.

Specific commodities in the industrial sector that produce CO_2 emissions include cement and other raw materials that contain calcium carbonate that is chemically transformed during the industrial process, which produces CO_2 as a by-product. Industrial processes that use petroleum products also contribute CO_2 emissions to the atmosphere. This includes the production of solvents, plastics, and lubricants that have a tendency to evaporate or dissolve over a period of time. The EPA has identified four principal industrial processes that produce significant CO_2 emissions:

- Production of metals, such as aluminum, zinc, lead, iron, and steel
- Production of chemicals, such as petrochemicals, ammonia, and titanium dioxide
- Consumption of petroleum products in feedstocks and other end uses
- Production and consumption of mineral products, such as cement, lime, and soda ash

Also important are industrial processes that produce other harmful greenhouse gases as by-products, such as methane, nitrous oxide, and fluorinated gases. These gases have a far greater global warming potential than CO_2. Methane's GWP is 23, nitrous oxide's is 296, and fluorinated gases range from 6,500 to 8,700, posing an even more critical negative effect.

Types of Carbon Sequestration

Carbon sequestration is the process through which CO_2 from the atmosphere is absorbed by various carbon sinks. Principal carbon sinks include agricultural sinks, forests, geologic formations, and oceanic sinks. Carbon sequestration or storage occurs when CO_2 is absorbed by trees, plants, and crops through photosynthesis and stored as carbon in biomass, such as tree trunks, branches, foliage, and roots, as well as in the soil. In terms of climate change and impacts to the environment, sequestration is very important because it has a large influence on the atmospheric levels of CO_2. According to the IPCC, carbon sequestration by forestry and agriculture alone significantly helps to offset CO_2 emissions that contribute to climate change. The amount of carbon that can be sequestered varies geographically and is determined by tree species, soil type, regional climate, type of topography, and even the type of land management practice used in the area. For example, in agricultural areas, if conservation tillage practices are used instead of conventional tillage, this limits the introduction of CO_2 into the atmosphere by sequestering larger amounts of CO_2 in the soil. According to the EPA, switching from conventional to conservation tillage can sequester 0.1–0.3 metric tons of carbon per acre per year.

Carbon sequestration does reach a limit, however. The amount of carbon that accumulates in forests and soils will eventually reach a saturation point where no additional carbon will be able to be stored. This typically occurs when trees reach full maturity or when the organic matter contained in soils builds up. According to the EPA, the United States landscape currently functions as an efficient carbon sink, sequestering more than it emits. The EPA does warn, however, that the overall sequestration amounts are currently declining because of increased harvests, land-use changes, and maturing forests. As far as global sequestration, the IPCC has estimated that 100 billion metric tons of carbon over the next 50 years could be sequestered through forest preservation; tree planting; and improved, conservation-oriented agricultural management. In the United States, Bruce McCarl (professor of Agricultural Economics at Texas A&M University) and Uwe Schneider (assistant professor of the Research Unit Sustainability and Global Change Department of Geosciences and Economics at Hamburg University in Germany) have determined that an additional 50–150 million metric tons of carbon could be sequestered through changes in both soil and forest management, new tree planting, and biofuel substitution (Schneider and McCarl, 2003).

Another positive aspect supporting carbon sequestration is that it also affects other greenhouse gases. In particular, methane (CH_4) and N_2O can also be sequestered in agricultural activities such as grazing and the growing of crops. Nitrous oxide can be introduced via fertilizers. Instead of using these fertilizers, which can have a negative environmental effect, other practices could be used instead, such as rotational grazing. In addition, if forage quality is improved, livestock methane emissions should be significantly reduced. Nitrous-oxide emissions could be avoided by eliminating the need for fertilizer. The EPA stresses that finding the right sequestration practices will help lessen the negative effects of all the greenhouse gases.

Other environmental benefits of carbon sequestration include the enhancement of the quality of soil, water, air, and wildlife habitat. For instance, when trees are planted, they not only sequester carbon, but they also provide wildlife habitat. When the rain forests are preserved, it keeps both plant and animal species from becoming endangered and helps control soil erosion. When forests are maintained it also cuts down on overland water flow, soil erosion, loss of nutrients, and improved water quality.

The continuation of climate change, however, can have an impact on carbon sequestration. According to the EPA: "In terms of global warming impact, one unit of CO_2 released from a car's tailpipe has the same effect as one unit of CO_2 released from a burning forest. Likewise, CO_2 removed from the atmosphere through tree planting can have the same benefit as avoiding an equivalent amount of CO_2 released from a power plant" (EPA, 2011).

The experts at the EPA also caution, however, that even though forests, agriculture, and other sinks can store carbon, the process can also become saturated and slow down or stop the storage process (such as traditional agricultural cultivation) or the sink can be destroyed and completely stop the process (such as complete deforestation). Carbon sequestration processes can naturally slow down and stop on their own as they age. In addition, carbon sequestration can be a natural or man-made process. Research is currently in progress to perfect the methodologies that enhance the natural terrestrial cycles of carbon storage that remove CO_2 from the atmosphere by vegetation and store CO_2 in biomass and soils. In order to accomplish this, research of biological and ecological processes are under way by the EPA. Specific technical areas that are currently being researched include:

- Increasing the net fixation of atmospheric CO_2 by terrestrial vegetation with emphasis on physiology and rates of photosynthesis of vascular plants,
- Retaining carbon and enhancing the transformation of carbon to soil organic matter,
- Reducing the emission of CO_2 from soils caused by heterotrophic oxidation of soil organic carbon; and
- Increasing the capacity of deserts and degraded lands to sequester carbon. Man-made processes include technologies such as geologic, mineral, and ocean sequestration.

In carbon sequestration, the main goal is to prevent CO_2 emissions from power plants and industrial facilities from entering the atmosphere by separating and capturing the emissions and then securing and storing the CO_2 on a long-term basis. Currently, the EPA is involved in research in an attempt to separate and capture the CO_2 from fossil fuels and from flue gases—these are both pre- and post-combustion processes. Underground storage facilities are also receiving large amounts of attention recently, and their potential is enormous. As an example, today more than 2.8 trillion liters of both hazardous and nonhazardous fluids are disposed of through a process called underground injection. The economics of carbon sequestration is discussed in greater detail in Chapter 12.

Impacts of Deforestation

Deforestation is occurring today at alarming rates, determined through the analysis of satellite imagery. It accounts for about 20 percent of the heat-trapping gas emissions worldwide. The Food and Agriculture Organization (FAO) estimates current tropical deforestation at 15,400,000 hectares per year. This equates to an area roughly the size of North Carolina being deforested annually (Figure 2.6). Tropical forests hold enormous amounts of carbon.

FIGURE 2.6
Deforestation, shown here in Madagascar, is a serious problem and has become so widespread that its connection to global warming cannot be overlooked. These huge resultant erosional features are called lavaka. The world's rain forests must be managed properly to avoid this kind of environmental damage. (From Rhett Butler, Mongabay.com.)

In fact, the plants and soil of tropical forests hold 460 to 575 billion metric tons of carbon worldwide. This equates to each acre of tropical forest storing roughly 180 metric tons of carbon.

When a forest is cut and replaced by pastures or cropland, the carbon that was stored in the tree trunks joins with the oxygen and is released into the atmosphere as CO_2. Because the wood in a tree is about 50 percent carbon, deforestation has a significant effect on climate change and the global carbon cycle. In fact, according to the Tropical Rainforest Information Center, from the mid-1800s to 1990, worldwide deforestation released 122 billion metric tons of carbon into the atmosphere. Currently 1.6 billion metric tons is released to the atmosphere each year. As a comparison, all of the fossil fuels (coal, oil, and gas) release about 5 billion metric tons per year. Tropical deforestation is the largest source of emissions for many developing countries (Tropical Rainforest Information Center, 2011).

According to Peter Frumhoff of the UCS, he and an international team of eleven research scientists found that if deforestation rates were cut in half by 2050, it would amount to 12 percent of the emissions reductions needed to keep the concentrations of heat-trapping gases in the atmosphere at relatively safe levels (Union of Concerned Scientists, 2007).

There are multiple impacts of deforestation. Many of the most severe impacts will be to the tropical rain forests. Even though they cover only about 7 percent of the earth's land surface, they provide habitat for about 50 percent of all the known species on earth. Some of these endemic species have become so specialized in their respective habitat niches, that if the climate changes, which then causes the ecosystem to also change, this will threaten the health and existence of multiple species, even to the point of forcing them to become extinct. In addition to the species that are destroyed, the ones which remain behind in the isolated forest fragments become vulnerable and sometimes are threatened with extinction themselves. The outer margins of the remaining forests become dried out, and are also subjected to die-off.

Two other major impacts in addition to the loss of biodiversity are the loss of natural resources, such as timber, fruit, nuts, medicine, oils, resins, latex, and spices; and the economic impact that causes along with the extreme reduction of genetic diversity. The loss of genetic diversity could mean a huge loss for the future health of humans. Hidden in the genes of plants, animals, fungi, and bacteria that may not have even been discovered yet could be the cures for diseases like cancer, diabetes, muscular dystrophy, and Alzheimer's disease. The United Nations FAO says, "The keys to improving the yield and nutritional quality of foods may also be found inside the rain forest and it will be crucial for feeding the nearly ten billion people the earth will likely need to support in the coming decades" (Lindsey, 2009).

Two of the largest climate impacts will revolve around rainfall and temperature. One-third of the rain that falls in the tropical rain forest is rain that was generated in the water cycle by the rain forest itself. Water is recycled locally as it is evaporated from the soil and vegetation, condenses into clouds, and falls back to earth again as rain in a repetitive cycle. The evaporation of the water from the earth's surface also acts to cool the earth. As climatologists continue to learn about the earth's climate and the effects of climate change, they are able to build better models. When the tropical rain forests are replaced by agriculture or grazing many climate models predict that these types of land-use changes will perpetuate a hotter, drier climate. Models also predict that tropical deforestation will disrupt the rainfall patterns outside the tropics, causing a decrease in precipitation—even to far-reaching destinations such as China, northern Mexico, and the south-central United States.

Predictions involving deforestation can get complicated in models, however. For instance, if deforestation is done in a "patchwork" pattern, then local isolated areas may actually experience an increase in rainfall by creating "heat islands" which increase the rising and convection of air that causes clouds and rainstorms. If rainstorms are concentrated over cleared areas, the ground can be vulnerable and susceptible to erosion.

The carbon cycle plays an important role in the rain forests. According to NASA, in the Amazon alone, the trees contain more carbon than ten years' worth of human-produced greenhouse gases. When the forests are cleared and burned, the carbon is returned to the atmosphere, enhancing the greenhouse effect. If the land is utilized for grazing it can also be a continual source of additional carbon. It is not certain today whether or not the tropical rain forests are a net source or sink of carbon. While the vegetation canopies hold enormous amounts of carbon, trees, plants, and microorganisms in the soil also respire and release CO_2 as they break down carbohydrates for energy. In the Amazon alone, enormous amounts of CO_2 escape from decaying organic matter in rivers and streams that flood huge areas. When tropical forests remain undisturbed, they remain essentially carbon neutral; but when deforestation occurs it contributes significantly to the atmosphere.

Rain-forest countries need to give serious consideration about what decisions need to be made for the future. Currently, Papua New Guinea, Costa Rica, and several other forest-rich developing countries are seeking financing from the global carbon market (such as the United States) to create attractive financial incentives for tropical rain forest conservation. Developed nations helping developing nations is one approach to help get a handle on the problem, but developed countries also need to cut back on their own emissions.

Anthropogenic Causes and Effects: Carbon Footprints

When looking for causes of climate change, it is easy to point fingers and put the blame on industry, other nations, transportation, deforestation, and other sources and activities, but the truth of the matter is that every person on earth plays a part and contributes to climate change. Even simple daily activities—such as using an electric appliance, heating or cooling a home, or taking a quick drive to the grocery store—contribute CO_2 to the atmosphere. Scientists refer to this input as a "carbon footprint." A carbon footprint is simply a measure of how much CO_2 people produce just by going about their daily lives. For every activity that involves the combustion of fossil fuels (coal, oil, and gas), such as the generation of electricity, the manufacture of products, or any type of transportation, the user of the intermediate or end product is leaving a carbon footprint. Of all the CO_2 found in the atmosphere, 98 percent originates from the burning of fossil fuels. Simply put, it is one measure of the impact people make individually on the earth by the lifestyle choices they make. In order to combat climate change, every person on earth can play an active role by consciously reducing the impact of their personal carbon footprint. The two most common ways of achieving this is by increasing their home's energy efficiency and driving less. A carbon footprint is calculated (carbon footprint calculators are available on the Internet), and a monthly, or annual, output of total CO_2 in tons is derived based on the specific daily activities of that person. The goal then is to reduce or eliminate their carbon footprint. Some people attempt to achieve "carbon neutrality," which means they cut their emissions as much as possible and offset the rest. Carbon offsets allow one to "pay" to reduce the global greenhouse total instead of making personal reductions. An offset is bought, for example, by funding projects that reduce emissions through restoring forests, updating power plants and factories, reducing the energy consumption in buildings, or investing in more energy-efficient transportation. In order to educate and make people more environmentally conscious, some companies are even advertising today what their carbon footprint is, drawing in additional business support because of their positive environmental commitments. Some commercial products today contain "carbon labels" estimating the carbon emissions that were involved in the creation of the product's production, packaging, transportation, and future disposal.

Carbon footprints are helpful because they allow individuals to become more environmentally aware of the implication of their own choices and actions and enables them to adopt behaviors that are more environmentally friendly—what is referred to as "going green." For example, transportation in the United States accounts for 33 percent of CO_2 emissions. Ways to make a difference include driving less, using public transportation, carpooling, driving a fuel-efficient car such as a hybrid, or bicycling. According to the EPA (2009), home energy use accounts for 21 percent of CO_2 emissions in the United States. Therefore, cutting down in these areas by increasing energy efficiency helps lessen the carbon footprint. There are several practical ways to do this, such as lowering the thermostat, installing double-paned windows, and installing good insulation, to name a few. Using compact fluorescent lamps and using energy efficient appliances (such as those listed on the Energy Star® program) also increases efficiency and lowers the carbon footprint. Carbon footprints can be a helpful measurement for those who want to take personal initiative and do their part to fight climate change.

The timing is right for individuals to become aware of their personal behavior and how it impacts the environment. The atmospheric concentration of CO_2 has risen by more than 30 percent in the last 250 years. Based on a study conducted by Michael Raupach of the Commonwealth Scientific and Industrial Research Organisation in Australia, worldwide carbon emissions of anthropogenic CO_2 are rising faster than previously predicted (Global Carbon Project, 2003). From 1990 to 1999, the increase in CO_2 levels averaged about 1.1 percent per year, but from 2000 to 2004, levels increased 3 percent per year. For this research, the world was divided into nine separate regions for analysis of specific factors such as economic factors, population trends, and energy consumption. The result of the study showed that the developed countries, which only account for 20 percent of the world's population, accounted for 59 percent of the anthropogenic global emissions in 2004. Developing nations were responsible for 41 percent of the total emissions in 2004, but contributed 73 percent of the emissions growth that year. Developing countries, such as India, are expected to become the major CO_2 contributors in the future. Today, the largest CO_2 emitter is China.

This study is significant because even the IPCC's most extreme predictions underestimated the rapid increase in CO_2 levels seen since 2000. The scientists involved in the study believe this shows that no countries are decarbonizing their energy supply and that CO_2 emissions are accelerating worldwide.

Also from Australia's Commonwealth Scientific and Industrial Research Organisation, Joseph G. Canadell (2007) calculated that CO_2 emissions were 35 percent higher in 2006 than in 1990, also a faster growth rate than expected. Canadell attributed this to an increased industrial use of fossil fuels and a decline in the amount of CO_2 being absorbed by the oceans or sequestered on land.

According to Canadell, "In addition to the growth of global population and wealth, we now know that significant contributions to the growth of atmospheric CO_2 arise from the slowdown of nature's ability to take the chemical out of the air. The changes characterize a carbon cycle that is generating stronger-than-expected and sooner-than-expected climate forcing" (Schmid, 2007).

In response to the study, Kevin Trenberth from the National Center for Atmospheric Research said, "The paper raises some very important issues that the public should be aware of: namely that concentrations of CO_2 are increasing at much higher rates than previously expected and this is in spite of the Kyoto Protocol that is designed to hold them down in Western countries" (MSNBC, 2007a).

According to the Brookings Institution, a research organization in Washington, DC, America's carbon footprint is expanding. As America's population grows and cities expand, they are building more, driving more, and consuming more energy, which means they are emitting more CO_2 than ever before. The Brookings Institution believes that the existing federal policies are currently limiting. They believe federal policy should play a more powerful role in helping metropolitan areas so that the country as a whole can collectively shrink their carbon footprint. They believe that besides economy-wide policies to motivate action, five targeted policies should be put in place that are extremely important within metro areas (Brown et al., 2008):

- Promote a wider variety of transportation choices
- Design more energy-efficient freight operations
- Require home energy cost disclosure when selling a home in order to encourage more energy-efficient appliances in homes
- Use federal housing policies to create incentives to build with both energy and location conservation in mind

Challenge metropolitan areas to develop (and reward them for developing) innovative solutions toward reducing carbon footprints.

Areas Most at Risk

About 634 million people—10 percent of the global population—live in coastal areas that are within just 10 meters above sea level, where they are most vulnerable to sea-level rise and severe storms associated with climate change. Three-quarters of these are in the low-elevation coastal zones in Asian nations on densely populated river deltas, such as India. The others most at risk are the small island nations. According to The Earth Institute at Columbia University, the ten countries with the largest number of people living within 10 meters of the average sea level in descending order are:

- China
- India
- Bangladesh
- Vietnam
- Indonesia
- Japan
- Egypt
- United States
- Thailand
- Philippines

The ten countries with the largest share of their population living within 33 feet (10 m) of the average sea level are:

- Bahamas (88%)
- Suriname (76%)
- Netherlands (74%)
- Vietnam (55%)
- Guyana (55%)
- Bangladesh (46%)
- Djibouti (41%)
- Belize (40%)
- Egypt (38%)
- Gambia (38%)

Fossil Fuels and Climate Change

The burning of fossil fuels (oil, gas, and coal) is one of the leading contributors to climate change, because fossil fuels are composed almost entirely of carbon and release CO_2 into the atmosphere when they are burned. Oil also contains toxic materials that when burned or when inhaled as fumes are known to cause cancer in humans.

In developed countries such as the United States, fossil fuels are the principal sources of energy used to produce the majority of fuel, electricity, heat, and air conditioning. In fact, more than 86 percent of the energy used worldwide originates from fossil-fuel combustion. Although for years fossil fuels have been readily available and convenient, they have also played a major role in climate change over the past few decades. According to the Center for Biological Diversity (CBD), fossil fuel use in the United States causes more than 80 percent of the greenhouse gas emissions and 98 percent of just the CO_2 emissions. This adds approximately 4.1 billion metric tons of CO_2 to the atmosphere each year and would even be greater if some of the earth's natural carbon sequestration processes such as carbon storage in the world's oceans, vegetation, and soils did not occur (CBD, 2008).

Even though scientists warn that climate change is already under way and will likely continue for the next several centuries because of the long natural processes involved, such as the long lifetimes of many of the greenhouse gases, there are ways the potential effects can be reduced. Because everyone uses energy sources every day, the biggest way to reduce the negative effects of climate change is by using less energy. By cutting back on the amount of electricity used, the use of fossil fuels can be reduced because most power plants burn fossil fuels to generate power. The most effective way to half the emission of greenhouse gases into the atmosphere is through the adoption of non-fossil-fuel energy sources, such as hydroelectric power, solar power, hydrogen engines, and fuel cells.

Former Vice President Al Gore, author of *An Inconvenient Truth,* and recipient of the 2007 Nobel Peace Prize (received jointly with the IPCC) said, "It is the most dangerous challenge we've ever faced, but it is also the greatest opportunity we have had to make changes" (MSNBC, 2007a).

Since the IPCC's last report was released identifying fuels as a principal cause of climate change with a virtual certainty (99% certainty) (IPCC, 2007), upgraded from its 2001 report of likely (66%), there have been several achievements and advancements made. The levels of research have grown, public awareness has increased, the subject has been incorporated into many school curriculums worldwide, and legislation—local, national, and international—has been passed and is being currently introduced into governments around the world. In addition to the 2007 Nobel Peace Prize shared by Al Gore and the IPCC, the film *An Inconvenient Truth* received two Oscar nominations at the 2007 Academy Awards in Los Angeles, California. Climate change issues are finally receiving the media attention they deserve, making the public more aware of the real issues and the reasons why they need to be addressed now.

Growing public education and awareness has not solved all the problems, however. Even though the public is becoming more educated, the opposition and skeptics are also raising their voices in protest, continuing to cloud the issues, making it important for the public to pay attention to the facts. In light of this, however, many cities worldwide, foreign countries, and individual states in the United States are taking action to curb fossil-fuel emissions. Arnold Schwarzenegger, former governor of California, for example, ordered the world's first low-carbon limits on passenger car fuels. The new standard will reduce the carbon content of transportation fuels at least 10 percent by the year 2020.

Climatologists at Lawrence Livermore National Laboratory in California have created a climate and carbon-cycle model to examine global climate and carbon-cycle changes. What they concluded was that if humans continued with the same lifestyles and habits to which they are accustomed today (commonly referred to as a business-as-usual approach), the earth's atmosphere would warm by 8°C if humans use all of the earth's available fossil fuels by the year 2300.

Their model predicted several alarming results: in the next few centuries, the polar ice caps will have vanished, ocean levels will rise by 7 meters, and the polar temperatures will climb higher than the predicted 8–20°C, transforming the delicate ecosystems from polar and tundra to boreal forest.

Govindasamy Bala, the Laboratory's Energy and Environment Directorate and lead author of the project, said, "The temperature estimate was actually conservative because the model did not take into consideration changing land use such as deforestation and build-out of cities into outlying wilderness areas" (TerraDaily, 2005).

Their model projected that by 2300, the CO_2 level will have risen to 1,423 ppm, roughly a 400 percent increase. The model identified the soil and biomass as significant carbon sinks, but according to Bala, the land ecosystem would not take up as much carbon dioxide as the model assumes, because they did not take land-use change into account.

The results of the model showed that ocean uptake of CO_2 starts to decrease in the twenty-second and twenty-third centuries as the ocean surface warms. It takes longer for the ocean to absorb CO_2 than it does for the vegetation and soil. By 2300, the land will absorb 38 percent of the CO_2 released from the burning of fossil fuels, and 17 percent will be absorbed by the oceans. The remaining 45 percent will stay in the atmosphere. Over time, roughly 80 percent of all CO_2 will end up in the oceans via physical processes, increasing its acidity and harming aquatic life.

According to Bala, the most drastic changes during the 300-year period will occur during the twenty-second century—when precipitation patterns change, when an increase in the amount of atmospheric precipitable water and a decrease in the size of sea ice are the largest, and when emission rates are the highest. Based on the model's results, all sea ice in the Northern Hemisphere summer will have vanished by 2150.

When referring to climate change skeptics, Bala says, "Even if people don't believe in it today, the evidence will be there in 20 years. These are long-term problems. We definitely know we are going to warm over the next 300 years. In reality, we may be worse off than we predict" (TerraDaily, 2005).

The New Carbon Balance: Summing It All Up

Carbon dioxide enters the air during the carbon cycle. Because it is so plentiful, it comes from several sources. Vast amounts of carbon are stored naturally in the earth's soils, oceans, and sediments at the bottoms of oceans. Carbon is stored in the earth's rocks and is released when the rocks are eroded. It exists in all living matter. Every time animals and plants breathe, they exhale carbon dioxide. The earth maintains a natural carbon balance. Throughout geologic time, when concentrations of CO_2 have been disturbed, the system had always gradually returned to its natural (balanced) state. This natural readjustment works very slowly.

Through a process called diffusion, various gasses that contain carbon move between the ocean's surface and the atmosphere. Because of this, plants in the ocean use CO_2 from the water for photosynthesis, which means that ocean plants store carbon, just as land plants do. When ocean animals eat these plants, they then store the carbon. Then when they die, they sink to the bottom and decompose, and their remains become incorporated in the sediments on the bottom of the ocean. Once in the ocean, the carbon can go through various processes: it can form rocks and then weather, and it can also be used in the formation of shells. Carbon can move to and from different depths of the ocean and also exchange with the atmosphere.

As carbon moves through the system, different components can move at different speeds. Scientists break these reaction times down into two categories: short-term cycles and long-term cycles. In short-term cycles, carbon is exchanged quickly. Examples of this include the gas exchange between the oceans and atmosphere (evaporation). Long-term cycles can take years to millions of years to occur. Examples of this would be carbon stored for years in trees or carbon weathered from a rock being carried to an ocean, being buried, incorporated into plate tectonic systems, and then later released into the atmosphere through a volcanic eruption.

Throughout geologic time, the earth has been able to maintain a balanced carbon cycle. Unfortunately, this natural balance has been upset by recent human activity. Over the past 150–200 years, fossil-fuel emissions, land-use change, and other human activities have increased atmospheric carbon dioxide by 30 percent (and methane, another greenhouse gas, by 150 percent) to concentrations not seen in the past 420,000 years.

Humans are adding CO_2 to the atmosphere much faster than the earth's natural system can remove it. Prior to the Industrial Revolution, atmospheric carbon levels remained constant at around 280 ppm. This meant that the natural carbon sinks were balanced between what was being emitted and what was being stored. After the Industrial Revolution began and carbon dioxide levels began to increase—315 ppm in 1958 to 383 ppm in 2007 to 392 ppm today—the "balancing act" became unbalanced, and the natural sinks could no longer store as much carbon as was being introduced into the atmosphere by human activities, such as driving and industry. In addition, according to Dr. Joseph Canadell of the National Academy of Sciences, 50 years ago for every ton of CO_2 emitted, 600 kg were removed by natural sinks (Center for Climate Change and Environmental Studies, 2011). In 2006, only 550 kg were removed per ton, and the amount continues to fall today, which indicates that the natural sinks are losing their carbon storage efficiency. This means that although the world's oceans and land plants are absorbing great amounts of carbon, they cannot keep up with what humans are adding. The natural processes work much more slowly than the human ones do. The earth's natural cycling usually takes millions of years to move large amounts from one system to another. The problem with human interference is that the impacts are happening in only centuries or decades, and the earth cannot keep up with the fast pace. The result is that each year the measured CO_2 concentration of the atmosphere gets higher, making the earth's atmosphere get warmer.

Levels of several greenhouse gases have increased by about 25 percent since large-scale industrialization began about 150 years ago. According to the National Energy Information Center, Energy Information Administration (EIA), 75 percent of the anthropogenic CO_2 emissions added to the atmosphere over the past 20 years are due to the burning of fossil fuels (Casper, 2009).

According to the EIA, natural earth processes can absorb approximately 3.2 billion metric tons of anthropogenic CO_2 emissions annually. An estimated

6.1 billion metric tons is added each year, however, creating a large imbalance, which is why there is a steady, continual growth of greenhouse gases in the atmosphere. In computer models, an increase in greenhouse gases results in an increase in the average temperature on the earth. The warming that has occurred over the past century is largely attributed to human activity. According to a study conducted by the National Research Council in May 2001: "Greenhouse gases are accumulating in earth's atmosphere as a result of human activities, causing surface air temperatures and sub-surface ocean temperatures to rise. Temperatures are, in fact, rising. The changes observed over the last several decades are likely mostly due to human activities" (Energy Information Administration, 2004). Table 2.2 shows the United States energy-related CO_2 emissions by fossil fuel type.

The EIA has determined that greenhouse gas emissions originate principally from energy use, driven mainly by economic growth, fuel used for electricity generation, and weather patterns affecting heating and cooling needs. Energy-related CO_2 emissions, resulting from petroleum and natural gas, represent 82 percent of the total U.S. human-made greenhouse gas emissions. According to the EIA, the graph represents historical and projected world CO_2 emissions by the top CO_2-emitting countries from 1990 to 2025.

Tables 2.3 and 2.4 illustrate CO_2 emissions caused by the use of fossil fuels by the world's top 20 emitters as well as major worldwide geographic regions as of 2006 based on data collected from the EIA and the Oak Ridge National Research Laboratory.

TABLE 2.2

U.S. Energy-Related Carbon Dioxide Emissions by
Fossil Fuel (Million Metric Tons CO_2)

Year	Petroleum	Coal	Natural Gas	Total
1995	2,206	1,893	1,192	5,301
1996	2,287	1,976	1,215	5,489
1997	2,309	2,025	1,225	5,570
1998	2,352	2,045	1,198	5,607
1999	2,414	2,046	1,198	5,669
2000	2,458	2,140	1,239	5,848
2001	2,469	2,084	1,190	5,754
2002	2,468	2,093	1,245	5,820
2003	2,513	2,130	1,216	5,872
2004	2,604	2,155	1,196	5,966
2005	2,621	2,163	1,179	5,974
2006	2,586	2,132	1,158	5,888
2007	2,583	2,154	1,234	5,984

Source: Energy Information Administration, *Annual Energy Review 2007*, U.S. Department of Energy, Washington DC, 2008.

TABLE 2.3

The Top Twenty Carbon Dioxide Emitters

Country	CO_2 Emissions (Million Metric Tons of Carbon Dioxide)
China (mainland)	6,017.69
United States	5,902.75
Russian Federation	1,704.36
India	1,293.17
Japan	1,246.76
Germany	857.60
Canada	614.33
United Kingdom	585.71
Republic of Korea	514.53
Iran	471.48
Italy (including San Marino)	468.19
South Africa	443.58
Mexico	435.60
Saudi Arabia	424.08
France (including Monaco)	417.75
Australia	417.06
Brazil	377.24
Spain	372.62
Ukraine	328.72
Indonesia	280.36

Developing new technologies that create energy by using fossil fuels more efficiently alone is not enough to control the emissions of greenhouse gases being ejected into the atmosphere. Without the increased use of renewable energy resources and the weaning off of the dependence on fossil fuels, it will not be possible to bring climate change under control

TABLE 2.4

World CO_2 Emissions by Geographic Region, 2001–2006

Geographic Region	2001	2002	2003	2004	2005	2006
North America	6,697.34	6,782.00	6,870.49	6,970.01	7,034.15	6,954.03
Central and South America	1,015.58	1,005.01	1,022.68	1,065.71	1,110.78	1,138.49
Europe	4,559.17	4,532.33	4,678.65	4,713.13	4,717.46	4,720.85
Eurasia	2,332.38	2,354.20	2,470.57	2,528.65	2,599.84	2,600.65
Middle East	1,118.75	1,175.37	1,240.40	1,330.10	1,444.16	1,505.30
Africa	922.55	924.10	974.71	1,024.82	1,061.61	1,056.55
Asia and Oceania	7,607.73	8,050.28	8,806.46	9,820.89	10,517.00	11,219.56
World total	24,253.49	24,823.30	26,063.96	27,453.30	28,485.00	29,195.42

in time to prevent irreversible damage to the environment and life on this planet.

The Concern about Coal Emissions

According to Andrew Revkin's article in *The New York Times* on December 4, 2007, titled "Stuck on Coal, Stuck for Words in a High-Tech World," whereas society of the twenty-first century has progressed technologically in many ways over the past century, society is still stuck in the old-fashioned, out-dated mode of popular energy choices (Revkin, 2007). In particular, even with new technology such as the growing cadre of renewable forms of energy, societies worldwide are still heavily dependent on coal and plan to remain that way despite the repeated pleas and warnings from climatologists about the future life-threatening consequences of climate change if changes are not made now.

Using coal-fired plants to generate electricity produces more greenhouse gases for each resulting watt than using oil or natural gas, but coal—often referred to as a "dirty fuel"—is attractive because coal is relatively inexpensive. In countries where there are no emission controls (such as China and India), the coal industry is booming today. The International Energy Agency projects that the demand for coal will grow by 2.2 percent a year until 2030, which is a rate faster than the demand for oil or natural gas. However, even with warnings to cut back on coal use and greenhouse gas emissions, according to the UCS, the United States currently has plans to increase its emissions by building many more coal-fired plants, with the majority of them lacking the carbon capture and storage technology—equipment that allows a plant to capture a certain amount of CO_2 before it is released and then store it underground (Union of Concerned Scientists, 2008).

One of the most serious contenders of coal use now and in the future is China. New figures from the Chinese government reveal that coal use has been climbing faster in China than anywhere else in the world. According to a report in *The Economist*, China opens a new coal-fired plant each week (*The Economist*, 2007). Their rising energy consumption is making it more difficult to effectively slow the climate change process. The International Energy Agency in Paris predicts that the increase in greenhouse gas emissions from 2000 to 2030 in just China will be comparable to the increase from the entire industrialized world (Bradsher, 2003).

China is currently the world's largest consumer of coal, and its power plants are burning it faster than the trains can deliver it from the mines in China. As a result, it is also importing coal from Australia to meet its rising demands. Concerning energy products, it has also become the world's fastest-growing importer of oil. The Chinese are using more energy in their homes than ever before, and with a population of one and a quarter billion people, energy usage is expected to skyrocket. This increase in energy usage will affect other energy-related sectors. China, for example, is now the world's largest market for television sets and one of the largest for other electrical appliances. It also has the world's fastest-growing automobile market. All of these commodities require the use of fossil fuels to manufacture and operate their industries, either directly or indirectly. Energy generated from coal and oil is expected to have a significant impact on climate change.

China is not the only country with a growing demand for energy, however. India, Brazil, and Indonesia are other countries and regions that are also showing a surging demand for energy. Power plants are burning increasing amounts of coal to meet the exploding new demands for electricity to serve both industry and private households. According to the EIA (2011), China's electricity generation and consumption, mostly from coal, has increased more than 110 percent since 2000. They also predict total net generation to be three times higher in 2035 than it was in 2009. China currently uses more coal than the United States, the European Union, and Japan combined (Bradsher and Barboza, 2006a).

If China's carbon usage keeps up with its current economic growth, their CO_2 emissions are projected to reach eight gigatons a year by 2030, an amount equal to the entire world's CO_2 production today. In 2000, steel production in China was reported at 127 million metric tons, and in 2006, they produced 380 million metric tons. In 2008, China's production led the world at 444 million metric tons, which is twice as much as the United States and Japan combined (Tang, 2010). In addition to new construction, steel is also being used for the manufacture of cars. In 1999, Chinese consumers bought 1.2 million cars. In 2006, 7.2 million cars were sold—an increase of 600 percent.

In a report in *The New York Times*, pollution from China's coal-fired plants is already affecting the world. In April 2006, a thick cloud of pollutants originating from northern China drifted airborne to Seoul, Korea, and then across the Pacific Ocean to the West Coast of the United States. Scientists were able to track the progression and route of this "brown cloud" via real-time satellite imagery. According to researchers in California, Oregon, and Washington, a coating of sulfur compounds, carbon, and other by-products of coal combustion were found on mountaintop detectors in the Pacific Northwest.

Steven S. Cliff, an atmospheric scientist at the University of California at Davis said, "The filters near Lake Tahoe in the mountains of eastern California are the darkest we've ever seen outside smoggy urban areas" (Bradsher and Barboza, 2006b).

The sulfur dioxide produced during coal combustion poses an immediate threat to China's population, contributing to roughly 400,000 premature deaths each year. In addition, it causes acid rain that poisons rivers, lakes, wetland ecosystems, agricultural areas, and forest ecosystems. The CO_2 coming from China will exist in the atmosphere for decades. According to the report, concerning China's newly growing economy, "Coal is China's double-edged sword—the new economy's black gold and the fragile environment's dark cloud" (Bradsher and Barboza, 2006a).

Health Issues Associated with Climate Change

One component of preparing for climate change involves planning for new or increasing threats to human health. As ecosystems change through intense weather events, shifting habitats, wildfires, heat waves, and other effects, humans must be prepared for inevitable changes. There is an overwhelming amount of evidence that rising levels of CO_2 in the earth's atmosphere are having a serious impact on the climate with secondary effects on the earth's physical systems and ecosystems, such as an increase in severe weather events, rising sea levels, migration and extinction of both plant and animal species, shifts in climate patterns, and melting of glaciers and permafrost. Another serious impact that has been clearly identified related to climate change and rising amounts of CO_2 in the atmosphere are the negative health effects being experienced by populations worldwide.

The segments of the population that are at the greatest risk of health problems from the atmospheric changes associated with climate change, such as increased air particulates, greenhouse gases, and pollution, are young children and the elderly of 65 years of age or older. Because young children are still developing physically and breathe faster than adults, they are more at risk to the adverse effects of air quality and extreme temperatures. According to the Environmental Defense Fund, there are four key health-related factors associated with climate change: heat waves, smog and soot pollution, food- and water-borne diseases, and stress from post-traumatic stress disorder (PTSD).

Infants and children four years old and younger are extremely sensitive to heat. When heat waves occur and when children are subjected to the urban heat-island effect—the situation in which urban areas are warmer than rural areas because asphalt pavements, buildings, and other human-made structures absorb incoming solar radiation and re-emit the energy as longwave (heat) radiation (see Chapter 11)—they face the risk of becoming rapidly dehydrated and can suffer the negative—sometimes fatal—effects of heat exhaustion or stroke. In addition, because young children's lungs are still developing, they can also suffer irreversible lung damage from being exposed to smog and soot pollution and breathing it into their lungs.

As climate change increases in intensity and food- and water-borne diseases spread into areas where they have never existed before, children are much more susceptible to becoming ill—especially those living in poverty. As extreme weather events occur—such as hurricanes and flooding—and families are left homeless, children are especially prone to complications of PTSD as they try to cope with the upheaval of their lives and often the loss of their homes.

The elderly, age 65 and older, are also at greater risk than the general population. By the year 2030, one-fifth of the U.S. population is projected to be older than 65. Because the elderly often have more frail health and less mobility, they are at greater risk to the negative effects of heat waves. If the income of an elderly person is limited, and they cannot afford air conditioning in their home, their health and safety can be in jeopardy because the elderly are more sensitive to changes in temperature and cannot physically adjust to extremes like younger people can.

Another sector of the population at risk are those with chronic health conditions. People, for example, with heart problems, respiratory illnesses, diabetes, or compromised immune systems are more likely to suffer serious health complications because of the effects of climate change than healthy people. Those who suffer from diabetes are at greater risk of death from heat waves. Stagnant hot air masses and areas with high ozone and soot concentrations pay a heavy toll on those with heat and respiratory illnesses. According to Dr. John Balmes of the American Lung Association of California, higher smog levels "may cause or exacerbate serious health problems, including damage to lung tissue, reduced lung function, asthma, emphysema, bronchitis, and increased hospitalizations for people with cardiac and respiratory illnesses" (Environmental Defense Fund, 2004).

Smog forms when sunlight, heat, and relatively stagnant air meet up with nitrogen oxides and various volatile organic compounds. Exposure to smog can do serious damage to lungs and respiratory systems. In the case of inflammation and irritation to the lungs, it can cause shortness of breath, throat irritation, chest pains, and coughing and lead to asthma attacks, hospital admissions, and emergency room visits. These consequences are even more severe if people are exposed while being active. More hot days mean better conditions for creating smog that can trigger asthma and other breathing problems.

Those with preexisting conditions of weakened immune systems, such as those suffering from AIDS or cancer, are also highly susceptible to catching diseases spread by mosquitoes or other vectors.

The Environmental Defense Fund has identified other groups as being potentially more vulnerable to the negative effects of pollution, illness, and climate change:

- Pregnant women and their unborn children: They may be unable to take specific medications or not get to properly air conditioned locations.
- People living in poverty: They may not have access to air conditioning or ready access to medical assistance.
- People living in areas of chronic pollution: They may have increased exposure to infectious diseases, and consistent exposure to unhealthy air conditions may be compromising.
- People living in geographic areas prone to harmful climate change: These areas may experience violent storms, coastal flooding and erosion, damage to buildings, and contamination of drinking water.

TABLE 2.5

Potential Regional Effects of Climate Change in the United States

Geographic Region	Potential Negative Effects
Southeast Atlantic and Gulf Coast	Violent storms, strong storm surges and flooding, coastal erosion, more damage to buildings and roads, contamination of drinking water
Southwest	Higher temperatures, less rainfall, hot, arid climate, increased wildfires, worsened air quality
Northwest	Heavy rainfall, flooding, sewage overflow, increased illness, spread of disease
Great Plains	Milder winters, scorching summers, decreased agricultural production, intense heat waves
Northeast	Higher temperatures, more allergies, spread of disease by insects and animals
Alaska	Melting permafrost, retreating sea ice, disturbed ecosystems, reduced subsistence hunting and fishing, milder temperatures, increase in insects and forest pests such as the spruce bark beetle

Source: Environmental Defense Fund, U.S. National Assessment of the Potential Consequences of Climate Variability and Change, regions and mega-regions, http://www.usgcrp.gov/usgcrp/nacc/background/regions.htm, 2003.

Table 2.5 shows the risks associated with potential climate change in different regions in the United States, according to a report issued by the U.S. Global Change Research Program (2009).

It is becoming more common today for individual cities to prepare emergency plans outlining actions that need to be taken if the negative effects of climate change and global warming affect their specific geographic area. According to the Environmental Defense Fund, several cities in the United States have already put together working Heat Health Watch/Warning System plans to make their populations prepared in order to deal with various climate issues when necessary. Cities already taking proactive action include: Dallas/Fort Worth, Texas; Cincinnati/Dayton, Ohio; Chicago, Illinois; Jackson, Mississippi; Lake Charles, Louisiana; Little Rock, Arkansas; Memphis, Tennessee; New Orleans, Louisiana; Philadelphia, Pennsylvania; Phoenix, Arizona; Portland, Oregon; St. Louis, Missouri; Shreveport, Louisiana; Seattle, Washington; and Yuma, Arizona (Trust for America's Health, 2009).

Local topography also plays a role in the pollution that an area experiences. Cities located in close proximity to mountain ranges experience unique patterns of recurrent pollution. Large cities that commonly experience photochemical smog—the brown air that often results from car, bus, and truck exhaust (composed of nitrogen oxide, oxygen, hydrocarbons, and sunlight to produce ozone, which can be deadly at lower elevations), as well as sulfurous fog (created from coal-burning plants), are often found with the negative consequences of air pollution and ill effects on health. Currently, there are some controls implemented on automobiles and coal-fired plants. Catalytic converters are used in some areas. They convert carbon monoxide, nitrous oxide, and hydrocarbons into CO_2, nitrogen, and water. Some cars recirculate exhaust along with a catalytic converter to reduce emissions.

In areas where the existence of mountain ranges is a factor, coal-fired plants may burn only low sulfur-content coal. They may also use a process that converts the coal directly to gas (gasification process) or use scrubbing technologies.

In areas where mountain ranges act as physical barriers and trap pollution over cities, air inversions are common, especially during the winter months. In valleys or on the lee side of mountains, if a warmer air mass moves above cooler air, it traps the cooler, denser air underneath and increases the severity of the air pollution. Los Angeles is an example of this, where the warm desert air from the east comes over the mountains to the east of Los Angeles and lies over the cooler Pacific Ocean air. The cooler air becomes trapped because it cannot rise through the less-dense warm air above it, and the pollution in the cold air accumulates. In mountain valleys, a similar situation occurs where warm air overlies the colder air that accumulates in the valleys. In cities, heat-island effects are common. Warm air filled with pollutants collects and then spreads out over the nearby suburbs. The greenhouse gases contributing the most to the anthropogenic greenhouse effect are listed in Table 2.6.

Based on research conducted at the Scripps Institution of Oceanography in San Diego, California, a new analysis of the pollution-filled "brown clouds" over southern Asia now offers hope that the region may be able to slow or stop some of the alarming retreat of glaciers in the region by reducing the existing air pollution (*Scripps News*, 2007).

Leading the research team, Dr. V. Ramanathan, a chemistry professor at Scripps, concluded from the research, "Concerning the rapid melting of these glaciers, the third-largest ice mass on the planet, if it becomes

TABLE 2.6

Gases Contributing to the Anthropogenic Greenhouse Effect

Gas	Rate of Increase (% per year)	Relative Contribution (%)
CO_2	0.5	60
CH_4	1	15
N_2O	0.2	5
O_3	0.5	8
CFC-11 (trichlorofluoromethane)	4	4
CFC-12 (trichlorofluoromethane)	4	8

widespread and continues for several more decades, will have unprecedented downstream effects on southern and eastern Asia" (*Scripps News*, 2007).

According to Achim Steiner, the United Nations undersecretary general and executive director of the United Nations Environment Programme, "The main cause of climate change is the buildup of greenhouse gases from the burning of fossil fuels. But brown clouds, whose environmental and economic impacts are beginning to be unraveled by scientists, are complicating, and in some cases aggravating, their effects. The new findings should spur the international community to even greater action. For it is likely that in curbing greenhouse gases we can tackle the twin challenges of climate change and brown clouds and in doing so, reap wider benefits from reduced air pollution to improved agricultural yields" (MSNBC, 2007b).

Jay Fein, program director in the National Science Foundation's Division of Atmospheric Sciences, remarked, "In order to understand the processes that can throw the climate out of balance, Ramanathan and colleagues, for the first time ever, used small and inexpensive unmanned aircraft and their miniaturized instruments as a creative means of simultaneously sampling clouds, aerosols, and radiative fluxes in polluted environments, from within and from all sides of the clouds. These measurements, combined with routine environmental observations and a state-of-the-science model, led to these remarkable results" (National Science Foundation, 2006).

What the study was successfully able to reveal was that the effect of the brown cloud was necessary to explain temperature changes that have been observed in the region over the last 50 years. It also clarified that southern Asia's warming trend is more pronounced at higher altitudes than closer to sea level. Ramanathan concluded, "This study reveals that over southern and eastern Asia, the soot particles in the brown clouds are intensifying the atmospheric warming trend caused by greenhouse gases by as much as 50 percent."

According to research done by the World Resources Institute, it was not until the late 1940s when air pollution disasters occurred on two separate continents that public awareness began to grow concerning outdoor air quality and its effects on human health (World Resources Institute, 2002). Both the 1948 "killer fog" in Denora, Pennsylvania, that killed 50 people and the London "fog" in 1952, where roughly 4,000 people died, spurred an investigation into the cause, and it was determined that the widespread use of dirty fuels was to blame. This began the concerted effort for governments to take the problem of urban air pollution seriously (Figure 2.7).

Since that time, many contaminants in the atmosphere have been identified as harmful, and serious efforts have been undertaken to clean up the atmosphere from harmful components. The most common and damaging pollutants include sulfur dioxide, suspended particulate matter, ground-level

FIGURE 2.7
A man guiding a bus with a flaming torch through thick fog during the London smog of 1952. (From Getty Images. Photo by Monty Fresco.)

ozone, nitrogen dioxide, carbon monoxide, and lead. All of these pollutants are tied either directly or indirectly to the combustion of fossil fuels. Even though major efforts are under way today to clean polluted air over cities, many cities worldwide still lack a healthy air quality. An inventory completed by the European Environment Agency determined that 70–80 percent of 105 European cities surveyed exceeded World Health Organization (WHO) air quality standards for at least one pollutant. In the United States, an estimated 80 million people live in areas that do not meet U.S. air quality standards, which are comparable with WHO standards. Other areas that do not meet WHO standards include Beijing, Delhi, Jakarta, and Mexico City. In these cities, pollutant levels sometimes exceed WHO air quality standards by a factor of three or more. Some of the cities in China exceed WHO standards by a factor of six. Worldwide, WHO estimates that up to 1.4 billion urban residents breathe air exceeding the WHO air guidelines and that the health consequences are considerable, with a mortality rate of 200,000–570,000 annually. In addition, the World Bank has estimated that exposure to particulate levels exceeding the WHO health standard accounts for roughly 2–5 percent of all deaths in urban areas in the developing world (World Resources Institute, 2002).

It is stressed, however, that these mortality estimates do not reflect the huge toll of illness and disability that exposure to air pollution brings on a global level. Health effects span a wide range of severity from coughing to bronchitis to heart disease and lung cancer. In developing cities alone, air pollution is responsible for approximately 50 million cases per year of chronic coughing in children younger than 14 years of age. Taking the health effects into consideration concerning climate change is critical because it affects the future of society.

Contributors to Climate Change and Pollution

The transportation sector is the largest single source of air pollution in the United States today. It causes almost 67 percent of the carbon monoxide (CO), a third of the nitrogen oxides (NO_x), and a fourth of the hydrocarbons in the atmosphere.

Cars and trucks pollute the air during manufacturing, oil refining and distribution, refueling, and, most of all, vehicle use. Motor vehicles cause both primary and secondary pollution. Primary pollution is that which is emitted directly into the atmosphere; secondary pollution is from chemical reactions between pollutants in the atmosphere.

There are six major pollutants: ozone (O_3), particulate matter (PM), NO_x, CO, SO_2, and hazardous air pollutants (toxics). The primary ingredient in smog is O_3. Particulate matter is particles of soot, metals, and pollen. The finest, smallest textured PM does the most damage, because it travels into the lungs easily. NO_x tend to weaken the body's defense against respiratory infections. CO is formed by the combustion of fossil fuels such as gasoline and is emitted by cars and trucks. When inhaled, it blocks the transport of oxygen to the brain, heart, and other vital organs, making it deadly. SO_2 is created via the burning of sulfur-containing fuels, especially diesel. It forms fine particles in the atmosphere and is harmful to children and those with asthma. Toxic compounds are chemical compounds emitted by cars, trucks, refineries, and gas pumps and have been related to birth defects, cancer, and other serious illnesses. The EPA estimates that the air toxics emitted from cars and trucks account for half of all cancers caused by air pollution.

According to the UCS, pollution from light trucks is growing quickly. This class of vehicles includes minivans, pickups, and sport-utility vehicles. Today, these vehicles represent one of every two new vehicles purchased. Because of this, California regulators and the EPA recently created new rules requiring light trucks to become as clean as cars over the next seven to nine years.

Even though for many years there have been air pollution control efforts, 92 million Americans still live in areas with chronic smog problems. Today, even with current control programs in effect, more than 93 million people live in areas that violate health standards for ozone (urban smog), and more than 55 million Americans suffer from unhealthy levels of fine particle pollution.

Trucks and buses are responsible for a large amount of toxic pollution. Although they account for less than 6 percent of the miles driven by highway vehicles in the United States, trucks and buses are responsible for:

- One-fourth of smog-causing pollution from highway vehicles
- More than half the soot from highway vehicles
- Six percent of the nation's climate change pollution
- More than one-tenth of America's oil consumption

Off-highway diesel equipment is another major contributor to pollution. All types of off-highway heavy diesel equipment, such as cranes, tractors, and combines, release more fine particulate matter than highway cars and trucks combined. Emissions from this equipment have continued to climb because this equipment has not had to meet the stricter standards that highway vehicles have had to face. Today, a typical tractor emits as much soot as 250 average cars (Off Highway Trucks, 2007).

Key Impacts

As the climate change issue has escalated in recent years, scientists have been able to observe the related impacts worldwide—across all continents and oceans. As illustrated in several examples, impacts have been described and future impacts have been predicted if current activity continues.

Because of the lifetimes of the various greenhouse gases, even if all activity were stopped immediately, the effects of climate change would continue to a certain extent, because past damage has already ensured those changes. There is still a time window, however, to avoid the most catastrophic of changes, if efforts, policies, and appropriate technologies are put into effect now. Figure 2.8 illustrates the key impacts as a function of increasing global average temperature change. It details the changes that would be expected with water sources, ecosystems, food, the status of the coasts, and major health issues and trends of an incremental temperature increase up to 5°C. Resultant impacts expected to a region will vary based on the extent they are able to adapt, the rate of temperature change, and their socioeconomic structure. For example, locations better equipped to handle emergencies and crisis management will fare better than locations that are not. But even being better equipped will not spare an area from feeling the ramifications of a changing climate.

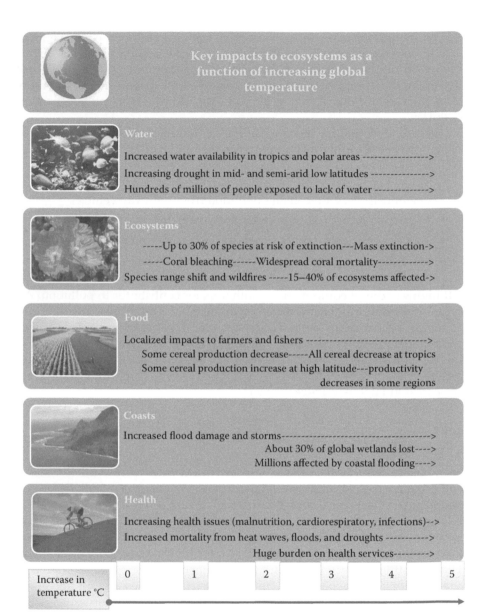

FIGURE 2.8
As temperatures increase, the impacts to the earth's various ecosystems will become more intensified, stressing natural resources to their limits and resulting in possible productive failure, loss of habitat, health disasters, disruption of food and water supplies, and extinction of life.

Conclusions

This chapter has served to illustrate the components and seriousness of the various greenhouse gases, their production, life cycles, and various contributions to climate change. Human activity has played a heavy hand in determining the atmosphere's composition and behavior since the onslaught of the Industrial Revolution, especially from 1850 until the present. Fortunately, Charles Keeling had the foresight to set up a monitoring station in Hawaii and diligently tracked atmospheric concentrations of CO_2. The results of his graph played an enormous hand in convincing the scientific community years later that climate change was real and deserved the world's attention.

As climate science entered its infancy, scientists began to see the critical role CO_2 and other greenhouse gases were playing in the larger picture. The bigger question was: could they convince the general public of what was going on and trigger what was necessary to solve the problem? Not just regulatory constraints, but major changes in mindset and the way people looked at the environment were needed. Through their subsequent research, they were able to unravel many mysteries and answer several questions, such as where carbon is sequestered (stored), how it is produced, the process by which it is released into the atmosphere, and the worldwide effects it was having.

Much progress has been made in solving and answering these problems. And finally, the public at last seems to be embracing this knowledge and choosing to take action. As new discoveries are made, it is hoped that the public will stay informed and adapt to a warming world with appropriate green responses, because those decisions and responses will shape the future landscape.

Discussion

Consider the following issues and their importance in the management of the environment:

1. Why is the integrity of data, data-collection techniques, and data analysis critical to final interpretation results of whether or not climate change is occurring for a particular geographic area or on a global scale? What are the extended ramifications of faulty techniques?
2. What is data amplification and how can it present a problem?
3. What do you think could be the root issues of why some claim climate change is not occurring and there is no need for alarm when multiple geographic locations present such overwhelming evidence to the contrary?

4. In addition to accurate scientific-process analyses, what other important factors do you see as playing a necessary role in the ultimate successful management of climate change for future generations?

5. If you were a business manager in a major energy production company and you had an opportunity to implement green energy production but were receiving pressure from traditional fossil fuel companies to maintain business with them, what would be your business decision and how would you handle it?

6. What are your views on carbon sequestration? If you were the owner of a major energy-producing company, would you consider justifying an investment in an expensive carbon sequestration project to help the environment long term if it meant sacrificing short-term revenue? If so, would your decision be based on your moral stand on the issue or the resulting public approval and attention? How would you respond to the guaranteed criticism of your nontraditional expenditure?

7. If you were the owner of a major energy corporation and were offered an inexpensive source of plentiful energy reserves at the expense of deforestation, would your priority be your business profit or protection of the rain forest ecosystem in order to mitigate long-term climate change? If you chose to protect the rain forest, how would you deal with potential angry stockholders?

8. If you functioned in the capacity of a policy maker for the EPA, which types of policies and practices would you foresee implementing concerning carbon sequestration for carbon management? What purposes would you put as priorities (e.g., forestry, farming, ranching)? Why? What are the main factors that would drive your decisions?

9. Do you think carbon sequestration technology should be funded through taxpayer revenue, even if it is experimental? Defend your reasons.

10. How could you reduce your personal carbon footprint?

11. Would you be more likely to do business with a company that advertises their greenness and carbon footprint? Why or why not?

12. Concerning promoting public awareness of climate change, what programs would you initiate to promote public awareness and encourage public involvement?

References

Bradsher, K. 2003. China's boom adds to global warming problem. *New York Times*, October 22. http://www.nytimes.com/2003/10/22/world/china-s-boom-adds-to-global-warming-problem.html (accessed September 14, 2010). Discusses China's intense industrialization efforts.

Bradsher, K., and D. Barboza. 2006a. Pollution from Chinese coal casts a global shadow. *New York Times*, June 11. http://www.nytimes.com/2006/06/11/business/worldbusiness/11chinacoal.html (accessed September 10, 2010). Discusses the environmental problems facing the world with China's unlimited use of coal.

Bradsher, K., and D. Barboza. 2006b. China's burning of coal casts a global cloud. *New York Times*, June 11. http://www.nytimes.com/2006/06/11/world/asia/11iht-coal.1947793.html (accessed August 31, 2011).

Brown, M. A., F. Southworth, and A. Sarzynski. 2008. Shrinking the carbon footprint of metropolitan America. http://www.brookings.edu/~/media/Files/rc/papers/2008/05_carbon_footprint_sarzynski/carbonfootprint_brief.pdf (accessed December 6, 2010). Discusses ways to lower the amount of CO_2 that is emitted into the atmosphere through several practical suggestions.

Canadell, J. G., C. Le Quéré, M. R. Raupach, C. B. Field, E. T. Buitenhuis, P. Ciais, et al. 2007. Contributions to accelerating atmospheric CO_2 growth from economic activity, carbon intensity, and efficiency of natural sinks. *PNAS*. 104(47): 18866–18870. doi: 10,1073/pnas.0702737104.

Casper, J. K. 2009. *Fossil Fuels and Pollution: The Future of Air Quality*. New York: Facts on File.

Center for Biological Diversity. 2008. Climate change solutions: Political and personal. http://www.biologicaldiversity.org/programs/climate_law_institute/solutions/index.html (accessed December 6, 2010). Discusses the use of fossil fuels and solutions to the problem.

Center for Climate Change and Environmental Studies. 2011. The global carbon cycle. http://www.center4climatechange.com/carboncycle.php (accessed August 31, 2011).

Cheshire, L. 1998. The dirt on carbon. NASA Earth Observatory, Oak Ridge National Laboratory. http://earthobservatory.nasa.gov/Features/DirtCarbon/ (accessed February 4, 2011). Discusses carbon stored in the boreal forests.

The Economist. 2007. Coal power still going strong: Efforts to curb greenhouse gas emissions have yet to dent enthusiasm for coal, November 15. http://www.economist.com/node/10145492 (accessed January 22, 2011). Discusses the current activity in China in building coal-fired plants.

Energy Information Administration. 2004. Greenhouse gases, climate change, and energy. http://www.eia.doe.gov/oiaf/1605/ggccebro/chapter1.html (accessed February 12, 2011). Discusses anthropogenic carbon emissions in the atmosphere.

Energy Information Administration. 2011. China energy data, statistics, and analysis. http://205.254.135.24/emeu/cabs/China/Full.html (accessed August 31, 2011).

Environmental Defense Fund. 2004. Hotter days mean unhealthier air. September 24. http://www.fightglobalwarming.com/page.cfm?tagID=242 (accessed September 15, 2010). Discusses the negative health effects of pollution.

Environmental Protection Agency. 2009. Buildings and their impact on the environment: A statistical summary. http://www.epa.gov/greenbuilding/pubs/gbstats.pdf (accessed August 29, 2011).

Environmental Protection Agency. 2011. Carbon sequestration in agriculture and forestry. http://www.epa.gov/sequestration/faq.html (accessed February 5, 2011). Provides general information and answers to commonly asked questions about climate change.

Global Carbon Project. 2003. Science framework and implementation. Earth System Science Partnership (IGBP, IHDP, WCRP, DIVERSITAS) Report No. 1. Canberra, Australia.

Hansen, J. 2007. Trial of the century—Act II. NASA Science Briefs, February. http://www.giss.nasa.gov/research/briefs/hansen_12/ (accessed March 2, 2011). Discusses the drastic buildup of CO_2 and the ramifications it has for life on earth.

Hansen, J. 2008. Twenty years later: Tipping points near on global warming. *The Huffpost Green*, June 23. http://www.huffingtonpost.com/dr-james-hansen/twenty-years-later-tippin_b_108766.html (accessed February 11, 2011). Discusses where the environment is now in reaching the point where there will be no recovery from global warming damage.

Hansen, J. E., A. A. Lacis, D. H. Rind, and G. L. Russell. Climate sensitivity to increasing greenhouse gases. *Sea Level Rise to the Year 2100*, Van Nostrand, New York, 1984. Published by U.S. EPA: http://epa.gov/climatechange/effects/downloads/Challenge_chapter2.pdf (accessed August 29, 2011).

Hopwood, N., and J. Cohen. 2008. Greenhouse gases and society. http://www.bokashicycle.com/library/GreenhouseGases.pdf (accessed August 31, 2011).

Hotinski, R. 2009. Stabilization wedges: Carbon mitigation initiative, November. http://cmi.princeton.edu/wedges/pdfs/teachers_guide.pdf (accessed February 24, 2011). Takes an innovative look at cutting back on the use of greenhouse gases.

Houghton, J. T., Meira Filho, L. G., Callander, B. A., Harris, N. Kattenberg, A., and Maskell, K., *Climate Change 1995: The Science of Climate Change. Contribution of WGI to the Second Assessment Report of the Intergovernmental Panel on Climate Change*. Cambridge, U.K.: Cambridge University Press. http://www.ipcc.ch/ipccreports/sar/wg_I/ipcc_sar_wg_I_full_report.pdf

Intergovernmental Panel on Climate Change. 2007. *WG I: The Physical Science Basis*. Cambridge, UK: Cambridge University Press.

Keeling, R. F. 2008. Recording earth's vital signs. *Science* 319: 1771–1772. Discusses the value of keeping accurate scientific data and where that can lead.

Lindsey, R. 2009. Tropical deforestation. Safe Rainforest for Our Children, March 18. http://tropical-rainforest-information.blogspot.com/2009/03/tropical-deforestation.html (accessed November 26, 2010). Provides answers to several questions about rain forest and climate change.

MSNBC. 2007a. Study: Warming is stronger. Happening sooner: Higher CO_2 emissions from fossil fuels, and weaker earth, cited as reasons, *MSNBC News*, October 22. http://www.msnbc.msn.com/id/21423872/ns/us_news-environment/ (accessed July 6, 2011). Discusses new discoveries in climate forcings.

MSNBC. 2007b. Haze added to warming in Asia, study finds: Particulates tied to energy usage—Himalayan glaciers tied to impacts. *MSNBC News*, August 6. Discusses how haze and soot are also discovered in aerosols on top of mountains, and all lead to warming.

MSNBC. 2007c. Gore, U.N. climate panel win Nobel Peace Prize. *MSNBC News*, October 12. http://www.msnbc.msn.com/id/21262661/ns/us_news-environment/ (accessed February 5, 2011). Provides feature coverage of the efforts and dedication Al Gore put forth in winning the Nobel Peace Prize.

National Science Foundation. 2006. *Autonomous Unmanned Aerial Vehicles Take to the Skies to Track Pollutants*, April 17. http://www.nsf.gov/news/news_images.jsp?cntn_id=106891&org=OLPA. Discusses the options of using light planes to collect large amounts of data remotely.

Off Highway Trucks. 2007. Off highway diesel equipment. *Heavy Equipment Machinery Information.* http://www.heavymachineryinfo.com/articles/524/1/ Off-Highway-Diesel-Equipment/Off-Highway-Diesel-Equipment.html (accessed January 16, 2011). Discusses pollution contributions from machinery.

Revkin, A. C. 2007. Stuck on coal, stuck for words in a high-tech world. *New York Times,* December 4. http://www.nytimes.com/2007/12/04/science/earth/04comm.html (accessed August 31, 2011).

Schmid, R. E. 2007. Carbon dioxide in atmosphere increasing. *Book Rags,* October 23. http://www.bookrags.com/news/carbon-dioxide-in-atmosphere-moc/ (accessed February 15, 2010). Discusses the rising trend of CO_2 in the atmosphere.

Schneider, U., and B. McCarl. 2003. The potential of U.S. agriculture and forestry to mitigate greenhouse gas emissions: An agricultural sector analysis. *Center for Agricultural and Rural Development.* 02-WP 300. http://www.card.iastate.edu/publications/DBS/PDFFiles/02wp300.pdf (accessed January 3, 2011). Discusses soil and forestry carbon sequestration.

Scripps News. 2007. Pollution amplifies greenhouse gas warming trends to jeopardize Asian water supplies, August 1. Scripps Institute of Oceanography, University of California San Diego. http://scrippsnews.ucsd.edu/Releases/?releaseID=830 (accessed August 14, 2010). Discusses the melting of the Himalayan and other Asian glaciers.

Tang, R. 2010. China's steel industry and its impact on the united states: Issues for Congress. CRS Report for Congress, 7-5700, R-41421. http://www.fas.org/sgp/crs/row/R41421.pdf (accessed August 31, 2010).

TerraDaily. 2005. Modeling of long-term fossil fuel consumption shows 14.5 degree hike in temperature, November 1. http://www.terradaily.com/news/climate-05zzzzzv.html (accessed March 13, 2011). Discusses projected warming scenarios.

Tropical Rainforest Information Center. 2011. Rain forest report card. http://www.trfic.msu.edu/rfrc/status.html (accessed October 8, 2010). Provides interesting information about tropical deforestation.

Trust for America's Health. 2009. How can we prevent and prepare for health issues in a changing climate? http://healthyamericans.org/assets/files/ClimateChangeandHealth.pdf (accessed December 15, 2010). Outlines a national health plan that takes climate change into account.

Union of Concerned Scientists. 2007. Tropical deforestation and global warming. May 15. http://www.ucsusa.org/global_warming/science_and_impacts/impacts/tropical-deforestation-and.html (accessed March 1, 2011). Discusses how deforestation must be stopped to avoid the worst effects from global warming.

Union of Concerned Scientists. 2008. Coal power in a warming world: A sensible transition to cleaner energy options, October. http://www.ucsusa.org/clean_energy/technology_and_impacts/energy_technologies/coal-power-in-a-warming-world.html (accessed November 23, 2010). Discusses coal technology and carbon capture and storage.

U.S. Global Change Research Program. 2009. *Global Climate Change Impacts in the United States.* Cambridge, UK: Cambridge University Press.

World Resources Institute. 2002. Health effects of air pollution. http://www.airimpacts.org/documents/local/HealthEffectsAirPollution.pdf (accessed October 6, 2010). Discusses pollution effects and health issues due to climate change.

Suggested Reading

Alley, R. B. 2004. Abrupt climate change. *Scientific American*. 291(5): 62–69.

Amos, J. 2007. Arctic summers ice-free by 2013. *BBC News*, December 12.

Biello, D. 2008. Out of sight, out of clime: Burying carbon in a vault of sea and rock. *Scientific American*, July 14. http://www.scientificamerican.com/article.cfm?id=undersea-carbon-capture-and-storage (accessed February 11, 2011).

Boyd, R. S. 2002. Glaciers melting worldwide, study finds. *National Geographic News*, August 21. http://news.nationalgeographic.com/news/2002/08/0821_020821_wireglaciers.html (accessed October 18, 2010).

Bridges, A. 2006. Greenland dumps ice into sea at faster pace. *Live Science*, February 16.

Environment News Service. 2009. Global Carbon Capture and Storage Institute launched in Canberra, April 17. http://www.ens-newswire.com/ens/apr2009/2009-04-16-02.html (accessed February 1, 2011). Maps out a global carbon capture and storage plan.

Gore, A. 2006. Finding solutions to the climate crisis. Speech presented to New York University School of Law, September 18. http://www.astrosurf.com/luxorion/climate-crisis-al-gore.htm (accessed October 15, 2010). Discusses the realities of climate change and the urgency to act.

Intergovernmental Panel on Climate Change. 2001. Climate change 2001: Synthesis report: Summary for policymakers. http://www.ipcc.ch/pub/un/syreng/spm.pdf (accessed December 9, 2010).

3

Climate Change and Its Effect on Ecosystems

Overview

As previously stated, one of the most misunderstood aspects of climate change is not the rise in air temperature by a few degrees—it is the sheer havoc that those "few degrees" will wreak not only on human society, but on the natural environment. One aspect of human nature is that if something is not directly affecting us, it probably will not make a lasting impression. For instance, consider this: Every time something upsets the balance in the Middle East and the price of gas skyrockets at the pumps as a result, people complain and drive less. They might put their SUVs up for sale, seriously consider buying a hybrid and adopting a greener lifestyle, and talk about the environment and recycling. Then, right after the prices fall again at the pump, people get comfortable in their "business as usual" ways and that is the end of it until the next panic at the pumps. What percentage of the general public do you suppose really takes it seriously and follows through for the environment?

If we had to trade places with someone or something that had to live with the permanent consequences of climate change every day, perhaps that would generate the interest and dedication necessary to make real and permanent changes. Knowledge is power, and that is the purpose of this chapter: to provide a basic, fundamental understanding of exactly how climate change is currently affecting—and will affect—the earth's varied ecosystems.

This chapter will discuss six unique ecosystems and the effect that climate change is currently having, what will happen as temperatures continue to rise, and what the future will look like for the inhabitants of these areas. First it will take a look at the world's forests—the temperate, boreal (northern), and tropical— so that you can get an idea of the destruction affecting these areas. Next it will focus on the world's grasslands and prairies. Following that, we will focus on the earth's most fragile ecosystem—the polar areas. Much of these areas are already feeling the disastrous effects of the warming temperatures. From there, we will travel to the world's great deserts, then to mountain areas, and finally to the diverse aquatic regions—lakes, oceans, and coastal areas—where many of the world's most heavily populated urban areas are located, so that you will see why the problem is much more than just a few-degree rise in temperature.

Introduction

The earth's ecosystems are each diverse and unique. They vary, sometimes drastically, from one another, and each offers its own blend of life forms, services, and purpose to this complicated planet. They are each rich in their own resources—irreplaceable if they are lost.

Currently, one of the biggest threats to their existence is climate change, caused principally by the action of human activity: chiefly the burning of fossil fuels but also other activities such as deforestation. Changes to ecosystems are being felt worldwide as climates are beginning to shift—some more subtle than others. Documented changes include the dying off of endemic vegetation, migration of vegetation northward (in the Northern Hemisphere), migration of animal species to cooler environments (northward in latitude and higher in elevation in mountainous areas), migration of grasslands and food belts, melting of ice caps and glaciers, loss of sea ice, loss of crucial animal habitat, increased incidences of drought and desertification, increased occurrence of heat waves, an upsurge in wildfire incidences, changes in mountain regimes, an increased lack of mountain water storage, sea-level rise and flooding in coastal regions, die-off of tropical coral reefs, and saltwater contamination in fragile wetland areas. Each one of these changes to the ecosystem they affect causes a negative impact of some type to the life that lives in them. For the birds that migrate, it might mean a two-week delay in flowering times of local plants and a subsequent lack of food supply. For a dry region suffering from the effects of drought or desertification, it may mean the lack of a food crop during a growing season desperately needed to support several villages. For the polar bear, it may mean the permanent melting of hunting and breeding grounds, restricting where they can travel and thrive—and to some it may spell death.

This chapter will illustrate the effects climate change is having, and will have, on the earth's major ecosystems, what to expect as temperatures steadily climb, and why it is so crucial that this process humans have set in motion be halted now before any more damage is done to fragile ecosystems. We may have no choices once this series of events has been set in motion.

Development

According to a special report issued by *Time* magazine on April 3, 2006, the Intergovernmental Panel on Climate Change (IPCC, 2001a) in their third report, released in 2001, had analyzed data from the past two decades representing properties such as air and ocean temperatures and the habitat characteristics and patterns of wildlife. Examples of observed changes included "shrinkage of glaciers; thawing of permafrost; later freezing and earlier break-up of ice on rivers and lakes; lengthening of mid- to high-latitude growing seasons; poleward and altitudinal shifts of plant and animal ranges; declines of some plant and animal populations; and earlier flowering of trees, emergence of insects, and egg-laying in birds. Associations between changes in regional temperatures and observed changes in physical and biological systems have been documented in many aquatic, terrestrial, and marine environments" (Kluger, 2006).

As a result of this comprehensive analysis, the IPCC determined that this steady warming has had a significant impact on at least 420 animal and plant species, as well as natural processes. Furthermore, this has not just occurred in one geographical location but worldwide. In the IPCC's fourth report, released in February 2007, they concluded that it is "very likely" (>90 percent) that heat-trapping emissions from human activities have caused "most of the observed increase in globally averaged temperatures since the mid-20th century."

Also, in the February 2007 report, they concluded that "Human induced warming over recent decades is already affecting many physical and biological processes on every continent. Nearly 90 percent of the 29,000 observational data series examined revealed changes consistent with the expected response to climate change, and the observed physical and biological responses have been the greatest in the regions that have warmed the most" (IPCC, 2007a).

In these studies, scientists have been able to break down the natural- and human-caused components in order to see how much of an effect humans have had. Human effects can include activities such as burning fossil fuels, certain agricultural practices (such as heavy use of fertilizers and other chemicals, tillage, mismanagement of livestock waste, irrigation erosion, introduction of invasive species, and soil compaction), deforestation, and local pollution from certain industrial processes (such as uncontrolled emissions, chemical use, and noncompliance of specific regulations), the introduction of invasive plant species, and various types of land-use change (such as urbanization, industrialization, and mismanaged recreational activities).

In many cases, scientists do not need to look very far to see the effects a warming world is having on the environment and the earth's ecosystems. Glaciers worldwide are melting at an accelerated rate never seen before (Figure 3.1). The cap of ice on top of Kilimanjaro is rapidly disappearing; the glaciers that have made Glacier National Park in the United States world renowned and that also extend across the border into Canada are melting and projected to be gone within the next few decades; and the glaciers in the European Alps are experiencing a similar fate.

In the world's tropical oceans, vast expanses of beautiful, brilliantly colored coral reefs are dying off as oceans slowly become too warm. Unable to survive the higher temperatures, the corals are undergoing a process called "bleaching" and are turning white and dying off at rapid rates. In the Arctic,

FIGURE 3.1
(See color insert.) This is the Muir Glacier located in Alaska. The top image was taken in 1941, and the bottom image was taken in 2004. The massive melting that has taken place is attributed largely to anthropogenic warming of the atmosphere. (Courtesy of National Snow and Ice Data Center; top photo by William O. Field; bottom photo by Bruce F. Molnia.)

as temperatures climb, ice is melting at accelerated rates, leaving polar bears stranded, destroying their feeding and breeding grounds, and causing them to starve and drown (Figure 3.2). Permafrost is melting at accelerated rates. As the ground thaws, it is disrupting the physical and chemical components of the ecosystem by causing the ground to shift and settle, toppling

FIGURE 3.2
Polar bears are in a precarious situation right now. As the Arctic ice retreats, polar bears are struggling to survive. Lack of ice takes away valuable hunting and breeding grounds as well as migration corridors. (From Publitek, Inc., Waukesha, Wisconsin.)

buildings and twisting roads and railroad tracks, as well as releasing methane gas (a potent greenhouse gas partially responsible for climate change) into the atmosphere.

Weather patterns are also changing. El Niño events are triggering destructive weather in the eastern Pacific (in North and South America). Extreme weather events and droughts have become more prevalent in some geographical areas, such as parts of Asia, Africa, and the American Southwest. Animal and plant habitats have been disrupted, and as temperatures continue to climb, there have been several documented migrations of individual species moving northward (toward the poles) or to higher elevations on individual mountain ranges. Migration patterns are also being impacted, such as those already documented of beluga whales, butterflies, and polar bears.

Spring, arriving earlier in some areas, is now influencing the timing of bird and fish migration, egg laying, leaf unfolding, and spring planting for agriculture. In fact, based on satellite imagery documentation of the Northern Hemisphere, growing seasons have steadily become longer since 1980.

Although species have been faced with changing environments in the past and have been able to adapt in many cases, the IPCC climate change scientists view this current rate of change with alarm. They fully expect the magnitude of these changes to increase with the temperatures over the next century and beyond. The concern is that many species and ecosystems will not be able to adapt as rapidly as the effects of climate change cause the environment to change. In addition, there will also be other disturbances along with climate change, such as floods, insect infestations, and spread of disease, wildfire, and drought. Any of these additional challenges themselves can destroy a species or habitat. In particular, alpine (high mountaintop) and polar species are especially vulnerable to the effects of climate change because as species move northward (poleward) or higher in elevation on mountains, these species' habitats will shrink, leaving them with nowhere else to go.

Climate zones could shift, completely disrupting land use practices. For instance, the current agricultural region of the Great Plains in the United States could be shifted instead toward Canada. The southern portion of the United States could become more like Central or South America. Siberia would no longer be a frozen, desolate landscape. Parts of Africa could become dry, desolate wastelands. If this were to happen, it would severely impact the production of agriculture and the agriculture-related industries. Areas currently equipped to produce agriculture would no longer be able to, and areas that could, based on the climate, might not have the financial resources or the proper soils. The ripple effect of these disruptions would be felt worldwide. Millions of people would be forced to migrate from newly uninhabitable regions to different areas where they could survive. There would be enough of a disruption that ecosystems worldwide would be thrown out of balance and altered.

If, however, the temperature rise tends toward the higher end of the estimate, the results on ecosystems worldwide would be disastrous. There would also be impacts to public health as a result. Rising seas would contaminate freshwater with saltwater; there would be more heat-related illnesses and deaths; and disease-carrying rodents and insects, such as mice, rats, mosquitoes, and ticks would spread diseases such as malaria, encephalitis, Lyme disease, and dengue fever.

Scientists of the IPCC agree that one of the most serious aspects of all this drastic change is that it is happening so rapidly. These changes are happening at a faster pace than the earth has seen in the last 100 million years, and although humans may be able to pick up and move to a new location, animals and their associated ecosystems cannot. Unfortunately, the choices humans make and actions they take today are determining the fate of other species and their ecosystems tomorrow.

The Results of Climate Change on Ecosystems

World Wildlife Fund scientists believe that climate change could begin causing extinction of animal species in the near future because the heating caused by accelerated climate change severely impacts the earth's many delicate ecosystems—both the land and the species that live within them. Worldwide, there are several species and habitats that have now been identified as being threatened and endangered because of the effects of climate change, such as those listed in Table 3.1. Because ecosystems can be altered to the point where the damage becomes irreversible and species must either adapt to survive or they will die, it is critical that the issue of climate change be addressed and acted upon now before it is too late.

TABLE 3.1

Some Species and Habitats Threatened and Endangered by Climate Change

---

Species Threatened and Endangered by Climate Change
Polar bear (*Ursus maritimus*)
Sea turtles
    Loggerhead (*Caretta caretta*)
    Leatherback (*Dermochelys coriacea*)
    Hawksbill (*Eretmochelys imbricata*)
    Kemp's ridley (*Lepidochelys kempii*)
    Olive ridley (*Lepidochelys olivacea*)
    Flatback (*Natator depressa*)
North Atlantic right whale (*Eubalaena glacialis*)
Giant panda (*Ailuropoda melanoleuca*)
Marine turtles (multiple *chelonian* species)
Pika (*Ochotona princeps*)
Multiple bird species (mountain, island, wetland, Arctic, Antarctic, seabirds, migratory birds) (*Avian*)
Wetland flora and fauna
Snowy owls (*Nyctea scandiaca*)
Salt wetland flora and fauna
Mountain gorilla (Africa) (*Gorilla beringei beringei*)
Cloud forest amphibians
Andes spectacled bear (*Tremarctos ornatus* or Andean bear)
Bengal tiger (*Panthera tigris tigris*)

Habitats Threatened and Endangered by Climate Change
Coral reefs
Mountain ecosystems
Coastal wetlands
Prairie wetlands
Mangroves
Ice edge ecosystems
Permafrost ecosystems

One of the most serious constraints to animal and plant adaptation is interrupted migration. Migration for many species, such as for geese, elk, salmon, leatherback turtles, wildebeest, and monarch butterflies, is a natural annual act. As climate change impacts ecosystems worldwide, many species unable to survive in the new climatic conditions of the geographic areas where they have always lived will need to migrate to areas with new climates in which they can survive. There are several potential problems with this, however. Under the effects of climate change, the environment may change faster than a species can adapt. In other cases, existing land use—such as heavily urbanized areas—may negatively impact wildlife habitat. Wide-ranging wildlife species need secure core habitats where human activity is limited, ecosystem functions are still intact, and wildlife populations are able to flourish. If they do not have this, their long-term health and survival will be negatively impacted. In this situation, corridors connecting core areas are important to have established before the effects of climate change are felt. According to the U.S. Fish and Wildlife Service, it is imperative to save species. Endangered species are like nature's call for help, because they often serve as an early warning sign for pollution and environmental degradation that can affect human health.

Impacts to Forests

Forests are products of hundreds of millions of years of evolution. During this time, natural climatic cycles have caused forests to adapt by changing the types of vegetation that live within them and by migrating to new habitats as conditions gradually change. When this has occurred in the past, it has happened at a much slower pace, allowing forests to successfully adapt. Today, changes in climate (temperature, precipitation, humidity, and air flow) are happening at a much more rapid pace and forests have not had the luxury of time to adapt.

Many human impacts have also left their harsh blow—clear cutting and urbanization and other activities have destroyed large sections of forest, leaving them fragmented. According to the IPCC, "it is in combination with these threats that the impacts of unprecedented rates of climate change can compromise forest resilience and distribution" (IPCC, 2001b). According to the special PBS presentation "Earth on Edge" by Bill Moyers, today nearly 9 percent of all known tree species are at some risk of extinction. Most of the world's forest loss has occurred in the last three decades—and most of this was caused by human activities (PBS, 2001).

Forests are extremely valuable ecosystems—they help to regulate rainfall and are also key sources of food and medicine. Forests provide abundant wildlife habitat, carbon storage, clean air and water, and recreational opportunities. They also provide a bounty of natural resources, such as wood, plants, berries, water, wildlife, and aesthetic value. The health and diversity of forests is largely influenced by climate. Native forests adapt to the local climate features. For example, in the far-northern boreal forests, cold-tolerant species, such as white spruce, have adapted there; in drier areas conifer and hardwood forests thrive because they need less water. Because air temperature affects the physiological processes of individual plants and productivity, the effects of climate change have a strong influence over the health of tomorrow's forests. Not all forests will have the same outcomes under the influence of climate change. Some forests will die back while others may extend their ranges. Amounts of CO_2 will vary, as well. Whatever the outcome of a particular forest, if the population cannot adapt or migrate with the changes, they will face extinction.

According to the IPCC, at least one-third of the world's remaining forests may be negatively impacted by climate change during this century. It may force plant and animal species to migrate or adapt faster than they physically can, disrupting entire ecosystems. The IPCC also predicts that forests will have changes in fire intensity and frequency (IPCC, 2007b) and increased susceptibility to insect damage and diseases.

Climate scientists are currently employing a variety of methods to predict the impacts of climate change on forests. On a global or large-regional scale, they are predicting shifts in ecosystems by combining biogeography models with atmospheric general circulation models (GCMs) that project changes in an environment where the CO_2 content has doubled (Malcolm et al., 2006). They also use biogeochemistry models to simulate the carbon cycle; flow of nutrients; and impacts of changes in precipitation, soil moisture, and temperature to study ecosystem productivity. They have also developed global models that simulate worldwide changes in vegetation composition and distribution. The concept of modeling is discussed in more detail in Chapter 13.

The IPCC also states that climate change both directly and indirectly affects the health of forests. The direct impacts of warmer temperatures, changing rainfall patterns, and severe weather events can already be seen in certain tree and animal species (IPCC, 2007b). Even small changes can affect forest growth and survival—especially the portions that lie along the outer edges of the ecosystem where the conditions are marginal. When it gets hotter, more water is lost through evapotranspiration, which causes drier conditions and decreases the vegetation's use of water. Water temperatures can also throw off the timing of flowering and fruiting for plants and adversely affect their growth rate. Forests will also be threatened when the seasonal precipitation patterns they have been used to change and water is not supplied when it is needed, causing drought conditions and stress; if the precipitation pattern supplies too much water when it cannot be assimilated, it can cause flooding and mudslides.

According to Natural Resources Canada (2002), the age and structure of the forest also plays an important role in determining how quickly a forest responds to changes in moisture conditions. Mature forests have well-established root systems, enabling them to tolerate drought better than younger forests or forests that have been disturbed in some other way, such as through disease infestation. Species type also plays a part—some species are more resistant than others.

Models have been run of forests simulating the impacts from elevated levels of CO_2. According to Natural Resources Canada (2002), the results of these models are not as straightforward or conclusive as other factors, tending to be more problematic. It was found that higher levels of CO_2 improve the efficiency of water use by some plants, but other plants have been shown to adjust to higher CO_2 levels, followed by absorption rates decreasing over time. It is also difficult to model CO_2 effects when other greenhouse gases may be involved, such as ozone (which works against CO_2) or nitrogen oxide (which can enhance tree growth), leading to problematic interpretation.

In the short term (50–100 years), changes caused by climate change will be focused on ecosystem function. In the long term, shift of forest types will be more significant. According to the IPCC, boreal forests will be impacted the most severely; their ecosystem will be greatly reduced because significant warming is expected in the polar regions (IPCC, 2007c).

Some of the most vulnerable temperate forests will be the "island" or isolated forest communities, such as the fragmented forests encroached upon by urban and agricultural areas. Island forests will also be at risk because there will be no migration options. Forests located in high elevations on mountain systems face a similar threat—as animals migrate upward, eventually there will be nowhere left to go. Individual species that are indigenous to small geographic areas or have limited seed dispersal will also be threatened and endangered. Under climate change conditions, surviving forests of the future are expected to look very different from those of today. Although changes will vary in degree from area to area, all forests will be affected in some manner. Impacts from climate change are already manifesting, such as forests in the United States and Canada. Over the past century, a 1–2°C increase in air temperature and changes in precipitation have already been documented. Experts believe that higher levels of CO_2 have already caused dieback of forested areas along the Pacific and Atlantic coasts.

It is expected that temperate forests in the United States will migrate northward from 100 to 530 kilometers over the course of the next century. If the air temperature warms 2°C over the same time period, models have predicted that tree species will have to migrate 1.6–4.8 kilometers per year, which is too fast for most temperate species, except for those whose seeds are carried by birds over greater distances. As a result, it is expected that grasslands will dominate many of these areas.

The effects of climate change are felt most at the poles, where the predicted rise in temperature could climb 5–10°C or higher over the next century. It is estimated that warming will negatively affect the species that live in the ecosystems and that approximately 24–40 percent of the species living in the boreal forests right now will be lost. Species that live in the north will be crowded out by species from the temperate regions that will be migrating northward in search of cooler climates. The species that will invade the present boreal forests will be today's temperate forest species and grasslands. According to the IPCC, as the boreal forest vegetation is forced out, it will migrate poleward 300–500 kilometers during the next century. Proof of this can already be seen in western Canada; their plant zones have already begun shifting poleward (IPCC, 2007a). Migrating vegetation will encounter severe challenges, however. For example, the soils in the tundra region are not fertile or conducive to high-density vegetation or tree growth. According to the IPCC, they lack the biota necessary for colonization. Specific seed dispersal rate and migration tolerance range are also important factors that could play a role in keeping trees from being able to survive the poleward migration rate dictated by climate change. The white spruce is one example; it can colonize at 100–200 kilometers over 100 years; Scots pine can migrate 4–8 kilometers every 100 years. According to A. Solomon and K. Jardine of Greenpeace, the rate of climate change will be about ten times faster than what is needed for successful species migration at an average migration rate of 25 kilometers each century through natural seed dispersal (Jardine, 1994).

There are other factors that will hurt species migration as well, such as habitat fragmentation (small isolated population clusters rather than one large cohesive unit) and competition from more hardy species. As temperatures change, it may also affect the timing and rate of seed production, which will affect the growth and strength of the trees. Trees that have limited seed dispersal mechanisms will also suffer. For instance, trees whose seeds are carried long distances by the wind will have a better chance of survival than those whose seeds fall closer to the tree.

The ability to adjust to a larger temperature range will also play an important role. Vegetation with narrow temperature tolerances will be vulnerable to extinction. The IPCC has stated that a drastic change in species composition and loss of habitat with even a 2°C warming near the poles will damage the ability of an ecosystem to function as species richness begins to be killed off.

According to Natural Resources Canada, an average rise in temperature of 1°C over Canada in the last century has had a negative impact on vegetation. At mid to high latitudes (45° N to 70° N), plant growth and the length of the growing season has increased. In portions of western Canada, there has been a decrease in rainfall as temperatures have risen, and this has hurt the growth of some tree species, such as aspen poplar. In Alberta, aspen are now blooming 26 days earlier than they were a hundred years ago (Natural Resources Canada, 2002).

Another major concern for boreal forests in a warmer climate is insect infestations. Insects commonly found in temperate forests, such as mountain pine beetle, will migrate north along with the forests and continue to infest and infect as they move northward, devastating industries such as logging and tourism.

As temperatures climb, drought-like conditions may develop. If this happens, there will also be greater incidence of wildfire. In the past 40 years, the trend has already been established that as climate warms, wildfires have become more frequent and are burning larger areas. This overall trend has been seen in places such as southern California and also in boreal locations in Canada and Russia. As climate change continues, longer fire seasons, drier conditions, and more frequent severe electrical storms are projected to increase, causing the fire season to become more problematic and devastating as the climate continues to change.

Although it is true that some forest species' seeds are actually dispersed by fire, which will aid their migration, and burnt litter will add nutrients to the soil, over time reoccurrences of wildfire will fragment established vegetation colonies and make it more difficult for them to migrate. In addition, as older trees burn, they will add carbon to the atmosphere, and as younger trees replace these burnt areas there will be less initial carbon storage capability.

The world's tropical forest ecosystems are very sensitive to disturbances such as overgrazing, logging, plowing, and burning. Converting natural ecosystems to agricultural and logging uses combined with climate change is the largest threat to the rain forests today. According to the Food and Agriculture Organization of the United Nations, a land area the size of Ireland or South Carolina (13 million hectares), is lost to these uses every few years in the rain forests. Developing countries are hit harder than developed countries. Unfortunately, it is often seen as necessary to farm the land in the pursuit of food or sell the land to logging companies in exchange for needed income (Food and Agriculture Organization, 2001).

According to Jose Antonio Marengo of Brazil's National Space Research Institute, if the issue of climate change is not addressed today, the rain forests will be negatively impacted. They will receive less rainfall in addition to experiencing higher temperatures; the effect could be enough to transform the Amazon—the world's largest remaining tropical rain forest—into a savanna (grassland) by the end of the century. The Amazon covers almost 60 percent of Brazil, and hosts one-fifth of the world's freshwater and nearly 30 percent of the world's plant and animal species. Jose Marengo is involved in research focused on the following two different scenarios:

1. If no action is taken to slow or halt climate change or deforestation, temperatures will rise 5–8°C by 2100, and rainfall will decrease by 15–20 percent. This is the scenario that would transform the Amazon into a savanna.

2. If action is taken now to slow climate change, temperatures would most likely rise 3–5°C by 2100, and rainfall will decrease by 5–15 percent.

Marengo stresses that "if pollution is controlled and deforestation reduced, the temperature would rise by about 5°C by 2100. Within this scenario, the rain forest will not come to the point of total collapse." Marengo also stated that he was optimistic that the worst-case scenario could be averted, but that it would require a major effort by industrialized nations to reduce emissions of greenhouse gases, such as carbon dioxide (Environmental and Energy Study Institute, 2007).

Unfortunately, many of the world's rain forests are being cut down at an accelerated rate for agriculture, pasture, mining, and logging. When these forests are cut down or burned, huge amounts of carbon are reintroduced back into the atmosphere. Tropical deforestation accounts for about 20 percent of the human-caused carbon dioxide emissions each year to the earth's atmosphere. This makes deforestation of the rain forests an important issue when dealing with the challenges of climate change.

In view of the impacts that may be felt by the world's forests—temperate, boreal, and rain forests—it is important to begin looking at adaptation strategies today. The World Wildlife Fund has suggested the following as viable adaptation options in order to reduce the threat on forest ecosystems from climate change:

- Reduce present threats that could harm the ecosystem, such as degradation and the introduction of invasive species.
- Manage large areas of land in a comprehensive way tied to the landscape. Focus on all the components that compose a large-scale area and plan for the adaptation migration of different species.
- Provide buffer zones and flexibility of land uses. The land must be managed to accommodate the movements of species migration.
- Protect mature forest stands. Mature trees are better able to withstand large-scale environmental changes and also provide a safe habitat for other species to adapt within.
- Maintain the natural fire regime. Different ecosystems have different fire ecologies, necessitating the development of fire-management plans.
- Actively manage pests. Because climate change is associated with invasions of insects, disease, and exotic species, healthy management practices must be set in place to protect the forests. This could include measures such as prescribed burning and nonchemical pesticides.

Different forested areas will require different monitoring, planning, and protection strategies. It is imperative that humans understand the unique forest ecosystems and their value (World Wildlife Fund, 2003).

Impacts to Rangelands, Grasslands, and Prairies

Native grasslands cover about one-quarter of the world's land surface, making them a significant component of the world's vegetation. Therefore, any impact on them could seriously impact a major global ecosystem. The extent and health of natural grasslands are typically controlled by rainfall and fire. These two factors limit the extent of their range, and it is already known that with an increase in climate change, there will be a decrease in rainfall in some areas and an increase in wildfire.

The world's grasslands stand to face serious consequences if climate change continues. These regions of the world provide grains and crops for the world's population, rangeland for cattle and sheep, and habitat for wildlife and play a role in carbon sequestration and maintaining an overall healthy ecological balance in nature.

Grasslands are important for carbon storage. Whereas carbon is stored in trees, plants, and organic matter on the surface of the soil in forests, in grasslands it is stored in the soil. Because of human development, most of the earth's grasslands suitable for agriculture have already been developed. According to the Food and Agriculture Organization of the United Nations, as much as 70 percent of the earth's 1.3 billion hectares of grasslands have become degraded, mostly because of overgrazing (Conant, 2010).

Temperate grasslands are located north of the Tropic of Cancer (23.5° N latitude) and south of the Tropic of Capricorn (23.5° S latitude). The four major temperate grasslands of the world are the plains of North America, the veldts of Africa, the steppes of Eurasia, and the pampas of South America. Dominant vegetation types in these areas are grasses. The principal places that trees, such as cottonwood, oak, and willow, are found are along river valleys, growing in areas near a reliable water source. The soil in temperate grasslands is nutrient-rich because of the growth and decay of deep, dense grass roots. The roots serve to hold the soil together as a cohesive unit. In fact, some of the world's most fertile soils are found in the grasslands of the eastern prairies in the United States, the steppes of Russia and Ukraine, and the pampas of South America.

Tropical grasslands are located near the equator, between the tropics of Cancer and Capricorn. These are the grasslands that occupy Africa, Australia, India, and South America. Tall grasses dominate these landscapes—and they are found in tropical wet and dry climates, where temperatures never fall below 18°C. These grasslands are usually dry, but there is a season of heavy rain, bringing an annual rainfall of 51–127 centimeters per year. One of the biggest threats to the tropical grasslands with increased climate change is desertification.

According to the IPCC, the structure and function of the world's grasslands makes them one of the most vulnerable to global climate change of any of the terrestrial ecosystems. Grasslands are vulnerable to vegetation change caused by changing temperatures and precipitation due to climate change (IPCC, 2007a).

According to an article in *Science Daily* on January 15, 2001, some experimentation has been done by scientists in grasslands to determine how much of a carbon sink they have the potential to be in light of climate change.

Dr. Shuijin Hu, assistant professor of plant pathology at North Carolina State University, was involved in a study that showed grasslands are able to act as carbon sinks when CO_2 concentrations in the atmosphere are increased. He discovered that grasslands actually sequester carbon in the soil. Soil microbes respond to the changes in carbon and nitrogen availability and store excess carbon as atmospheric levels increase. Sequestration occurs when the amount of carbon dioxide absorbed by vegetation is greater than the amount of gas released by decomposing plant material (Science Daily, 2001).

In the study, Dr. Hu, along with Dr. F. Stuart Chapin III at the University of Alaska, Dr. Christopher B. Field at the Carnegie Institution of Washington, DC, and Dr. Harold A. Mooney of Stanford University, focused for 5 years on a study site at Stanford University's Jasper Ridge Biological Preserve in California. From 1992 to 1997 two study plots were maintained: one at a CO_2 level of 360 ppm and the other at double that amount (720 ppm).

Hu explained that "other scientists have proposed that grasslands can act as carbon sinks when atmospheric carbon dioxide is elevated." The research they conducted showed that grassland soils can sequester carbon and also found a trend toward increased soil carbon under elevated carbon dioxide conditions.

At the end of the experiment, they found that the soil core sample from the plot that had received double the normal CO_2 had stored more carbon in it. Their results presented an interesting relationship in grasslands: When CO_2 was increased in the atmosphere, grassland plants grew more rapidly while drawing nitrogen from the soil. This nitrogen depletion restricted it from the soil microbes, which in turn reduced their ability to decompose dead plant material (which releases CO_2 to the atmosphere). There did seem to be a tipping point, however, where only so much carbon could build up. Once it reached that level, the lack of nitrogen kept additional plants from growing.

Dr. Hu concluded that, "Forests may be of greater potential as a long-term carbon sink than annual grasslands because trees can sequester carbon in above-ground biomass." Grasslands do, however, play a role in carbon storage.

According to the IPCC, GCMs predict that grassland ecosystems will experience climatic changes such as higher maximum (daytime) and minimum (nighttime) temperatures and more intensive precipitation events. Recent studies of climate change processes reveal a surprising situation: Daytime high temperatures are not the problem; it is the nighttime highs that are causing problems. During late winter and early spring is when the situation is most pronounced. Nighttime high temperatures have risen, changing the temperature regime enough that the last frost date occurs on average two weeks earlier now than it did 20 years ago. In addition, native grasses are now germinating later. Invasive and noxious plants that have invaded grasslands over the years are germinating early, taking the moisture out of the soil and using the nutrients that would have been available for the grasses. In addition, cattle grazing occurs on much of the grassland areas, and the animals do not ingest the weeds, which further promotes their invasion.

Based on the results of the IPCC studies, in the world's grasslands that are expected to receive an increase in precipitation, such as the western United States, these areas may not be faced with the drying out resulting from lack of water and rising temperatures, but these areas may suffer from accelerated nutrient cycling, which in turn could encourage the spread of more invasive species. In the hot desert grasslands (such as in Australia, in the Sonoran and Chihuahua deserts of Mexico, and in the United States), GCMs predict an increase in the frequency of intense rainfall and flash floods. These areas are expected to have increased erosion and nutrient loss. In ecosystems that have already been disturbed in other ways, such as by wildfire or with desertification, they will be impacted even more and have a more difficult time surviving under climate change conditions.

There are several threats to grasslands in the face of climate change. Marginal grasslands could be converted into deserts as rainfall decreases. The threat of desertification is all too real in many parts of the world. As areas struggle to find water to support populations, cultivate crops, and support livestock, grassland ecosystems stand to face further degradation as temperatures rise during climate change. Therefore, as climate changes, it is important for land managers to adapt to the changes as they happen and manage the land in a responsible way. One of the biggest obstacles they face with climate change is the threat of invasive species encroaching on native grassland species. Invasion of species can happen via seed deliverance by livestock, off-road vehicles, road-maintenance crews, and outdoor recreationalists. Invasive species need to be dealt with immediately before they become established. Land managers will also have to put management plans in place in order to be prepared for the changes that climate change will bring.

Impacts on Polar Ecosystems

The effects of climate change will be felt most strongly in the earth's polar regions. Already temperatures have climbed by 2.4°C in some areas in polar regions, whereas the global average is 0.6°C over the past century. Weather conditions are so harsh that polar ecosystems must maintain a delicate balance to survive. In addition, because of their sensitive nature, the polar regions are sometimes the first ecosystems to show warning signs of climate change, such as acceleration in the melting of glaciers. In the Arctic, climate change is expected to be rapid and extensive. As temperatures warm up and ice melts, Arctic ecosystems will be impacted heavily over the next century. There is extensive sea ice at the periphery of the Arctic Ocean that forms and melts each year. These waters are important to the fishing industry, accounting for almost half of the total global production. Glacial decline, melting sea ice, rising sea levels, impacts to wildlife habitat, melting permafrost, impacts to native tribal inhabitants, and changes to plant life are all being affected in the Arctic today because of climate change.

Arctic ice is getting thinner and more prone to breaking. The Ward Hunt Ice Shelf, located on the north coast of Ellesmere Island, Nunavut, Canada, is one example—it was the largest single block of ice in the Arctic. About 3,000 years old, it began cracking in the year 2000. By 2002, it had broken all the way through. Today it is breaking into pieces. The breakup of the Ward Hunt Ice Shelf has already caused several impacts: polar bears, whales, and walrus have changed their migration and feeding patterns, native people are having to change their hunting territories, and coastal villages are being flooded (Figure 3.3).

There are about 2,000 valley glaciers in Alaska. Today less than 20 are still advancing, according to Bruce Molnia, a geologist at the United States Geological Survey (USGS). Instead nearly all of the glaciers in Alaska are melting. In the past 7 years, the melting rate is roughly twice what it has been previously. According to Molnia, a striking example of glacial retreat is found at Glacier Bay, a popular destination for Alaskan cruise ships. "Ironically, the climate event that made cruising into Glacier Bay not only possible, but popular, could ultimately take away its top attraction as many tidewater glaciers now retreat out of the water," Molnia points out (UniSci, 2001).

According to a report in the *Washington Post*, scientists at the National Oceanic and Atmospheric Administration say that the Arctic ice cap is melting faster than scientists had originally expected and will most likely shrink 40 percent by 2050 in most regions, which will devastate wildlife populations, such as polar bear, walruses, and other marine animals (Struck, 2007). They also predict that Arctic sea ice will retreat hundreds of kilometers farther from the coast of Alaska in the summer. Although that could be good news for fisherman and open shipping routes and new areas for oil and gas exploration, it will spell disaster for the wildlife species that inhabit the region. In 2001, the IPCC predicted that there will be "major ice loss by 2100." Then, in 2007, when they issued their next report, they remarked that "without drastic changes in greenhouse emissions, Arctic sea ice will 'almost entirely' disappear by the end of the century."

According to a report published May 1, 2007, in *The New York Times*, a study shows that climate scientists may have significantly underestimated the power of climate change from human-generated heat-trapping gases to shrink the sea ice in the Arctic Ocean. Dr. Julienne Stroeve, a researcher at the National Snow and Ice Data Center in Boulder, Colorado, discovered that since 1953, the area of sea ice in September has declined at an average rate of 7.8 percent per decade. "There are huge changes going on," Dr. Stroeve says. "Just with warm waters entering the Arctic, combined with warming air temperatures, this is wreaking havoc on the sea ice" (Revkin, 2007).

The warmer it gets, the more the glaciers melt, which adds freshwater to the ocean and upsets the earth's energy balance, ocean circulation patterns, and ecosystems. Some estimates of global sea-level rise from Arctic melt

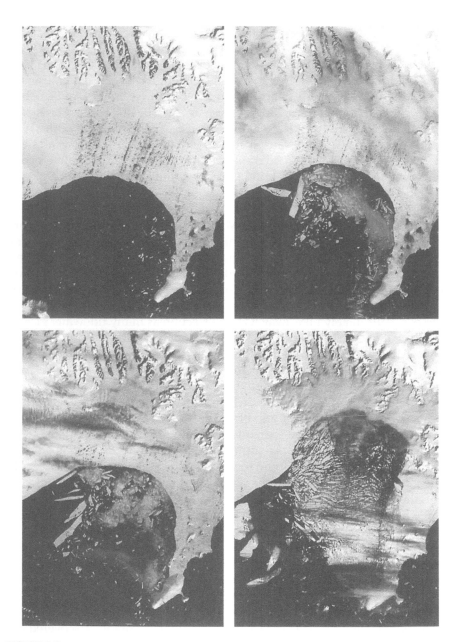

FIGURE 3.3
A large portion of the Larsen B Ice Shelf disintegrated in 2002, as evidenced in these MODIS satellite images from January 31, 2002 (upper left), February 17, 2002 (upper right), February 23, 2002 (lower left), and March 5, 2002 (lower right). Earth scientists predict that if global warming continues, incidences like this will become more common. (Courtesy of National Snow and Ice Data Center, Larsen B Ice Shelf collapses in Antarctica, http://nsidc.org/news/press/larsen_B/2002.html, 2002.)

have been placed at 1 meter by 2100. According to the U.S. Environmental Protection Agency, this increase would drown roughly 58,016 square kilometers of land along the Atlantic and Gulf coasts of the United States—specifically Texas, Florida, North Carolina, and Louisiana.

Warming in the Arctic will also affect weather patterns and food production worldwide. An example of this is given by the National Aeronautics and Space Administration (NASA): in one of their computer models they calculated that Kansas would be 2.4°C warmer in the winter without Arctic ice, which usually sends cold air masses into the United States. Without these Arctic-induced cold air masses, winter wheat cannot be grown, and soil would be 10 percent drier in the summer, wreaking havoc on summer wheat crops.

In 2005, the ice cap covering the Arctic Ocean melted to a size smaller than it had been since records began to be kept a century ago. In Greenland, glaciers are melting and traveling faster to the ocean, adding freshwater. These changes in the Arctic have the potential to impact the entire earth. One of the most serious consequences could be the disruption of the thermohaline circulation, with the ramifications discussed in Chapter 1.

The Arctic Council, an intergovernmental forum comprised of eight Arctic nations (Canada, Denmark/Greenland/Faroe Islands, Finland, Iceland, Norway, Russia, Sweden, and the United States) and six indigenous people's organizations, had an assessment prepared by the Arctic Monitoring and Assessment Programme, the Conservation of Arctic Flora and Fauna, and the International Arctic Science Committee of impacts to the Arctic as a result of climate change. They support the notion that the Arctic is extremely vulnerable to observed and projected climate impact and that it is currently facing some of the most rapid and severe climate change on earth. They report that over the next hundred years, climate is expected to change, and these changes in the Arctic will be felt worldwide. Although some of these changes are those caused by nature, the overwhelming trends and patterns are results of human influence, resulting from the emission of greenhouse gases.

In many Arctic locations winter temperatures are rising faster than summer temperatures. In fact, according to the Arctic Council's Arctic Climate Impact Assessment, in Alaska and western Canada, winter temperatures have increased as much as 3–4°C in the last 50 years. Using a moderate emissions model, annual temperatures over the next 100 years are projected to rise 3–5°C over land and 7–10°C over the oceans (Arctic Climate Impact Assessment, 2004).

One of the biggest key climate change indicators in the Arctic is the declining sea ice. Researchers actually use the quantitative amounts of Arctic sea ice as an early warning system of climate change. According to the International Arctic Research Center, over the last 30 years, the average sea ice extent has decreased by 18 percent, which is the equivalent area of roughly 1 million square kilometers (SIMBA, 2005). As a comparison, this represents an area roughly the size of Egypt. Of even more concern, the melting rate is accelerating. The most pronounced season of melting is in the late summer. It is projected that by 2100, late-summer sea ice will have decreased in a range somewhere from 50 percent to complete disappearance.

Another major global impact that will occur as a result of ice sheet melting in the Arctic is sea-level rise. The Greenland Ice Sheet is the largest ice sheet in the Arctic. Its surface-melt area increased by 16 percent from 1979 to 2002, which is an area about the size of Sweden. In 2002, the total area of surface melt on the ice sheet broke all records kept to date—melting reached 2,000 meters in elevation. Data from satellites are used to monitor the surface of the ice sheet, and the sophisticated instruments on board can accurately measure slight changes in elevation, enabling scientists to accurately assess melt rates. Satellite data have shown an increasing melt rate since 1979. The only year the rate slowed down was in 1992. This was due to the eruption of Mt. Pinatubo. The violent volcanic eruption ejected massive amounts of particles into the atmosphere, which reduced the amount of sunlight able to reach the earth's surface, thereby cooling global temperatures that year.

The sea level rose roughly 8 centimeters over the past 20 years, and as temperatures rise, the rate is increasing. This is due to two factors: melting ice and thermal expansion of seawater. The global average sea level is projected to rise anywhere from 10 to 90 centimeters by 2100. Some climate models have predicted that the Greenland ice sheet will completely melt, causing sea levels to rise 7 meters. As a comparison, a 50-centimeter rise in sea level will typically cause a shoreward retreat of coastline of 50 meters if the land is relatively flat. This flooding would cause significant environmental and economic impacts (Ritter and Hanley, 2011).

Impacts to Desert Ecosystems

The world's principal deserts are found in two major zones: at 25–35° latitude north and south of the equator. Desert ecosystems are defined as regions that are very arid and dry, receiving less than 25 centimeters of rain a year. By definition, deserts are not confined to just hot, dry areas—they can also exist in the cold, dry areas within polar regions.

Because these regions are so physically harsh, the plants and animals that live within them have learned to adapt and survive in an extremely hostile environment. Covering about one-fourth of the earth's land surface (54 million square kilometers), they are dominated by bare soil and scarce vegetative cover.

As climate change continues, current deserts will be impacted as temperatures rise, water becomes more scarce, and food sources become problematic. As desertification spreads because of climate change, ecosystems worldwide will be impacted in some way. Understanding the fragile balances in desert ecosystems and how climate change will relate to them is important as land managers look at future management options.

Climate change is expected to cause intensification in the hydrologic cycle, with a marked increase in evaporation over land and water. The higher the evaporation rate, the greater the drying will be of soils and vegetation. In areas where there is a decrease in rainfall and an increase in evaporation, droughts will occur, be more severe, and last longer.

Drought already threatens the lives of millions of people on earth. As the atmosphere continues to warm under the effects of climate change, the negative effects and hardships brought on by drought will spread across the globe as areas heat up and dry out. According to a study conducted by the Hadley Centre for Climate Prediction and Research, as reported by *The Independent* of the United Kingdom in 2006, "Extreme drought, in which agriculture is in effect impossible, will affect about a third of the planet" (McCarthy, 2006).

The climate research center also added that their prediction may be an underestimate. It is also recognized that the worst hit will be the large populations of developing countries—those who can least afford to deal with the unfortunate consequences.

A National Oceanic and Atmospheric Administration measure of drought from a climate model, called the Palmer Drought Severity Index, developed in the 1960s by Wayne Palmer, predicts there will be a notable increase during this century with predicted changes in rainfall and heat around the world because of climate change. The Palmer Drought Severity Index figure for "moderate drought" is currently at 25 percent of the earth's surface. It is predicted that by the year 2100, this figure will rise to 50 percent. The figure for "severe drought" is currently at 8 percent; it is predicted to be 40 percent by 2100. The figure for "extreme drought" is currently 3 percent, and it is predicted to rise to 30 percent by 2100, based on the study conducted by the Hadley Centre (McCarthy, 2006).

Mark Lynas, author of *High Tide*, in a response to the model, said, "We're talking about 30 percent of the world's land surface becoming essentially uninhabitable in terms of agricultural production in the space of a few decades. These are parts of the world where hundreds of millions of people will no longer be able to feed themselves" (Douglas, 2006).

For a realistic glimpse into the future, predictions are that valleys once fertile will turn dry and brown. For inhabitants of desert valleys that rely on a short rainy season each year in order to be able to grow crops and graze their animals, they will wait in anticipation for rains that, instead, will never come. Year after year, the situation will repeat itself as millions of people near starvation. Nomadic herders' animals will die; their cattle are emaciated to skin and bones now. Bleached skeletons of cows, goats, and sheep will litter the barren, dusty landscape. Nomadic herders will set out and walk for weeks without finding a watering hole or a riverbed. As people begin dying of starvation, inhabitants of different geographic areas will start fighting each other for what slim, meager resources are still in existence.

As brutal as this may sound, if climate change continues, this scenario is projected to become commonplace. According to the Hadley Centre, the number of food emergencies in Africa each year has almost tripled since the 1980s. Across sub-Saharan Africa, one in three people today is undernourished (McCarthy, 2006).

A news report from the World Wildlife Fund on January 13, 2003, from Sydney, Australia, says human-induced climate change was a key factor in the severity of the 2002 drought in Australia. The report, entitled "Global Warming Contributes to Australia's Worst Drought," warns that higher temperatures and drier conditions have created greater bush fire danger than previous droughts. The drought has also negatively affected their agricultural areas. Professor David Karoly, co-author of the report, stated, "The higher temperatures experienced throughout Australia last year are part of a national warming trend over the past 50 years which cannot be explained by natural climate variability alone. Most of the warming is likely due to the increase in greenhouse gases in the atmosphere from human activity such as burning fossil fuels for electricity and transport and from land clearing" (Karoly et al., 2002). Dr. Karoly believes this is the first drought in Australia where the impact of human-induced climate change can be directly observed.

According to MSNBC in a release issued in June of 2006, the United Nations said "The world's deserts are under threat as never before, with climate change making lack of water an even bigger problem for the parched regions; those areas are facing dramatic changes." The report said that most of the twelve desert regions whose future climate was studied faced a much drier future. Experts predicted that rainfall would decrease as much as 20 percent by the end of the century due to human-induced climate change (Associated Press, 2006).

Increasing drought will also lead to the possibility of more wildfires. Some models that simulate climate change predict drier summers at northern high latitudes. According to the IPCC, there are already documented decreases in precipitation in parts of Africa, the Caribbean, and tropical Asia. The 1999 drought in the eastern United States may be a glimpse at what future conditions may be like there as the earth's atmosphere heats up (IPCC, 2007a).

Desertification

One of the impacts that climate change may have on the surface of the earth is to exacerbate the worldwide progression of desertification. If there is a significant decrease in the amount of rainfall reaching an arid or semi-arid area, it could increase the extent of the dryland areas globally, destroying both vegetation and soils. Desertification is a degradation process inflicted on arid and semi-arid landscapes caused by human activities, climate change, or a combination of both. Although desertification has always existed, it has become more prominent and a much bigger concern in recent years as populations have rapidly expanded across previously untouched landscapes and as arable land has been cultivated and grazed.

The first most well-known incidence of desertification in the United States was the Dust Bowl of the 1930s, caused by a combination of drought and improper farming and land-management practices. During this period, millions of people were forced to leave their homes and abandon their farms in the biggest migration in U.S. history. The Great Plains were eventually restored over time through the practice of good land-management plans, improved agricultural methods, and responsible conservation efforts, initiating the birth of the Soil Conservation Service. In other areas of the world, however, population explosions and increasing livestock pressure on marginally healthy land has worsened the problem of desertification, speeding it up. As the problem intensifies, the productive capability of the land decreases, destroying the original biodiversity. Different plant species produce physically and chemically different litter compositions in the ground. The litter, along with the natural biologic decomposers in the ground, helps form the soil complex and plays an active role in nutrient cycling. All the vegetation supports the primary production that provides food and wood that works together to sequester carbon, which plays an ultimate role in global climate. When these connections are broken, desertification is triggered, and habitats can be jeopardized and lost. Biodiversity loss affects the health of the habitat. Climate change increases evapotranspiration and adversely affects biodiversity.

The loss of vegetation in the food chain impacts life along it. In addition, when native vegetation dies off, invasive species, such as cheat grass (*Bromus tectorum*) in the southwestern United States, moves in and takes over. Invasive species have a better probability of survival because they lack the native species' predators. It does not take long for invasive species to overrun a landscape, pushing native species out. Once this happens, it may be that only a few invasive species are supported in an area, where at one time dozens of species once existed. In areas where this occurs, the biodiversity is significantly lowered, and unless rehabilitation efforts are put into effect, it may never be able to re-establish itself to a natural, ecologically balanced state. Even if native vegetation is reintroduced, it may not be able to survive if desertification degrades the land to the point where nutrients and water supplies are no longer available.

Biodiversity, which contributes to many of the benefits provided to humans by dryland ecosystems, is greatly diminished by desertification. For example, diversity of vegetation is critical in soil conservation and in the regulation of surface water and local climate. If these delicate balances are disrupted, it can threaten the health and existence of habitats. Desertification affects global climate change through both soil and vegetation losses. Another problem that exists involves the use of land for livestock grazing. When desertification becomes an issue and invasive weeds move into the area, the livestock may consider the new vegetation species unpalatable and refuse to graze.

It is not just one human activity that is responsible for desertification—there are many. Overgrazing of livestock and cultivation are two of the most common reasons, but deforestation is also a major contributor. If groundwater is overused or irrigation of farming areas is not done properly, they can introduce the desertification process. Desertification is also encouraged if soils collect more salt, raising their salinity levels. Climate change is being largely blamed for worldwide desertification in the earth's arid, semi-arid, and subhumid areas. This means that desertification is the result of a combination of social, economic, political, physical, and natural factors, which vary from region to region.

Currently, lands that seem to be most prone currently to desertification include the areas at the fringes and outskirts of deserts. These transition areas have fragile, delicately balanced ecosystems, usually operating under various microclimates. Already delicate, once these ecosystems are stressed to their limit of tolerance, they cannot recover on their own. Grazing by livestock is especially harmful because the animals pack down the soil. The more the subsoil gets packed into an impermeable layer, the less water is able to percolate down into the ground. Because the scarce water is not able to penetrate the ground's subsurface, it flows off the surface, eroding the land with rills and gullies. South Africa, for example loses 262 million metric tons of topsoil each year. In addition, as the surface dries out, the soil is carried away by the wind.

According to the USGS, there are many things scientists still do not know about desertification of productive lands and the expansion of deserts. To date, there is no consensus among researchers as to the specific causes, extent, or degree of desertification. Desertification is a subtle and complex process.

From a global context, since the mid-1980s, satellites have helped scientists study climate change by providing global imagery with which to study the effects of desertification. The existence of satellite imagery, such as LANDSAT, SPOT, Quickbird, and Digital Globe, has made it possible to monitor areas over time and determine the susceptibility of the land to desertification. The problem is a global one. It is predicted that by 2100, climate change will increase the area of desert climates by 17 percent, meaning more areas at risk of desertification. Worldwide about 12 million hectares becomes useless for cultivation each year—an area equal to about 87 percent of the area of agricultural land in the United States.

Desertification needs to be monitored and managed as a worldwide effort as climate change intensifies. Like global uplifting, it does not stop at international borders. If it is not controlled, biodiversity will be negatively impacted. Another way desertification contributes to climate change is by allowing the carbon that has been stored in dryland vegetation and soils to be released to the atmosphere as it dries out and dies. This could have significant consequences for global climate. According to Greenfact's desertification synthesis report in 2005, it is estimated that every year 272 million metric tons of carbon are lost to the atmosphere as a result of desertification. This equates to around 4 percent of global emissions from all sources combined (Greenfact, 2005).

The relationship between climate change and desertification is not straight-forward; its variabilities are extremely complex. The best way to deal with desertification is to prevent it from happening in the first place. The most appropriate way to do this is through proper management at local, regional, and global levels. Besides being environmentally better, prevention is also more cost effective than rehabilitation. The best form of prevention requires a change in the attitudes of those living on and working with the land and employing sustainable, environmentally friendly agricultural and grazing practices. For areas that have already been degraded, rehabilitation and res-toration measures can help to restore lost ecosystems, habitat, and services the drylands originally supplied. One of the most important reasons to avoid desertification is to avoid extreme poverty and hunger—when drylands become too degraded in developing countries, it leaves populations with little access to food and clean water.

According to the Ecosystems and Human Well-being: Desertification Syntheses, which is part of the report published in the 2005 Millennium Ecosystem Assessment, the following actions can be employed to prevent desertification (Corvalan and Hales, 2005):

- Implementation of a land and water management plan to protect soils from erosion, salinization, and degradation.
- Creation of economic opportunities outside of dryland areas, taking the stress off of drylands.
- Protection of vegetative cover so that it will stabilize the soil under-neath and keep it from being eroded by wind and water.
- Becoming involved in alternative livelihoods that do not depend on intensive land use. These types of activities include greenhouse agriculture, tourism, and aquaculture.
- Combining areas of farming and grazing in order to centrally man-age natural resources more effectively.
- Empowering local communities to effectively manage their own resources and combining traditional practices with local ones.

In areas where desertification has already become established, it is important to rehabilitate and restore the lands in order for them to return to their previ-ous conditions. Successful restoration must be done at the local level. Several methods are commonly used, such as:

- Reintroduction of the original, native species that used to live there.
- Combating erosion through the systematic terracing of steep areas so that water does not run down slopes, eroding the land's surface.
- Establishing seed banks to ensure that species do not become endan-gered or extinct. Then when climate conditions exist for the plant to survive, seeds can be planted.

- Enriching the soils with nutrients, making them more fertile and conducive to vegetative growth.
- Planting additional reserves of trees.

THE STRAIGHT FACTS ABOUT DESERTIFICATION

For land managers, it is critical that appropriate, effective, and well-thought-out decisions be made about the land and its resources. The following facts reflect the shocking realities of desertification facing today's world:

- Roughly 3.6 billion hectares of the world's 5.2 billion hectares of dryland used for agriculture has been degraded by erosional processes.
- One out of every six people is directly affected by desertification.
- Desertification has forced many farmers to abandon farming and look for urban employment.
- Each year 51,800 square kilometers of land is destroyed by desertification.
- Desertification affects almost 75 percent of the land in North America.
- Dust from deserts can be blown great distances into cities. Dust has been blown from Africa to Europe and the United States. During the Dust Bowl, it was blown from Oklahoma and out across the Atlantic Ocean.
- Desertification destroys the topsoil of an area, making it unable to grow crops, support livestock, or provide suitable habitat for humans.
- Climate change and poor land-management practices can trigger desertification.
- The destruction caused by desertification carries a hefty price tag. More than $40 billion per year in agricultural goods are lost, causing an increase in agricultural prices, negatively impacting the consumer.
- According to a United Nations study, roughly 30 percent of the earth's land is affected by drought. Every day, approximately 33,000 people starve to death.
- Desertification makes the environment more likely to experience wildfires.

Desertification is a condition that can be stopped, but it usually is not brought to the public eye until it is well under way, making rehabilitation of the landscape much more difficult and expensive. If it were brought to the public's attention sooner and not allowed to reach a critical point, it would make recovery much more simple. One way to ensure the success of this is through outreach and public education. By keeping people informed and educated about the results of their behavior on the environment, land degradation can be avoided altogether.

According to the United Nations Environmental Programme, desertification causes crop losses of approximately $42 billion a year. The Sahel countries have been experiencing drought-like conditions for the past 35 years with up to 20 percent less rainfall than they previously had. Semi-desert regions have already advanced 100 kilometers southward. Continental interior locations are expected to become drier (Dregne et al., 1992).

According to the IPCC, many deserts will face a decrease in rainfall from 5 to 15 percent. The Colorado Great Basin region of the southwestern United States is expected to be one of the hardest-hit drylands (Arizona, New Mexico, Nevada, and Utah). In order to fight against desertification, action must be taken now. Restorative measures may take 10 or more years to take effect. Low-intensity agriculture will have to be controlled in recovery areas. Not only does the scientific and technical community need to stand behind positive action, but today's and tomorrow's leaders do as well (Intergovernmental Panel on Climate Change, 2007a).

In a study featured in *The Independent*, in a news release in the United Kingdom on July 23, 2006, it was suggested that the Amazon rain forest could become a desert, which could then speed up the process of climate change with "incalculable consequences" (McCarthy, 2006).

In studies conducted by the Woods Hole Research Centre in the Amazon, the rain forest cannot withstand more than two consecutive years of drought without extreme ecological impacts and destruction. The research pointed to the fact that drought in the Amazon would trigger drought in the Northern Hemisphere and could accelerate climate change, creating a process that could get so far out of control that it could make the earth uninhabitable (Rohter, 2005).

The results of drought in the rain forest were that the trees managed fairly well after the first year. In the second year, they sunk their roots deeper in the ground to find moisture and still survived. But by the third year, the trees began dying. The tallest trees were impacted first. As they toppled to the forest floor, the lower canopies were exposed to the direct sunlight. The forest floor was also exposed to direct sunlight and dried out quickly.

After the end of the third year, the biomass had released more than two-thirds of the stored CO_2 to the atmosphere. Where once the forests operated as a carbon sink, they were now a major carbon source. What has scientists concerned is that the Amazon currently holds about 81.6 billion metric tons of carbon, enough in itself to increase the rate of climate change by 50 percent. On top of that, if a wildfire were to start in these remote locations and destroy the vegetation, the rain forests would be transformed into a desert. Dr. Deborah Clark from the University of Missouri, a renowned forest ecologist, says that the research shows that "the lock has broken" on the Amazon system. The Amazon is headed in a terrible direction (Real Climate, 2006).

Heat Waves

Temperatures are obtained from monitoring stations worldwide in order to calculate global mean temperature rise. When temperatures are taken near cities, they must be corrected to eliminate the specific effect that the presence of the urban area has on the temperature reading. Because urban areas have so many dark surfaces—such as asphalt-covered roads and dark roofs on buildings, they absorb more heat than natural surfaces such as grass, prairie, and woodlands. This absorbed heat is reradiated by the buildings and roads, and the resultant increase in temperature from these sources, as well as heat released from industry, cars, and other sources of burning fossil fuels, adds to the increased temperatures.

In order to use a reliable temperature value instead of the skewed value created by artificial inputs collectively referred to as the urban heat island effect (discussed in greater detail in Chapter 11), this contribution to the temperature must be accounted for and removed. In addition, the various instruments and methods used worldwide must be calibrated so that the temperatures collected are directly comparable.

According to the Union of Concerned Scientists, with all of these conditions taken into account, the global mean temperature has increased approximately 0.3–0.6°C over the past 150 years. More importantly, since 1975, the increase of the 5-year mean temperature has been calculated at 0.5°C, which is a rate that is faster than any other 5-year period in the past 150 years. Fourteen of the earth's warmest recorded years have occurred since 1990. In addition, reconstruction of past climates through the use of proxies (items that present evidence of past climatic conditions) such as fossil pollen, tree rings, coral, ice cores, and sediment cores reveal that the twentieth-century warming is significantly different than the past 400 to 600 years (Union of Concerned Scientists, 2007).

The only reliable explanation scientists have been able to come up with this far in their models is human interference. The temperature rises in the models accurately reflect the actual temperature rises the earth has experienced so far when the effects of greenhouse gas levels from the burning of fossil fuels and deforestation are entered into the mathematical model equations. If the human interference factor is not added, the models do not work—they underestimate actual temperatures. The polar latitudes have been identified in models as those areas on earth that are being impacted the fastest and most significantly. In addition to this, nighttime temperatures have increased much more than daytime temperatures, keeping the earth's atmosphere warmer overall. This is significant because when the earth stays warmer at night, it retains the heat that was generated during the day in the atmosphere, starting the next day off warmer than normal.

The Union of Concerned Scientists believes that by the year 2100 there will be an increase in average surface air temperature of around 2.5°C. The result of this will be an increase in temperature at many scales, such as days, seasons, and years. When looking at significant local levels of temperature increases, this means that some areas will succumb to more "extremely hot summer days" during the summer and "killer heat waves" will occur.

According to A. Kattenberg, temperate climates (such as the United States) would experience a doubling of extremely hot days, resulting in a 2–3°C increase in average summer temperature. D. J. Gaffen and R. J. Ross of the R. J. Ross Company have researched increased summer heat stress in the United States and have determined that the number of days per year that temperature thresholds for death have been passed have increased significantly over the past 50 years for many U.S. cities (Gaffen et al., 1998).

As heat wave incidents increase, more people will be negatively affected, and many of them could die. The sick, very young, elderly, and those who cannot afford indoor air conditioning are at the most risk of dying. As an illustration, in 1995 in Chicago, nearly 500 people died during a significant heat wave. In July and August of 1999, another heat wave hit Chicago. Because emergency response network specialists learned some valuable lessons in

1995, they were better prepared for the second heat wave. Even though the response was better, 103 heat-related deaths still occurred.

Chicago is not alone. In 1980, a heat wave in the United States killed 1,700 people in the East and Midwest, and again in 1988 one killed 454 people. In 1998, more than 120 people died in a Texas heat wave. Europe experienced a deadly heat wave in 2003, so hot, in fact, it was considered the hottest summer in 500 years. During this period, 27,000 people died from heat-related problems. Some of those who did not die suffered irreversible brain damage from advanced fevers as a result of the intense temperatures (Environmental Defense Fund, 2011). The heat wave in the United States in 2006 was one of the worst it had ever experienced. It held the entire country in its grip and lasted for almost a month. The effects and costs of this wave were enormous—hundreds of people died, massive power outages were triggered, and unmanageable wildfires burned large areas of ground. Tens of thousands of people in New York went without electricity for over a week.

One thing that scientists cannot do with the overall issue of climate change is to blame a single weather event—like a heat wave, a hurricane, a blizzard, or a tornado—on climate change, because weather fluctuates naturally. However, what they can relate to climate change are trends. Based on the fact that climate models predict more wild and unpredictable weather as a future trend, expecting an increase in heat waves does fit into the future climate scenario. One thing that both developed and developing countries need to be aware of is that continued urbanization will increase the number of people vulnerable to these urban heat islands and heat waves.

Wildfire

The likelihood of a disastrous wildfire occurring increases significantly during periods of drought. The world saw proof of this in 1997 through 1998 when an El Niño episode caused extremely dry conditions in many areas across the globe. Large forest fires occurred in Brazil, Central America, Florida, eastern Russia, and Indonesia. Wildfires can easily occur during drought periods from both natural and human-caused factors. Lightning strikes commonly cause them, as well as campers or hikers. Regardless of the cause, during drought conditions, the vegetation is so dry that it does not take much to start a wildfire, and once they start, they burn fast and intense.

According to a report on CNN on November 2, 2000, climate change may be a significant contributor to accelerating the fire cycle in desert ecosystems of North America. According to the report, which was published in the journal *Nature*, rising CO_2 levels from burning fossil fuels can negatively alter the delicate desert ecosystem (Environmental News Network Staff, 2000).

Stan Smith, a professor of biology at the University of Nevada in Las Vegas, as well as lead author of the study, said, "This could be a real problem for land managers." Smith and his team of scientists ran an experiment in the desert where they increased the CO_2 emissions by 50 percent and then monitored the resultant impact on four plant communities in the Mojave Desert ecosystem. With the elevated CO_2 levels, all the plants' densities and biomasses increased. The scientists likened this result to what would happen out in nature during high precipitation years and increased CO_2. Then, when drought occurred during an El Niño, this extensive biomass became dried out tinder, highly susceptible to wildfire.

Based on a report from the Natural Resources Defense Council, the 2006 wildland fire season set new records in both the number of reported fires as well as acres burned. Close to 100,000 fires were reported, and nearly 4 million hectares burned—125 percent above the 10-year average. If warming continues to spur wildfire seasons, it could be economically devastating. Fire fighting expenditures recently have totaled $1 billion per year.

In 2007, California was subjected to a series of devastating wildfires that began burning across southern California in October. More than 1,500 homes were destroyed, and more than 202,000 hectares of land were burned from Santa Barbara County southward to the United States–Mexico border. Nine people died, and eighty-five were injured. These fires occurred at the end of a very dry summer and were made even worse by the Santa Ana winds, which were blowing 97 kilometers per hour, fueling the fire. In addition, temperatures were consistently in the 32°C range, setting the stage for destructive, dangerous wildfires to do the most damage. When the last of the fires had been extinguished, the number of displaced residents that had been evacuated totaled above 900,000. It was the largest evacuation in California's history.

Air condition and quality also suffered because of the concentrations of particulate matter 10 micrometers and smaller (referred to as PM10) that reached unhealthy levels in the atmosphere, affecting those with breathing problems. It reached such unsafe levels that residents were requested to consider a voluntary evacuation from the city until the particulates were cleared from the air. As climate change continues, drought, desertification, and wildfire are a combination expected to become even more of a problem.

Mountain Ecosystems in Danger Worldwide

According to *National Geographic News* on February 1, 2002, the United Nations University in Tokyo has determined that climate change along with other issues such as pollution, deforestation, and overuse and development are heavily impacting and stressing mountain environments today (Institute for Global Environmental Strategies, 2010).

According to Dr. Jack Ives, a professor of geography and environmental science at Carleton University in Ottawa, Canada, as these mountain systems become more taxed, "water shortages, landslides, avalanches, and catastrophic flooding will become major consequences" (Ives, 2002).

Water shortages alone present a critical issue that warrants close attention. Mountainous regions occupy roughly 25 percent of the earth's land surface and provide a home to 10 percent of the world's population. These regions function as the principal source of water for half of the world's population. If climate change negatively impacted mountain ecosystems, the repercussions could be significant.

According to Dr. Ives, "the most severe examples of environmental and socioeconomic degradation—now near total disaster—are the Hindu Kush in Afghanistan, the Karakorum and western Himalaya, and the disputed territory of Kashmir" (Ives, 2002).

Other mountain ranges in trouble included in the United Nation's report as "ecologically endangered" are the European Alps, the Sierra Chincua in central Mexico, the Amber Mountains in Madagascar, the Snowy Mountains in Australia, and the Rocky Mountains in North America.

The problem consistent with all the mountain ranges is the melting of their glaciers and ice caps. The world's glaciers have been shrinking faster than they have been growing; there is a net loss worldwide because of climate change (Jowit and McKie, 2008). Losses during 1997–1998 have been identified as some of the most "extreme" and predict that up to 25 percent of the earth's glacier mass could completely melt by 2050, and up to 50 percent by 2100. If this were to occur, the only areas left on earth that would have any remaining glacial deposits would be Patagonia, the Himalayas, and Alaska.

The United Nations reported that the ice on Mount Kilimanjaro in Tanzania has lost 82 percent of its mass over the last 100 years. In addition, the glaciers in Montana's Glacier National Park are melting so rapidly that they could be completely gone in 30 years (MSNBC, 2006). Losing these ice-covered surfaces would have a negative impact on the earth's energy balance. According to Lisa Mastny of the Worldwatch Institute, "when the ice melts, newly exposed land and water surfaces retain heat, leading to even more melt and creating a feedback loop that accelerates the overall warming" (National Geographic News, 2010).

A serious consequence of these changing mountain ecosystems is the loss of species. As the climate warms, the unique plants and animals that cannot adapt will be forced to extinction. The United Nations reports that just in the Snowy Mountains of Australia, the warmer temperatures and lack of snowfall of the past few years threaten more than 250 species of plants. Vegetation has already begun to migrate in response. Subalpine trees are now growing at altitudes 40 meters higher than they were 25 years ago.

Another issue documented in the United Nations report was that of insect infestations. Warmer temperatures encourage insects to migrate further north into territories where it previously was too cold for them to survive. The Rocky Mountain chain of North America has experienced several insect infestations of the past decade in the Canadian Rockies, and the consistently warm winters of the past few years have triggered a pine-beetle infestation that has now affected the trees of more than 5,000 square kilometers of the pristine forested areas of British Columbia. Mountain forest stands in Alaska are facing the same problem.

Lack of Water Storage

In a report issued by *Space Daily* in 2004, based on a recently developed climate change model, climate change is expected to reduce the amount of water stored as snow in the western United States by up to 70 percent over the next 50 years. They warn that a reduction in snowfall means much more than a shortage of drinking water for the next summer. If there is a reduction in the snowfall that covers the Sierra Nevada that feeds California and that which reaches the Cascades of the Pacific Northwest, it will lead to increased fall and winter flooding because the snow pack will not stay frozen. Without sufficient snow pack, there will not be enough water stored in the mountains to supply an adequate amount to residents along the populated West Coast areas. Wildlife indigenous to the region will also suffer. In drought situations, water is often rationed and residents are restricted as to how much personal water they can consume for household use and landscaping. Significant amounts of water are also needed for industry and commercial businesses. All this must be budgeted and accounted for during years of water shortage. Some of the biggest impacts in the West will be felt with West Coast fisheries, agriculture, and hydropower generation (Space Daily, 2004). L. Ruby Leung, a scientist at the Department of Energy's Pacific Northwest National Laboratory, said that this prediction concerning water storage along the West Coast is "a best case scenario."

The model developed jointly by the Department of Energy and the National Center for Atmospheric Research is on the conservative end compared with most climate prediction models. This model assumes a 1 percent increase each year in the rate of greenhouse gas concentrations through the year 2100, little change in precipitation, and an average temperature increase of 1.5–2°C through 2050 (Pacific Northwest National Laboratory, 2004). The model's result was that most of the precipitation would fall as rain instead of snow: 0.5–1.3 centimeters each day, which would push the snowline in the mountains from 914 meters to 1,219 meters and higher. This would mean that instead of snow melting in late April, by 2050, the snow would melt off much earlier. In fact, over the past 50 years, the coastal mountain ranges have lost 60 percent of their normal snow pack. Leung says, "The change in the timing of the water flow is not welcome. The rules we have now for managing dams and reservoirs and irrigation schedules cannot mitigate for the negative effects of climate change" (Pacific Northwest National Laboratory, 2004).

Glaciers and Flooding

One of the prominent concerns in mountain ecosystems in light of climate change is the melting of glaciers and flooding of valley bottoms and other lowlands upon the rupture of huge melt-water lakes that have built up behind glaciers. As glaciers begin to melt and become unstable, it does not take much force from the water backed up behind them to push its way forward, escape its temporary dam, and rush downhill, destroying every-thing in its path. Martin Beniston, a climate scientist at Freiburg University in Switzerland, says, "In the Himalayas, some glaciers are up to 70 km long. In Bhutan alone, there are at least 50 lakes in this category, and a similar number in Nepal, as well. Towns and villages in their path could be hit by a tsunami" (Agence France-Presse, 2005).

Not only has the climate warmed up, but these areas have also received less snowfall. According to Heinz Slupetzky of the University of Utrecht in the Netherlands, all of the glaciers could be melted as early as 2080. In another part of the world, Yao Tandong, a glacier expert from China's Institute of Tibetan Plateau Research at The Chinese Academy of Sciences, said that up to 64 percent of China's high altitude glaciers could be com-pletely melted by 2050 (Agence France-Presse, 2005). As a measure of the volume of ice that represents, he compared it with an amount equivalent to all the water in the Yellow River. One of the most crippling side effects of this is that 23 percent of China's 1.3 billion people currently depend on the glacial runoff as their source of drinking water.

Challenges in Alpine Regions

The mountain regions are experiencing some of the most profound changes due to climate change. In addition to the obvious loss of glaciers (which is easily quantifiable), hotter temperatures are driving the alpine zones farther toward the mountain summits. Alpine zones are usually found at an altitude of about 3,048 meters or higher. They lie just below the snowline of a moun-tain. Before long, under the constant impact of rising temperatures, the alpine plants and animals will have nowhere else to go. Species that cannot adapt will not survive. Already in the American West, this is being seen with the destruction of pika habitat in many portions of the Rocky Mountains today.

Vegetation is facing the same dilemma. On Mount Rainier and in Olympic National Park, mountain hemlock and subalpine fir have been documented moving into areas that have traditionally served as alpine meadows. In Yellowstone National Park, a similar migration is occurring with white bark pine. It is now migrating toward the mountain summits. If the white bark cannot find suitable habitat, it will affect the grizzly bear population, because it is a significant food source for the grizzly. Therefore, climate change can jeopardize entire food webs.

In Europe, the concerns scientists have regarding alpine regions are for the biodiversity and survival of the fragile life forms that exist, such as the delicate mosses and flowers. According to Georg Grabherr at the University of Vienna, if climate change continues, the alpine vegetation in Europe will migrate northward and eastward. He predicts that vegetation that is cur-rently found in the Mediterranean will eventually be found in the central European countries. Studies he has already conducted support his predic-tions: he has confirmed that alpine vegetation is already beginning to shift.

Plants are not only migrating north, but they are also moving upward in elevation on mountain ranges (SwissInfo, 2003). One of the big concerns he has expressed is whether or not vegetation will be able to migrate northward because so much of the vegetation's natural habitat has been taken for urbanization and other human uses.

As stated by Vera Markgraf of the Institute of Arctic and Alpine Research at the University of Colorado at Boulder, "You reduce the capacity of plants to propagate if the habitat is already occupied or not available, as the situation is today where most habitats have been altered by human intervention" (SwissInfo, 2003).

Challenges in Marine Environments

Temperate Marine Environments

There are several components of the temperate marine environment that climate change could impact, such as changes in temperature, shoreline ecology, and major storm tracks. Two elements that are the least complicated to predict are temperature and sea-level rise; others are much more complex depending on scale and degree of interaction between other components in the environment.

It is difficult to use existing climate models to address these issues because they produce outputs that generally depict broader geographical regions than the scales at which local storms occur. Models are not refined enough at this point for their spatial resolution to be able to model areas as small as a specific bay or coast.

The IPCC predicts that by 2100, the earth's near-surface temperature averaged worldwide will increase by 1.4–5.8°C from the 1990 levels. This means that sea-surface temperatures will also rise. The IPCC believes that this will be the highest temperature rise seen in the past 10,000 years. The aspect of the temperature rise that will cause the most significant ecological change in estuary and marine ecosystems is the rapid speed with which it is expected to happen, leaving species little time to adapt (IPCC, 2007a).

Based on data from the IPCC, globally averaged sea level rose between 10 and 20 centimeters during the twentieth century. They predict the oceans will rise another 9–88 centimeters between now and 2100. The rise will be the result of both thermal expansion of the current ocean water (the warmer the temperature, the more water expands) and melting from land-based glaciers and ice sheets.

The Effects of Climate Change on Coastal Locations

The Pew Center on Global Climate Change is a nonprofit organization that brings together business leaders, policy makers, scientists, and other experts worldwide to create a new approach to managing the problem of climate change, what they refer to as an extremely "controversial issue." Not slanted in any particular way politically or economically, they "approach the issue objectively and base their research and conclusions on sound science, straight talk, and a belief that experts worldwide can work together to protect the climate while sustaining economic growth."

In January 2007, the Pew Center was one of the inaugural members of the U.S. Climate Action Partnership—an alliance of major businesses and environmental groups that calls on the federal government to enact legislation requiring significant reductions of greenhouse gas emissions.

According to the Pew Center on Global Climate Change, in terms of climate change, the biggest impact on estuarine and marine systems will be temperature change, sea-level rise, the availability of water from precipitation and runoff, wind patterns, and storminess. In these often-fragile systems, temperature has a direct and serious effect. For the sea life living within the ocean, temperature directly affects an organism's biology, such as birth, reproduction, growth, behavior, and death.

Temperature differences can also influence interaction between species, such as predator-prey, parasite-host, and other competitions that may develop over the struggle for limited resources. If temperatures change the distribution patterns of organisms, it could also change the balance of predators, prey, parasites, and competitors in an ecosystem, completely readjusting food chains, behaviors, and the equilibrium of the ecosystem.

Climate change can also change the way that species interact by changing the timing of physiological events. One of the key changes that it could alter is the timing of reproduction for many species. Rising temperature could interfere with the timing of the birth being correlated with the availability of food for that species. This can be a problem, for example, for bird species that migrate and depend on a specific food source to be available when they reach their breeding grounds. If warmer temperatures have changed the timing of when the food will be available by a few weeks, and it is no longer synchronized with the migrating birds, it could leave the birds without available food, threatening their survival.

Sea-Level Rise

Melting glacial and polar land ice will add to sea-level rise. The effects of sea-level rise will not be constant but will vary with location, how fast the sea level rises, and the biogeochemical responses of the individual ecosystems involved. The South Atlantic and the Gulf of Mexico have been identified as susceptible areas.

As sea levels rise, ocean water will submerge and erode the shorelines. In natural areas covered with marshes and mangroves, as sea level rises, water will flood the wetlands and waterlog the soils. Because the plants that live in the wetlands, which do not contain salty water, are not accustomed to salt, the salty ocean water would kill them. Because wetlands provide habitat for wildlife, including several migrating birds, this would also destroy their habitat.

The areas that would be hardest hit are those where species cannot migrate inland because urbanization has resulted in building right up to the shoreline, effectively removing any possible potential wetland habitat. This presents a serious impact to the environment because wetlands are an important part of the biological productivity of coastal systems. Marshes provide many critical services; they function not only as habitats for wildlife but as nurseries for breeding and raising young, and as refuges from predators. Wetlands function as part of an integrated system. If they are jeopardized, their loss will affect the availability and transfer of nutrients, the flow of energy, and the availability of natural habitat needed by the multitudes of organisms already living there. One of the most unfortunate losses will be those areas where rare, threatened, or endangered plant and animal species, such as the American alligator, Florida black bear, West Indian manatee, Florida panther, southern bald eagle, snowy egret, and roseate spoonbill, live. If invading salt water destroys these habitats, they could become extinct.

The Effects of Climate Change on Open Oceans

Because the oceans are so vast, much of the predictions made about the earth's open oceans in light of climate change have been generated by computer models. The two most important functions that govern the behavior of the ocean are temperature and circulation, and these calculations are fairly straightforward in computer models and readily calculable. Another thing that makes it possible to model the oceans using computers is that human impact has not had as large an impact on the oceans as it has on land.

Tropical Marine Environments

The world's tropical and subtropical marine environments represent some of the most diverse habitats on earth. Often characterized by reef-building corals, the complex arrays of marine inhabitants that occupy these waters have developed many strategies for survival. Their ecosystems are so complex that a delicate balance is needed to protect these numerous marine resources, while also accommodating an economy centered on commercial fisheries and recreation.

The tropical marine environment is also subjected to many of the same environmental concerns as the temperate marine environment, such as temperature, sea-level rise, wind circulation, and algae blooms. An additional risk the tropical habitat encounters is the negative impact on reefs and corals. These ecosystems represent some of the most fragile on earth and some of the hardest hit with the negative effects of climate change.

Fragile Ecosystems: Reefs and Corals

As climate change causes the earth's tropical oceans to heat up and more acidic and strong storms become more common, the world's coral reefs are taking a beating. Rod Fujita of the Environmental Defense Fund says, "Coral reefs may prove to be the first ecological victims of unchecked global warming" (Environmental Defense Fund, 2007b).

According to the United States Commission on Ocean Policy (2004), the loss of coral reefs would translate into huge economic losses in coastal regions dependent on reefs, which currently provide about $375 billion each year in food and tourist income.

Destruction of reef systems also represents an ecological disaster. Coral reefs are sometimes referred to as the rain forests of the ocean because they provide habitat to a rich diversity of marine life, including reef fish, turtles, sharks, lobsters, anemones, sponges, shrimp, sea stars, sea horses, and eels.

Coral reefs attract scuba divers from around the world each year to swim among the beautiful, otherworldly shapes and color combinations. Corals actually obtain their food and color from tiny algae called zooxanthellae that live in them. Corals have a very narrow temperature tolerance range. In fact, if the water increases only 1.1°C above the typical maximum summer temperature, it can cause coral to expel their algae and turn white, through a process called "bleaching." If bleaching continues for a prolonged period of time, the corals will die.

Unfortunately, there have been several highly destructive or fatal bleachings to the world's coral. In 1997–1998, during one of the warmest 12-month periods ever recorded, there was massive documented bleaching of the world's corals. Of these, 16 percent of the earth's reefs suffered severe damage. The warming caused ancient, thousand-year-old corals to die off. Doug Rader, a scientist with the Environmental Defense Fund, says, "Within a century, very large portions of coral reefs could be gone. It seems hard to believe that it is happening—and happening on our watch" (Environmental Defense Fund, 2007b).

In an article in *The New York Times*, it was confirmed that increasing acidity in the oceans hurts the reef algae as well as corals. Ilsa B. Kuffner of the USGS was involved in experimentation that illustrated that when seawater absorbs CO_2 it leads to lower saturation levels of carbonate ions, which reduces calcification. Calcification is the process that corals use to make their hard skeletons. Ocean acidification also negatively impacts crustose coralline algae, another important reef builder, which acts as a cement-like substance that bonds the reefs and helps the reef ecology. The result of the experiment was that the specimens subjected to higher heat and acidity became soft, unhealthy, resistant algae (Fountain, 2008).

Climate Change Stress to Coral Reefs

The World Wildlife Fund reports that coral reefs around the world have been severely damaged by unusually warm ocean waters. They predict that less than 5 percent of Australia's Great Barrier Reef will remain by 2050 if the world fails to reduce carbon dioxide emissions. They project that "if the present rate of destruction continues, a good proportion of the world's coral reefs could be killed within our lifetime" (World Wildlife Fund, 2011).

The World Wildlife Fund has identified both the Seychelles Islands and American Samoa as locations under high stress for coral bleaching. The United Nations Educational, Scientific, and Cultural Organization recognizes the Seychelles Islands as a natural World Heritage Site. They have a high coral diversity and support rare land species, such as the giant tortoise. In addition to increasing ocean temperatures, these areas are also threatened by climate change because of more frequent tropical storms (which could break up the coral) and more frequent rains, flooding, and river runoff (which deposits sediments in the ocean) (Figure 3.4).

According to NASA, reef habitats are so complex and warrant so much exploration "that for marine biologists, the destruction of the reefs has proven to be as frustrating as it is heart breaking. At the rate the reefs are disappearing, they may be beyond repair by the time a comprehensive plan to save reefs can be put into place" (NASA, 2001). Even in view of unstoppable damage to reefs in the future, conservationists say that given the complex factors affecting coral health, there is still a lot that can be done to help reefs recover if action is taken now.

Freshwater Environments

Only 2.58 percent of the water on earth is fresh. The largest share of that, 1.97 percent, occurs as ice in the world's glaciers and ice caps. The remaining

FIGURE 3.4
Coral reefs are often referred to as the rain forests of the ocean because they provide habitats for an extremely diverse collection of life forms. Climate change is threatening their existence. (Courtesy of National Oceanic and Atmospheric Administration, Photo Library, http://www.photolib.noaa.gov/htmls/reef2559.htm, 2010.)

freshwater, 0.61 percent, is stored in lakes, rivers, and groundwater. The earth's climate—the processes of precipitation, evaporation, and water vapor transport—determine the amount and distribution of freshwater at any given time on earth.

IPCC Assessment

According to the IPCC, whereas growing populations are already overtaxing freshwater supplies in many parts of the world, climate change will make freshwater supply even scarcer in certain areas. Warming temperatures will also have consequences. The winter ice cover of streams and lakes will decline, and there will be a trend toward later freeze and earlier ice breakup, as is already present in Europe. The timing and duration of freeze and the breakup of ice is important because it affects both biological and ecological processes. If precipitation decreases, the water flow rates will drop, and this could have impacts on lakes and streams. This could lead to changes in habitat and breeding locations of aquatic flora and fauna. The IPCC has also determined that hydrologically isolated systems, such as wetlands and topographical depressions, would be the most vulnerable areas to global climate change. Areas along larger rivers and lake shores would be impacted the least.

Because freshwater supplies diminish with increasing temperatures and in areas that have a lack of precipitation, there will also be increased competition for diminishing freshwater resources. Even if precipitation may increase in some areas during the winter, if it cannot be stored as groundwater and runs off the surface of the ground when it melts in the spring, it will be unusable. Summers are predicted to become warmer and drier, which will also lead to a deterioration of freshwater ecosystems.

Conclusion

As evidenced from the impacts currently happening to ecosystems, the effects are not minor, nor are they localized. On the contrary, they are serious, large-scale impacts that have endangered species, degraded living conditions, destroyed habitats, impacted the health and livelihood of societies, caused starvation and extensive misery, and threatened the future of both humans and wildlife. So far. Yet step back and reflect a bit.

Scientists have been warning policy makers for the past two decades of the changes humans are making to the environment—changes largely caused by lifestyle preferences. Only recently has the political community been listening—Europe for the past decade has been on board, but countries such as the United States have stubbornly turned their heads the other way in almost defiant response, instead demanding that the rest of the world cut back their greenhouse gas emissions while they do nothing.

The temperatures and parts per million (ppm) carbon content in the atmosphere have just begun to rise toward what climate scientists predicted a decade ago, with a delayed response time in the earth's atmosphere, just as they warned. It is happening just like clockwork.

What do you suppose it will be like when the temperature climbs another degree? Or two? What about the people in distant, undeveloped countries that never even contributed significantly to the problem who may be suffering the most, while people in developed countries are still trying to live their "business as usual" lifestyle?

Sometimes it takes the people who can listen to the truth, read the facts, and take a good look around their environment—the global environment—taking leadership and making their voices heard in order to make a real difference. We all have personal choices to make, and that includes the way we treat the environment. The damage being done right now to the earth's ecosystem is oftentimes irreversible, and what our direct or indirect actions are causing to be lost will not only impact those of us who inhabit the earth right now but will also impact our children and grandchildren. For all those future generations who did not have a say and have not yet seen the beauty that this planet has to offer, the time to act is now.

Discussion

Consider the following issues and their importance in protecting the health of ecosystems:

1. How do you see "climate change" as one of the most pressing problems of the twenty-first century, as many prominent scientists worldwide have labeled the issue?

2. What are your views on species endangerment and extinction? How much research have you personally done, or are you willing to do, to educate yourself on the issue? Where would you go for information?

3. If you were a land or resource manager, how would you handle current species survival and habitat suitability projections and balance urban expansion issues? If you had to sacrifice key urban areas for wildlife habitat conservation, which types of urban-use areas would you convert?

4. What kind of price do you put on the environment you leave for future generations? If you were in control of key budgets, what

percentage of funding would you justify to protect ecosystems? What funding items would you omit in order to address climate change, despite extreme pressure from public-interest groups?

5. In order of priority, which natural resources would you deem most important to protect if you were a land manager and your funding was limited?

6. If you were a resource manager for a national government, how would you allocate future budgets to accommodate climate change and why?

7. If you were a key political figure in your country's government, what would be your agenda on climate change and what programs would you implement to address climate change issues in order to solve the problem in both the short and long term?

8. As a multiple-use land manager, you are tasked to balance wildlife habitat management with other land uses, such as tourism, oil exploration (one of the largest sources of income for your area), farming, outdoor recreation, and conservation measures. You are offered a multimillion-dollar budget to develop a major recreation project right in the middle of an area needed as a corridor by wildlife—one essential for their survival in a warming world. What will you do? Is there latitude for compromise? If so, would such a plan involve an incremental or phased introduction approach? Justify your decision to environmentalists, land investors, developers, taxpayers, recreational enthusiasts, land use planners, etc.

9. What methods can you think of to implement positive action toward the prevention and/or remediation of desertification? What public outreach and educational programs would you develop and launch to increase public awareness and involvement? How would you justify funding of tax dollars spent? How would you handle protests and petitions?

References

Arctic Climate Impact Assessment. 2004. *Impacts of a Warming Arctic.* Boston, Massachusetts: Cambridge University Press. http://amap.no/acia/ (accessed May 1, 2011). Discusses the research done on climate change in the polar ecosystem.

Agence France-Presse. 2005. Global warming: Mountains face tsunami risk. United Nations. http://www.un.org/special-rep/ohrlls/News_flash2005/13%20 Feb%20Global%20warming-Mountains%20face%20tsunami%20risk.htm (accessed March 15, 2011). Discusses the effects of climate change on mountain lakes.

Associated Press. 2006. Deserts drying and widening, UN report finds. MSNBC, June 6. http://www.msnbc.msn.com/id/13147504/ns/world_news-world_environment/ (accessed October 6, 2010). Discusses climate change in deserts.

Conant, R. T., 2010. Challenges and opportunities for carbon sequestration in grassland systems: A technical report on grassland management. *Integrated Crop Management*, Vol. 9-2010.

Corvalan, C., and S. Hales. 2005. *Ecosystems and Human Well-Being*. Geneva, Switzerland: World Health Organization. http://www.who.int/globalchange/ecosystems/ecosys.pdf (accessed March 27, 2011). Discusses climate change and the various effects it has on health.

Douglas, J. 2006. Global warming to cause massive drought over next 100 years, say climatologists. *Natural News.com*, October 6. http://www.naturalnews.com/020649_drought_global_warming_climate_change.html (accessed October 15, 2010). Discusses the impacts that will be faced if changes are not made today.

Dregne, H. E., and N. T. Chou. 1992. *Global Desertification Dimensions and Costs*. Lubbock: Texas Tech. University.

Environmental Defense Fund. 2007a. Mother nature signals climate warnings. *Environmental Defense Fund*, August 16. http://www.edf.org/article.cfm?contentID=4823 (accessed March 11, 2011). Discusses the damage to the natural environment as a result of climate change.

Environmental Defense Fund. 2007b. Threats: Coral reef bleaching. August 16. http://www.edf.org/article.cfm?contentID=4709 (accessed February 22, 2011). Offers information on how the world's coral reefs are being endangered and how they can be saved.

Environmental Defense Fund. 2011. Deadly heat waves more likely. http://www.fightglobalwarming.com/page.cfm?tagID=251 (accessed March 21, 2011). Discusses the deadly health effects of heat waves associated with climate change.

Environmental and Energy Study Institute. 2007. Amazon could become grassland under climate change. *Climate Change News*, January 5. http://www.eesi.org/ccn?page=199 (accessed September 21, 2010).

Environmental News Network Staff. 2000. Vicious cycle: Global warming feeds fire potential. CNN, November 2. http://archives.cnn.com/2000/nature/11/02/global.warming.enn/index.html (accessed October 5, 2008). Discusses the correlation between global warming and wildfires.

Food and Agriculture Organization of the United Nations. 2001. Pan-tropical survey of forest cover changes 1980–2000. *Global Forest Resources Assessment 2000*. Rome, Italy: Food and Agriculture Organization of the United Nations. http://www.fao.org/docrep/004/y1997e/y1997e1f.htm (accessed September 17, 2010).

Fountain, H. 2008. More acidic ocean hurts reef algae as well as corals. *The New York Times*, January 8. http://www.nytimes.com/2008/01/08/science/earth/08obalga.html (accessed August 23, 2008). Explains that as more carbon dioxide is added to the atmosphere, the oceans are becoming acidic, and scientists believe this is damaging not only coral but also reef algae.

Gaffen, D. J., and R. J. Ross. 1998. Increased summertime heat stress in the U.S. *Nature*. 396: 529–530. Touches on the realities of global warming and the effects that will only get worse as temperatures continue to climb.

Greenfact Millennium Ecosystem Assessment. 2005. Scientific facts on desertification. http://www.greenfacts.org/en/desertification/index.htm (accessed September 8, 2011).

Institute for Global Environmental Strategies. 2010. 2010 top news on the environment in Asia (Provisional Version), December 20. http://www.iges.or.jp/en/news/topic/1012topnews.html Discusses the damage to mountain systems in Japan.

Intergovernmental Panel on Climate Change. 2001a. Climate change 2001: Synthesis report: Summary for policymakers. http://www.ipcc.ch/pub/un/syreng/spm.pdf (accessed December 9, 2010).

Intergovernmental Panel on Climate Change. 2001b. *Impacts, Adaptations and Vulnerability. Working Group II, Third Assessment Report.* Cambridge, UK: Cambridge University Press.

Intergovernmental Panel on Climate Change. 2007a. *WG1: The Physical Science Basis.* Cambridge, UK: Cambridge University Press.

Intergovernmental Panel on Climate Change. 2007b. Food, fibre and forest products. *Impacts, Adaptation, and Vulnerability: Working Group II, 275–303.* http://www.ipcc-wg2.gov/AR4/website/05.pdf

Intergovernmental Panel on Climate Change. 2007c. *The Regional Impacts of Climate Change: Executive Summary.* http://www.ipcc.ch/ipccreports/sres/regional/index.php?idp=43

Ives, J. 2002. Mountain disasters. *International Year of the Mountain Conference.* November 15 and 16. University of Colorado at Boulder. Discusses the future of mountain ecosystems.

Jardine, K. 1994. The Carbon bomb: Climate change and the fate of the northern boreal forests. *Greenpeace International.* Amsterdam: the Netherlands.

Juliette, J., and R. McKie. 2008. Glaciers melt at fastest rate in 5000 years. *The Observer,* March 16. http://www.guardian.co.uk/environment/2008/mar/16/glaciers.climatechange1.

Karoly, D., J. Risbey, and A. Reynolds. 2002. Global warming contributes to australia's worst drought. World Wildlife Fund. http://www.wwf.org.au/publications/drought_report/ (accessed November 23, 2010). Discusses the effects of climate change in Australia.

Kluger, J. 2006. Global warming heats up. *Time,* March 26. http://www.time.com/time/printout/0,8816,1176980,00.html. Discusses several environmental aspects of the environment that point to the effects of global warming.

Malcolm, J. R., C. Liu, R. P. Neilson, L. Hansen, and L. Hannah. 2006. Global warming and extinction of endemic species from biodiversity hotspots. *Conservation Biology.* 20(2): 538–548. http://ddr.nal.usda.gov/bitstream/10113/28647/1/IND44194501.pdf.

McCarthy, M. 2006. The century of drought. *The Independent,* October 4. http://www.commondreams.org/headlines06/1004-02.htm (accessed November 22, 2008).

MSNBC. 2006. Africa's tallest mountains nearly bare of ice: UN warns of human disaster If kilimanjaro, kenya run dry in 50 years. October 12. http://www.msnbc.msn.com/id/15238801/ns/world_news-world_environment/ (accessed March 28, 2011). Discusses the melting of glaciers worldwide due to global warming.

National Geographic News. 2002. Mountain ecosystems in danger worldwide, UN says. February 1. http://news.nationalgeographic.com/news/2002/02/0201_020201_wiremountain.html (accessed January 4, 2009). Addresses water shortages brought on by global warming.

Natural Resources Canada. 2002. *Climate Change Impacts and Adaptation: A Canadian Perspective.* Climate Change Impacts and Adaptation Directorate. http://adaptation.nrcan.gc.ca/perspective/pdf/report_e.pdf (accessed January 11, 2009). Discusses global warming impacts on forests.

NASA Earth Observatory. 2001. Mapping the decline of coral reefs. http://earth observatory.nasa.gov/Features/Coral/ (accessed March 15, 2011). Discusses the heartbreak of the vast majority of dying reefs worldwide because of climate change.

Pacific Northwest National Laboratory. 2004. Global warming to squeeze western mountains dry by 2050. http://www.scienceblog.com/community/older/2004/3/20042513.shtml (accessed February 16, 2011). Discusses the effects of climate change in mountain environments.

PBS. 2001. *Bill Moyers Reports: Earth on Edge.* http://www.pbs.org/earthonedge/program/index.html.

Real Climate. 2006. Amazonian drought. http://www.realclimate.org/index.php. archives/2006/08/amazonian-drought/ (accessed July 8, 2011). Discusses the future of the Amazon under the conditions of severe desertification.

Revkin, A. C. 2007. Arctic Sea ice melting faster, a study finds. *New York Times,* May 1. http://www.nytimes.com/2007/05/01/us/01climate.html (accessed September 30, 2010). Discusses the impacts from climate change in the Arctic.

Ritter, K., and C. J. Hanley. 2011. Seas could rise up to 1.6 meters by 2100: New report. *Coastal Care News,* May 3. http://coastalcare.org/2011/05/seas-could-rise-up-to-1-6-meters-by-2100-new-report/ (accessed May 3, 2011). Discusses the latest results on climate change's effects on the coasts.

Rohter, L. 2005. A record Amazon drought, and fear of wider ills. *The New York Times,* December 11. Discusses the problems associated with desertification.

Science Daily. 2001. Scientists find that grasslands can act as "carbon sinks." January 15. http://www.sciencedaily.com/releases/2001/01/010111073831.htm (accessed November 9, 2010).

SIMBA. 2005. *Sea Ice Mass Budget of the Arctic: Bridging Regional to Global Scales: Executive Summary.* International Arctic Research Center, Fairbanks, AK. http://www.iarc.uaf.edu/workshops/SIMBA_2005/SIMBAreport.pdf.

Space Daily. 2004. Global warming to squeeze western mountains dry by 2050. February 18. http://www.spacemart.com/reports/Global_Warming_To_Squeeze_Western_Mountains_Dry_By_2050.html (accessed January 11, 2009). Addresses the effects global warming will have on future snow pack and how that will impact the use and demand of water.

Struck, D. 2007. NOAA scientists say Arctic ice is melting faster than expected. *Washington Post,* September 7. http://www.washingtonpost.com/wp-dyn/content/article/2007/09/06/AR2007090602499.html.

SwissInfo. 2003. Global warming threatens alpine plants. September 9. http://www.swissinfo.ch/eng/Home/Archive/Global_warming_threatens_alpine_plants.html?cid=3501194 (accessed March 28, 2011). Discusses the fragile alpine ecosystems in the Swiss Alps and the challenges faced there with climate change.

Union of Concerned Scientists. 2007. IPCC Fourth Assessment Report: Climate Change Mitigation. *IPCC Highlights Series: Findings of the IPCC Fourth Assessment Report.* Presents the IPCC's findings for other scientists.

UniSci. 2001. Many Alaskan glaciers are thinning, USGS study says. *Daily University Science News*. http://www.unisci.com/stories/20014/1211011.htm (accessed October 6, 2010). Discusses the fate of Alaska's glaciers currently affecting the tourist industry.

U.S. Commission on Ocean Policy. 2004. *An Ocean Blueprint for the 21st Century*. Washington, DC: U.S. Commission on Ocean Policy.

World Wildlife Fund. 2003. *Buying Time: A User's Manual for Building Resistance and Resilience to Climate Change in Natural Systems*. World Wildlife Fund Climate Change Program.

World Wildlife Fund. 2011. Coral reefs bleach to death. http://www.worldwildlife.org/what/wherewework/coraltriangle/coral-reefs-bleaching-to-death.html. Presents critical information on the fate of the world's coral reefs.

Suggested Reading

Bridges, A. 2006. Greenland dumps ice into sea at faster pace. *Live Science*, February 16.

Britt, R. R. 2005. Scientists put melting mystery on ice. *LiveScience*, June 30. http://www.msnbc.msn.com/id/8421342/40041707 (accessed December 2, 2010).

Christianson, G. 1999. *Greenhouse: The 200-Year Story of Global Warming*. New York: Walker.

Gore, A. 2006. *An Inconvenient Truth*. Emmaus, PA: Rodale.

Houghton, J. 2004. *Global Warming: The Complete Briefing*. New York: Cambridge University Press.

Karl, T., and K. Trenberth. 1999. The human impact on climate. *Scientific American*, 281(6): 100–105.

United Nations Environmental Programme/Arctic Monitoring and Assessment Programme. 2011. Climate change and POPS: Predicting the impacts. http://arctic-council.org/climate_change_and_pops_predicting_the_impacts (accessed April 1, 2011). Discusses the latest research results in the Arctic ecosystem.

4

The Inception of Climate Change Management

Overview

Because climate change is a global phenomenon, it will take the concerted effort of every person on earth to solve the problem. Not just a select few, or one region, or even one country can slow and reverse the processes that have been put in motion. The most efficient way to get results from a global effort is to have it well thought out, organized, and mandated. Voluntary, disorganized efforts do not work. Humans have a natural tendency to "just let someone else take care of the problem" and think that it will simply go away.

Climate change, however, is one problem that is not going to just go away if only a few dedicated people join together and give it their full attention while the rest of humanity goes on with "business as usual." This situation, for many, will be an entirely new paradigm shift. This is a time when organization is critical. Because of the highly technical nature of the problem, it is imperative that there be organized scientific involvement to keep abreast of changes as they occur and then keep leaders, policy makers, and the general public aware of the facts so that proper action can be taken along the lines of safety, preparedness, health, education, planning, and lifestyle adjustments. People need to be kept apprised so that they can plan ahead and react wisely rather than panic or be caught unprepared. Likewise, government leaders and policy makers need to be kept appraised so that planning can occur and the proper infrastructure can be put in place.

This chapter introduces the inception of climate change management. It begins by looking at the working template that it was partly modeled after: the Montreal Protocol, the document put in place in 1987 to globally lower the chemicals being added to the atmosphere that were destroying the stratospheric ozone. The Protocol had been so rapidly adopted and implemented worldwide, meeting such success, that it was hoped the same could be done for the greenhouse gases being pumped into the atmosphere by industrialized nations. We will find out what happened. Next, the chapter looks at the United Nations Framework Convention on Climate Change and the creation of the Kyoto

Protocol, as well as the infamous reaction by the United States and subsequent international reactions towards "the country who wouldn't play on the team."

The chapter then explores the international G8 advisory group and the creation of the Intergovernmental Panel on Climate Change (IPCC) and the critical, worthwhile work they are involved in. This chapter then sums up the vital climate conferences that have been held over the past decade, whose purpose has been to put a working long-term plan in place to keep future temperature rise below the critically designated 2°C, a threshold long ago identified by the world's most notable climate scientists. Pivotal climate conferences include the Bali Conference in 2007, the Copenhagen Climate Change Conference in 2009, and the Cancun Convention in 2010. The chapter discusses the expected results and unexpected surprises that occurred, as well as what policy makers and the public alike learned—all made possible because of organized climate change management.

Introduction

Putting a working scheme of international climate management together is not an easy task. In fact, meshing different political systems, economic situations, sociological philosophies, and vastly different lifestyles and expectations is extremely difficult and to cut through all the differences up front and still be able to put a global climate policy together is nothing short of outstanding. It is not only outstanding, but also critically necessary. Without someone leading climate change management, the outcome would most likely be a disarray of opinions, pieces of partial information, errors of miscommunication, and straight lack of communication. In any kind of a critical issue—especially one this large and serious—there would be no way to even deal with it without organized management.

Fortunately for climate science, the United Nations had the resources available to be able to offer an avenue in which to manage the research, assemble regular international meetings, make international decisions, implement and enforce treaties and agreements, and act as mediators. Today, they represent, through groups such as the IPCC, some of the largest, most well-represented governing bodies dedicated to the understanding, control, and solution of scientific issues. Providing insight on climate change issues and directing those who are in a position to make sound decisions is one of the greatest assets they have to offer.

Development

The Montreal Protocol: A Working Model

The Montreal Protocol on Substances That Deplete the Ozone Layer is a protocol to the Vienna Convention for the Protection of the Ozone Layer, which is an international treaty designed to protect the ozone layer through the phasing out of the production of several substances that are believed to be responsible for ozone depletion. The purpose of the treaty is to recognize that worldwide emissions of certain substances can significantly deplete and modify the ozone layer in a way that is likely to result in adverse effects on human health and the environment. Therefore, the treaty is to protect the ozone layer by taking precautionary measures to control global emissions of ozone-depleting substances, with the ultimate objective being their elimination from use. The treaty was opened for signature on September 16, 1987, and entered into force on January 1, 1989. It has been ratified by 196 nations.

Since that time, the Protocol has evolved through seven revisions, and it is believed that if the international agreement is strictly followed, the ozone layer will fully recover by 2050. Because of its widespread adoption and implementation, it is often referred to as an outstanding example of desirable international agreement and cooperation. In fact, Kofi Annan, the seventh Secretary General of the United Nations, said it is "perhaps the single most successful international agreement to date."

The history leading up to its implementation was relatively short. In 1973, chemists Frank Sherwood Rowland and Mario Molina of the University of California at Irvine began studying the impacts of chlorofluorocarbons (CFCs) in the atmosphere. They discovered that CFC molecules remained stable in the atmosphere until they reached the middle of the stratosphere (10–50 kilometers above the earth's surface) where they would finally, after an average of 50–100 years, be broken down by ultraviolet radiation releasing a chlorine atom. Molina and Roland (1974) then proposed that these chlorine atoms might be expected to cause the breakdown of large amounts of ozone (O_3) in the stratosphere. The two chemists were awarded the Nobel Prize for Chemistry for their discovery.

The problem lay in the fact that there was a tremendous environmental consequence. Because the stratospheric ozone absorbs most of the ultraviolet B radiation reaching the earth's surface, destruction of the ozone layer through use of CFCs would lead to an increase of ultraviolet B radiation reaching the earth's surface and causing an increase in skin cancer, damage to crops, and detrimental impacts to marine phytoplankton.

Naturally, this discovery and public announcement of it led to strong objections from the aerosol and halocarbon industries, which used the chemicals in their products. Rowland and Molina testified at a hearing before the U.S. House of Representatives in 1974, however, which led to government funding being appropriated to research of the problem in order to confirm their findings. In 1976, the U.S. National Academy of Sciences released a report that confirmed the scientific credibility of the ozone-depletion hypothesis.

Following this activity, in 1985, British Antarctic Survey scientists Farman, Gardiner, and Shanklin shocked the world when they published results of a study showing an ozone "hole" in the journal *Nature*. Shortly after, in the same year, 20 nations, including most of the major

CFC producers, signed the Vienna Convention, which established the framework for negotiating international regulations on ozone-depleting substances. Following that, as part of the phase-out solution, a fund, called the Multilateral Fund for the Implementation of the Montreal Protocol, was set up; it provides funds to help developing countries to phase out the use of ozone-depleting substances.

The Montreal Protocol provides for international regulation of ozone-depleting substances in order to protect public health and the environment from the potential negative effects of the depletion of stratospheric ozone. The Protocol was negotiated under the United Nations Environment Program (UNEP), representing a historic agreement because it was a voluntary international agreement for the purpose of protecting a vital global resource. Since its implementation, the atmospheric concentrations of the most prevalent CFCs and related chemicals have either leveled off or decreased. A 2006 statement by the UNEP says, "The Montreal Protocol is working: There is clear evidence of a decrease in the atmospheric burden of ozone-depleting substances and some early signs of stratospheric ozone recovery" (World Meteorological Organization, 2006).

Because of its success, the Montreal Protocol has often been called the most successful international environmental agreement to date. It was what many had in mind for a similar solution with the Kyoto Protocol and the ultimate international cooperation and solution to climate change. It had worked so smoothly that it was initially expected—and hoped—that the solution of an even bigger problem would follow suit, but the climate change issue proved to be a much tougher beast to tame, and the entire discussion was fraught with problems right from the beginning, whether because it was a bigger global issue; generated so much controversy between industry, government, and private industry; involved more people directly; lacked the support of one of the world's most developed nations; required long-term universal commitment; or a combination of all these reasons. The negotiations, which began in 1992 at the Framework Convention, were just the beginning of what would turn out to be a long, long road of controversy and debate, yet the problem has not lacked appropriate, sound scientific study and evidence.

In addition, as an interesting side note to the Montreal Protocol, several nations suggested in 2010 that its scope be expanded to also address climate change. The United States, Canada, Mexico, and Micronesia are currently proposing to phase out a family of industrial chemicals called hydrofluorocarbons (HFCs), which have a minimal effect on the ozone layer but are extremely potent greenhouse gases. If this move were taken, it would represent a very unique political move with an existing international agreement. It has received opposition based on technical, procedural, and financial grounds, but the idea is appealing to 91 countries so far. To support the notion, scientists say that by eliminating HFCs, it could slow climate change by a decade. The objectors to the plan include China, India, and Brazil, who are the world's major producers of HFCs.

According to Durwood Zaelke, president of the Institute for Governance and Sustainable Development, "Phasing down HFCs under the Montreal Protocol is a brilliant and necessary climate mitigation strategy. This may be the only climate strategy with support from industry, environmental groups, and a majority of parties. We owe it to the world to take advantage of this unique opportunity" (Broder, 2010).

What is interesting here is that a protocol adopted for another environmental issue is garnering support to play a role as policy to deal with climate change in the hopes that it will be more effective and better received than previous policy and suggested mediation already proposed or implemented specifically for climate change. The unintended foresight and dual nature of the Montreal Protocol is also an unexpected surprise.

The United Nations Framework Convention on Climate Change and the Kyoto Protocol

The international policy response to climate change began with the negotiation of the United Nations Framework Convention on Climate Change (UNFCCC), which eventually led the way for the creation and establishment of the Kyoto Protocol, the legal framework for global action to cut greenhouse gas (GHG) emissions.

The UNFCCC

The UNFCCC is an international environmental treaty that was produced at the United Nations Conference on Environment and Development, which is also known as the Earth Summit, the Rio Summit, and Eco '92. It was held in Rio de Janeiro from June 3 to 14, 1992. The purpose of the treaty was to stabilize greenhouse gas concentrations in the atmosphere in order to prevent climate change. As it was set up, the treaty was nonbinding, because it did not set any mandatory limits on greenhouse gas emissions for individual countries.

THE IMPACTS OF WARMING ON THE UNITED STATES AND CANADA

According to the IPCC, the United States and Canada will not escape the effects of climate change. In their report issued on April 6, 2007, they confirm that climate change is already affecting the environment. When the atmospheric temperature rises a little higher—even a few degrees—what may merely be uncomfortable heat now may become dangerous to the point of causing death. This will be felt all the way from Florida and Texas to Alaska and Canada's Northwest Territories.

According to Achim Steiner, executive director of the UNEP, "Canada and the United States are, despite being strong economies with the financial power to cope, facing many of the same impacts that are projected for the rest of the world." Chicago and Los Angeles will likely face increasing heat waves. Chicago is expected to see a 25 percent increase in heat waves later this century, and dangerously hot days in Los Angeles are projected to increase from a dozen per year to between 44 and 100. North American wood and timber production could suffer huge economic losses of $1 to $2 billion a year during the twenty-first century if climate change triggers diseases, insect infestations, and wildfires. Groundwater aquifers in Texas, South Dakota, Nebraska, Wyoming, Colorado, Kansas, Oklahoma, and New Mexico could see a lessening of recharge of 20–40 percent, causing problems for farmers and population centers (UNEP, 2007). Winter recreation in eastern North America may disappear by the 2050s, striking a hard blow to the recreation industry. Costs to replenish Florida's beaches with new sand after sea-level rise may cost upward of $9 billion.

The IPCC also cautions that severe storm surges could hit Boston and New York City. Cities that rely on melting snow for water, such as those in the drainage basins of the Rocky Mountains and Sierra Nevada, may experience serious water shortages. In particular, increased tension over water availability will result. As rainfall patterns shift, temperatures rise, and glaciers melt around the world, the demand for dwindling supplies of water will likely increase tensions across cultural and political borders.

The IPCC predicts that as temperatures rise, summer flows will drastically reduce, leaving huge areas without adequate water. As an example, they report that "A warming of a few degrees by the 2040s is likely to sharply reduce summer flows. As population increases, by then Portland, Oregon, alone will need over 26 million cubic meters of additional water due to climate change and population growth. The Columbia River's water supply is expected to be much lower, however: about 5 million cubic meters lower."

The IPCC also warns of storm surges and high tides and predicts that by the 2090s, levels of flooding that are currently only expected to occur once every 500 years could start happening once every 50 years as flooding intensity increases (IPCC, 2007).

What the treaty did include was provisions for updates (called protocols) that would set mandatory emission limits. The principal update is the Kyoto Protocol, the international agreement that sets binding targets for 37 industrialized countries and the European community for reducing GHGs.

The UNFCCC was opened for signature on May 9, 1992, and entered into force on March 21, 1994. Its principal objective was "to achieve stabilization of greenhouse gas concentrations in the atmosphere at a low enough level to prevent dangerous anthropogenic interference with the climate system."

One of the UNFCCC's first major achievements was that it set into place a "National Greenhouse Gas Inventory," which serves as a tabulation of GHG emissions and removals. All countries that signed the treaty must submit a greenhouse gas record on a regular basis. Nations that signed the treaty are divided into three groups:

1. Annex I countries, which are the industrialized countries
2. Annex II countries, which are the developed countries that pay for costs of developing countries
3. Developing countries

The UNFCCC has been ratified by the United States, Canada, France, Germany, the United Kingdom, Australia, Austria, Denmark, Finland, and virtually the entire international community.

The Annex I countries agree to reduce their GHG emissions to levels that are below their 1990 levels. If industries exceed their allotted limits, they must buy emission allowances or offset their excesses through a mechanism that is agreed upon by the UNFCCC. The Annex II countries (which are a subgroup of the Annex I countries) also participate as Organisation for Economic Co-operation and Development members. The developing countries are not expected to cut back on their carbon emissions unless developed countries provide them with the necessary funding and technology to accomplish it. Developing countries can become Annex I countries once they have become developed.

There have been opponents to the treaty who believe that not requiring developing countries to control their emissions is not fair politically, economically, or socially. They feel that all countries should have to reduce emissions equally. Some developing countries have said they cannot afford the costs of compliance. Other countries have countered that, saying that the Stern Review calculates that the cost of compliance is actually less than the cost of the consequences of doing nothing.

At the Earth Summit on June 12, 1992, 154 nations signed the UNFCCC, which, when ratified, committed those countries to a voluntary agreement to reduce atmospheric concentrations of greenhouse gases with the goal of "preventing dangerous anthropogenic interference with the earth's climate system." The actions were targeted mostly at industrialized nations to get them to stabilize their emissions of GHG at 1990 levels by the year 2000.

From the United States' perspective, on September 8, 1992, the U.S. Senate Foreign Relations Committee approved the treaty, reporting it through Senate Executive Report 102-55 on October 1, 1992. The Senate then consented to ratification on October 7, 1992. President George Bush signed the instrument of ratification on October 13, 1992, and deposited it with the UN Secretary-General. The treaty became effective on March 21, 1994, once it received the ratification of fifty countries. Since that time, the participating nations have met once a year at the Conference of the Parties in order to assess the progress being made in dealing with climate change. In the mid-1990s, negotiations began on the drafting of the Kyoto Protocol to establish legally binding obligations holding participating developed countries responsible for reducing their greenhouse gas emissions.

The Kyoto Protocol

The first Conference of the Parties took place in March 1995, and a ruling was given concluding that the present commitments under the United Nations Framework Convention on Climate Change were not adequate. Under the Framework Convention, developed countries pledged to take measures aimed at returning their greenhouse gas emissions to 1990 levels by the year 2000. Because this goal was lacking in scope and enforceable measures, the Berlin Mandate was created, which established a process that would enable the Parties to take appropriate action for the period beyond 2000, including a strengthening of developed-country commitments, through the adoption of a protocol or other legal instruments. In addition, the Berlin Mandate is what launched the talks that led to the adoption of the Kyoto Protocol.

FIGURE 4.1
The Kyoto Protocol enacted on February 16, 2005. (Courtesy of IISD/Earth Negotiations Bulletin.)

The Kyoto Protocol established legally binding commitments for reduction of four principal GHGs: carbon dioxide (CO_2), methane (CH_4), nitrous oxide (NO_x), and sulfur hexafluoride (SF_6), and two groups of gases: hydrofluorocarbons and perfluorocarbons. These gases are produced by the Annex I countries (industrialized nations). At the UNFCCC Conference of the Parties held in Kyoto, Japan, in 1997, the Kyoto Protocol was adopted for use after long, intense negotiations. The majority of industrialized nations and some central European countries agreed to the Protocol, and it was enacted on February 16, 2005 (Figure 4.1). It is significant that the Protocol was the first legally binding agreement enforcing reductions in greenhouse gas emissions. It called for reductions of 6–8 percent below 1990 levels to occur between 2008 and 2012, a time period referred to as the first emissions budget period. At that time, the United States would be required to reduce its total emissions by an average of 7 percent below 1990 levels. Neither the Bill Clinton nor George W. Bush administrations sent the Protocol to Congress for ratification. The Bush administration completely rejected the Protocol in 2001.

The objective of the Protocol was to stabilize greenhouse gas concentrations so that they remain below a level that causes climate change. There are five principal concepts of the Kyoto Protocol, which are as follows:

- Commitment: The Protocol establishes a legally binding commitment by the Annex I countries to reduce their GHG emissions.

- Implementation: Official policies and measures must be prepared by each participating country concerning how they will meet their objectives. Each country must also implement and use all mechanisms possible to absorb GHGs in order to be awarded credits that would allow for additional emissions.

- Assistance: The impacts on developing countries will be minimized through the establishment of an adaptation fund for climate change. This will facilitate the development and deployment of techniques that can help increase resilience to the impacts of climate change.

- Reporting: Each country is held responsible for accounting, reporting, and review to ensure they are strictly abiding by the terms of the Protocol. They submit annual emission inventories and national reports at regular intervals.

- Compliance: A compliance committee is established to ensure that individual countries are in strict compliance with their commitments under the Protocol.

One of the provisions of the Protocol is the manner in which it sets up an understanding of responsibility. The UNFCCC agreed to what they referred to as "common but differentiated responsibilities." The participating parties agreed on three terms.

The developed (industrialized) countries are currently (and have been historically) the largest emitters of GHGs. Per capita emissions in developing countries are still relatively low, and the share of global emissions originating in developing countries will grow as their social and developmental needs grow. What is so critical about this is that China, India, and other developing countries were not included in the original GHG restrictions of the Kyoto Protocol because they were not among the main contributors when the treaty was negotiated. Today, however, both China and India are developed nations. China is developing so rapidly that it is opening an average of one new coal-fired power plant each week, adding enormous amounts of GHGs to the atmosphere, unimpeded by the terms of the Kyoto Protocol. These rapidly developing countries' unaccountability was a principal reason why the Bush administration did not ratify the Kyoto Protocol. Also agreed upon in the original Protocol were financial commitments. It stipulates that it is the responsibility of developed countries to invest billions of dollars and supply the proper technology to developing countries to finance climate-related studies and projects.

In addition, the Protocol also allows an environmental policy tool called cap and trade (for a more detailed discussion, see Chapter 13, this volume). What this means is that there are caps (or limits) set on developed countries (Annex I group) as to the volume of GHGs they can legally emit. On average, the cap requires countries to reduce their emissions 5.2 percent below their 1990 baseline over the 2008–2012 period. Although the caps apply to the country itself, in practicality they are then divided within the country to the various industrial entities—power plants, car and computer manufacturers, and so forth. If a particular industry—for example, a power plant—knows it is going to exceed its allotted quota, it is allowed to purchase credits elsewhere to offset the overage. The credits (or excess allowances) are often purchased through a broker or an exchange set up expressly for that—a global carbon market. As a business venture, the Protocol allows groups of Annex I countries to join together to create a market within a market. Several exist today, such as in the European Union (EU), which created the EU Emissions Trading System (ETS). The EU ETS uses EU allowance units, which are each equivalent to a Kyoto assigned amount unit. The United Kingdom uses the UK ETS.

The sources of Kyoto credits are what are called the clean development mechanism and joint implementation projects. The clean development mechanism allows the creation of new carbon credits by developing emissions reduction projects in non–Annex I countries. Under the Protocol, countries' actual emissions have to be monitored, and precise records have to be kept of the trades carried out. Registry systems trace and record transactions by countries under the mechanisms. The UN Climate Change Secretariat, based in Bonn, Germany, keeps an international transaction log to verify that transactions are consistent with the rules of the Protocol. The enforcement branch was created and given the responsibility of ensuring compliance. If it is determined that an Annex I country is not in compliance with its emissions limitation, then the country is required to make up the difference plus an additional 30 percent. In addition, that country is then suspended from making transfers under an emissions trading program. Since the Protocol's inception, it has become apparent that in order to meet the original objective of stabilizing GHG emissions to control climate change, even larger emissions reductions will need to be achieved than those originally required by Kyoto. Table 4.1 illustrates the changes in GHG emissions of some prominent countries.

When the United Nations met at their annual climate conference in December 2005 in Montreal, participating nations began negotiations for a second set of targets for the period beginning in 2013 (once the original period ended in 2012). At that time, 2009 was viewed as a crucial year in the international arena of finding a workable solution to climate change. In 2007, the parties agreed to create an ambitious and effective international response to climate change to be agreed on at the climate conference held in Copenhagen in December 2009.

The U.S. Response and International Reactions

Although the bulk of the world's countries agreed to Kyoto, the United States took a different stance, choosing to approach the issue on its own terms. The former vice president Al Gore was a main participant in putting the Kyoto

TABLE 4.1

Greenhouse Gas Emissions of Prominent Countries

Country	Change in GHG Emissions (1992–2007)
India	+103%
China	+150%
United States	+20%
Russian Federation	–20%
Japan	+11%
Worldwide total	+38%

According to estimates from the Netherlands Environmental Assessment Agency (PBL), in the second half of 2008 there was a halving of the annual increase in global CO_2 emissions from fossil fuel use and cement production. Emissions increased by 1.7 percent in 2008 against 3.3 percent in 2007. Since 2002, the overall worldwide annual increase has averaged 4 percent. Besides high oil prices and financial crises, the increased use of renewable energy resources (such as biofuels for highway transportation and wind energy for electricity generation) has caused a noticeable mitigating impact on CO_2 emissions. CO_2 emissions in the United States fell 3.12 percent in 2008. In 2007, for the first time, CO_2 emissions in the United States were surpassed by those from China. There was a small absolute decline in the European Union as a whole, with declines also reported in Australia and Japan. Emissions in the Eastern European/CIS region increased 1.72 percent in 2008. Emissions from the large developing nations of Brazil, China, and India grew 6.9, 6.6, and 7.2 percent, respectively—together these nations accounted for 27.6 percent of the world total in 2008.

Protocol together in 1997. President Bill Clinton actually signed the agreement on November 12, 1998, but the U.S. Senate refused to ratify it, citing potential damage to the U.S. economy if the nation were forced to comply. The Senate also objected because the Protocol excluded certain developing countries, including China and India, from having to comply with new emission standards.

On March 29, 2001, the Bush administration withdrew the United States from the 1997 Kyoto Protocol on Climate Change. From a statement released by the U.S. Embassy in Vienna, Austria, it said that although the U.S. government was committed to developing an effective way to address the problem of climate change, it believed that the Kyoto Protocol was "fundamentally flawed" and therefore "is not the best approach to achieve a real environmental solution." The administration stated that, "The Kyoto Protocol does not provide the long-term solution the world seeks to the problem of climate change. The goals of the Kyoto Protocol were established not by science, but by political negotiation, and are therefore arbitrary and ineffective in nature. In addition, many countries of the world are completely exempted from the Protocol, such as China and India, who are two of the top five emitters of greenhouse gases in the world. Further, the Protocol could have potentially significant repercussions for the global economy" (U.S. Embassy, 2001).

President Bush commented on the treaty:

> This is a challenge that requires a 100 percent effort; ours, and the rest of the world's. The world's second-largest emitter of greenhouse gases is the People's Republic of China. Yet China was entirely exempted from the requirements of the Kyoto Protocol. India and Germany are among the top emitters. Yet India was also exempt from Kyoto … America's unwillingness to embrace a flawed treaty should not be read by our friends and allies as any abdication of responsibility. To the contrary, my administration is committed to a leadership role on the issue of climate change . . . Our approach must be consistent with the long-term goal of stabilizing greenhouse gas concentrations in the atmosphere.

(Note that as of 2006, China has become the largest GHG emitter in the world.)

Therefore, 10 days after taking office, Bush established a cabinet-level working group to find a more practical method to work with global climate change. The result of the working group was an energy policy that reflected the seriousness of the future of U.S. environmental policy. Bush announced the Clear Skies and Global Climate Change Initiatives in February 2002. The initiatives cover the following goals for managing global climate change:

- By 2018, emissions of the three worst air pollutants will be cut by 70 percent.
- In the next 10 years, the United States will cut greenhouse gas intensity by 18 percent.
- Goals similar to those of the Kyoto Protocol will be achieved, using market-based approaches.

These solutions differ from Kyoto in that they are based on free-market solutions. A free market is a market in which there is no economic intervention and regulation, except to enforce taxes, private contracts, and the ownership of property. There are four recommendations:

1. Ensuring continuing economic growth: It is in no country's best interest to sacrifice economic growth. With market-based incentive structures to spur innovation, it will be possible to move forward in the field of environmental conservation. Provisions under the Kyoto Protocol would rely on inflexible regulatory structures that would distort investment and waste billions of dollars on pollution permits, accomplishing no real change for the environment.

2. Finding global solutions: Addressing this issue must be as comprehensive as possible. All nations including developing nations must participate. For those that already account for a majority of the world's greenhouse gas emissions, it would be irresponsible to absolve them from shouldering some of the shared obligations.

3. Using the most modern technology: The United States is committed to investing heavily in research and development and encouraging private companies to do the same through market-based incentives (market-based incentives are measures intended to directly change relative prices of energy services and overcome market barriers). Since 1990, the United States has spent more than all of the countries of the European Union on research in new energy and environmentally friendly technology.

4. Focusing on bilateral relations to provide assistance: The United States has already worked with more than fifty-six countries on their energy and environmental policies.

According to Bush, "The United States fully acknowledges the problem of climate change, and is committed to pursuing a practical and sustainable plan to address this grave situation. The United States hopes to find a workable solution to this serious problem that affects all of us in the global community" (U.S. Embassy, 2001).

The international reaction to Bush's response to climate change was heated. Although there was faint support from some sectors that the administration finally acknowledged climate change as a problem worthy of attention and committed U.S. involvement toward finding a feasible solution, most reactions were negative. Accusing the administration of trying to create a new ad hoc process—separate from the official framework established by the United Nations—critics stated that the U.S. response would do nothing more than distract from the progress the rest of the world was trying to make toward stabilizing climate change. If anything, they felt it would actually hamper any progress being made to reduce greenhouse gas emissions and slow global warming. Great Britain and Germany especially criticized the United States, stating that all international climate agreements should logically stay within the jurisdiction of the United Nations.

German chancellor Angela Merkel said, "For me, that is nonnegotiable. In a process led by the United Nations, we must create a successor to the Kyoto agreement, which ends in 2012. But it is important that they flow from the United Nations" (West, 2001). Hilary Benn, Britain's international development secretary, remarked, "I think it is very important that we stick with the framework we've got. In the end, we have to have one framework for reaching agreement. I think that is very clear" (West, 2001). Leaders from environmental groups also had strong opinions. Philip Clapp, president of the National Environmental Trust, said, "This is a transparent effort to divert attention from the president's refusal to accept any emissions reductions proposals at next week's G8 summit" (West, 2001). David Doniger, the climate policy director for the Natural Resources Defense Council, commented, "There is no more time for longwinded talks about unenforceable long-term goals. We need to get a serious commitment to cut emissions now and in the G8" (West, 2001).

The Bush administration offered an alternative environmental plan on June 11, 2001, promising increased environmental research and commitment from the United States. Bush announced that he was "committing the United States of America to work within the United Nations framework and elsewhere to develop an effective science-based response to the issue of climate change." Bush also stated that, "The rest of the world emits 80 percent of all greenhouse gases, and many of those emissions come from developing countries. The world's second largest emitter of greenhouse gas is China, yet China was entirely exempted from the requirements of the Kyoto Protocol" (West, 2001). Bush committed his administration to fully fund high-priority areas for scientific research into climate change over the next 5 years and help developing nations to match the U.S. commitment. According to CNN News, former president Clinton signed the Kyoto Protocol but also said he would not send it to the Senate for ratification until several changes were made.

One country that did not seem to be up in arms over the U.S. stand was Australia. The Australian Prime Minister John Howard supported Bush's plan. According to Howard, "We are a net exporter of energy, and unless you have the developing countries involved we would be hurt. Our position . . . is much closer to that of the United States than the attitude of the European countries. I do think what the president indicates in his speech will lead to an alternative to simply saying 'no' to the Kyoto Protocol, and I welcome that" (West, 2001). Pia Ahrenkilde-Hansen, the EU spokeswoman, remarked, "It is positive that the U.S. administration is realizing that there needs to be something done about climate change but we feel that the multilateral approach is the best way to face up to this tremendous challenge" (CNN, 2002).

Many environmental groups opposed Bush's voluntary plan, however, saying that it ultimately would do nothing to curb U.S. emissions. According to a December 4, 2003, _New York Times_ report, "The 1997 Protocol had many flaws, but it represented the only international response to the climate change problem thus far devised, and at the very least it provides a plausible framework for collective international action" (CNN, 2002).

The international community was not alone in disagreeing with the Bush administration's stand. Several U.S. cities rose to the occasion, and dozens of mayors—representing more than 25 million Americans—pledged that their cities would cut greenhouse gases by 7 percent by 2010.

Greg Nickles, Seattle's mayor who spearheaded the event, said, "This campaign has clearly touched a nerve with the American people. The climate affects Democrats and Republicans alike. Here in Seattle, we rely on the snow for our drinking water and hydroelectricity but it is disappearing" (Brown, 2005).

Nickles also warned that each city had a tough target of cutting emissions by 7 percent, and each mayor would choose a different way to accomplish that goal. He also said, "There are changes we will have to make but there are many opportunities to create employment and make for a better life. In any event, the costs of doing nothing are greater than doing something."

Some of the specific proposals for cities include using hybrid cars, investing in renewable energy, improving public transportation, planting trees, promoting carpooling, and providing cycling lanes.

The G8

The G8, or Group of Eight, is a forum that was created by France in 1975 for the governments of eight nations of the Northern Hemisphere. The participating members are Canada, France, Germany, Italy, Japan, Russia, the United Kingdom, and the United States. The European Union (EU) is also represented, but it cannot host or chair. Table 4.2 lists the current members.

Each year, the G8 holds a conference in the country of whoever is currently serving as president. The number of participating countries has evolved over the years since 1975, and just recently it has been proposed that the group be expanded to include five developing countries, referred to as the Outreach Five (O5): Brazil, China, India, Mexico, and South Africa. These countries have attended as guests in the past. It has been proposed that the name be changed to the G8+5.

The G8 is an informal forum that began in 1973 after the oil crisis and global recession that followed it. The object of the gathering is to discuss issues of mutual or global concern, such as energy, the environment, terrorism, economics, health, trade, and so forth. At the Heiligendamm Summit held in 2007, the G8 addressed the issues of energy efficiency and climate change.

The group agreed, along with the International Energy Agency, that the best way to promote energy efficiency was on an international basis. As a result, on June 8, 2008, the G8 and China, India, South Korea, and the European Community jointly established the International Partnership for Energy Efficiency Cooperation. The G8 finance ministers agreed to the "G8 Action Plan for Climate Change to Enhance the Engagement of Private and

TABLE 4.2

The G8 Leaders

Country	World Leader
Canada	Prime Minister Stephen Harper
France	President Nicolas Sarkozy
Germany	Chancellor Angela Merkel
Italy	Prime Minister Silvio Berlusconi
Japan	Prime Minister Naoto Kan
Russia	President Dimitry Medvedev
United Kingdom	Prime Minister David Cameron
United States	President Barack Obama
Also represented: European Union	Commission President José Manuel Barroso
	Council President Herman Van Rompuy

Public Financial Institutions." They also initiated the climate investment fund by the World Bank, which is put into place to help existing efforts until a new framework under the UNFCCC is implemented after 2012, when Kyoto expires.

The Intergovernmental Panel on Climate Change

In order to make meaningful management decisions to minimize the negative impacts of climate change, it is necessary to have an organized body of professionals working together toward the common goal of understanding the science of climate change. This way they can advise political leaders who can then develop regulations that enforce positive human response to that change. The IPCC is a scientific organization established by UNEP and the World Meteorological Organization in 1988. The IPCC is comprised of the world's top scientists in all relevant fields who review and analyze scientific studies of climate change and provide authoritative assessments on the state of knowledge regarding climate change. The IPCC was established to provide decision-makers and others interested in climate change with an objective source of information. The IPCC itself does not conduct any research. Its key role is "to assess on a comprehensive, objective, open, and transparent basis the latest scientific, technical, and socio-economic literature produced worldwide relevant to the understanding of the risk of human-induced climate change, and its observed and projected impacts and options for adaptation and mitigation." The reports they produce are of a high scientific and technical standard, meant to reflect a range of views and expertise and encompass a wide geographical area.

The IPCC produces reports at regular intervals. The flowchart in Figure 4.2 illustrates the process followed to produce the reports. To date there have been four major assessments: 1990, 1995, 2001, and 2007. The IPCC is comprised of about 2,500 of the world's top climate scientists and is chaired by Dr. Rajendra Pachauri of India (Figure 4.3). Once the reports are released, they become standard works of reference that are widely used by policy makers, experts, and others. For example, in 1990, the findings of the First Assessment Report played a critical role in establishing the UNFCCC. The Second Assessment Report, released in 1995, provided key input for the negotiations of the Kyoto Protocol in 1997. The Third Assessment Report in 2001 was used in the development of the UNFCCC (IPCC, 2001).

Currently, the IPCC has three working groups and has undertaken the National Greenhouse Gas Inventories Programme in collaboration with the Organisation for Economic Co-operation and Development and the International Energy Agency. Each working group has its own agenda and is assisted by a technical support unit and the working group or task force bureau. Working Group I (WGI) is titled *The Physical Science Basis*. Working Group II (WGII) is called *Impacts, Adaptation and Vulnerability*. Working Group III (WGIII) is called *Mitigation of Climate Change*.

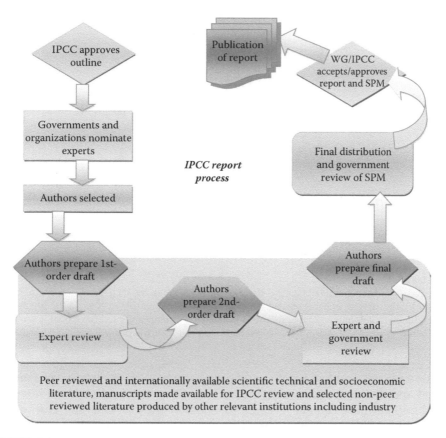

FIGURE 4.2
The review process to which the IPCC reports are subjected.

The main objective of the Greenhouse Gas Inventories Programme is to develop and refine a methodology for the calculation and reporting of national GHG emissions and removals. In addition, there is a provision written into the agreement where further task groups and steering groups may be established for a duration of time to consider specific topics or concerns.

Working Group I

WGI assesses the physical scientific aspects of the climate system and climate change. Their latest report, published on February 2, 2007, was released in Paris. This report covers information on changes in greenhouse gases and aerosols in the atmosphere and the role they play in determining the behavior of the climate. The report provides specific details in the changes of air, land, and ocean temperatures, glaciers, rainfall, and ice sheets. It takes into account enormous amounts of satellite-derived data for broad global coverage.

FIGURE 4.3
Dr. Rajendra Pachauri has been the chair of the IPCC since 2002. He is an environmentalist and also the director general of the Energy and Resources Institute in New Delhi, involved in sustainable development. On December 10, 2007, Dr. Pachauri accepted the Nobel Peace Prize on behalf of the IPCC, along with co-recipient Al Gore. (Courtesy of IISD/Earth Negotiations Bulletin.)

In addition to the current status of the atmosphere, it also focuses on the past and includes a paleoclimatic review of the earth's glacial and interglacial periods, the evidence left behind, and how the past can offer clues about the future. This working group also looks at how climate change interacts and affects geochemistry and the biosphere. Complex climate models are evaluated, and the driving factors—or climate *forcings*—are analyzed so that projections can be made as to what the future climate may be like both globally and locally.

Working Group II

WGII assessed the vulnerability of socioeconomic and natural systems to climate change, the negative and positive consequences of climate change, and options for adapting to climate change. Their most recent report was released on April 6, 2007, in Paris, and was entitled *Impacts, Adaptation and Vulnerability*. It provides a detailed analysis of how climate change is affecting natural and human systems, what its future impacts will be, and to what extent adaptation and mitigation can reduce these impacts. It analyzes how adaptation and mitigation work together and how societies can make the best use of the resources they have so that they can maintain a *sustainable development*.

This report looks at specific natural earth systems, such as ecosystems, water resources, coastal systems, oceans, and forests. It also analyzes human-controlled sectors, such as industry, agriculture, and health. It examines these issues on a geographical basis, breaking the data into subregions such as North America, Latin America, Polar Regions, Africa, Asia, Australia and New Zealand, Europe, and small islands.

Working Group III

WGIII is responsible for assessing practical options for mitigating climate change through limiting and preventing greenhouse gas emissions (Figure 4.4). They also focus on identifying methods that remove greenhouse gas emissions from the atmosphere. Their fourth report was released by the IPCC on May 4, 2007, summing up their meeting in Bangkok. The report analyzes the world's GHG emission trends and analyzes various mitigation options for the main economic sectors from up to 2030. It provides an in-depth analysis of the costs and benefits of various mitigation approaches and also looks at short-term strategies and projects how effective they would be in the long term. The report focuses on policy measures and instruments available to governments and industries to mitigate climate change and stresses the strong relationships between

FIGURE 4.4
The IPCC Working Group III focusing on their interest in the Fourth Assessment Report, *Mitigation of Climate Change*, at the ninth session in Bangkok, Thailand, held April 30 to May 4, 2007. (Courtesy of IISD/Earth Negotiations Bulletin.)

mitigation and sustainable development. The Task Force on National Greenhouse Gas Inventories was established by the IPCC to oversee the National Greenhouse Gas Inventories Programme.

IPCC Reports

The IPCC's Fourth Assessment Report, released in 2007, represents the work of more than 1,200 authors and 2,500 scientific expert reviewers from more than 130 countries. The terminology the IPCC uses when they make projections is very specific. When discussing their degree of confidence, the following terminology applies:

- Very high confidence – at least a 90 percent chance
- High confidence – ~80 percent chance
- Medium confidence – ~50 percent chance

In terms of likelihood of occurrence, the terms are as follows:

- Extremely likely – >95 percent
- Very likely – >90 percent
- Likely – >66 percent
- More likely than not – >50 percent
- Less likely than not – <50 percent
- Unlikely – >33 percent
- Very unlikely – 10 percent
- Extremely unlikely – >5 percent

Working Group I Report: The Physical Science Basis

This report contains the strongest language yet of any of the IPCC's reports, and it found that it is very likely (>90 percent probability) that emissions of heat-trapping gases from human activities have caused "most of the observed increase in globally averaged temperatures since the mid-20th century." The report concludes that it is "unequivocal" that earth's climate is warming, "as is now evident from observations of increases in global average air and ocean temperatures, widespread melting of snow and ice, and rising global mean sea level."

The report also verifies that the current atmospheric concentration of CO_2 and methane "exceeds by far the natural range over the last 650,000 years." Since the beginning of the industrial revolution, concentrations of both gases have increased at a rate that is "very likely to have been unprecedented in more than 10,000 years."

The report also identified the following findings:

- Eleven of the last 12 years were among the 12 hottest years on record.
- Over the past 50 years, cold days, cold nights, and frost have become less frequent, while hot days, hot nights, and heat waves have become more frequent.
- The intensity of hurricanes in the North Atlantic has increased over the past 30 years, which correlates with increases in *tropical* sea surface temperatures. They are likely to become more intense.
- Between 1900 and 2005, the Sahel, the Mediterranean, southern Africa, and parts of southern Asia have become drier, adding stress to water resources in these regions.
- Droughts have become longer and more intense and have affected larger areas since the 1970s, especially in the tropics and subtropics.
- Since 1990, the Northern Hemisphere has lost 7 percent of the maximum area covered by seasonally frozen ground.
- Mountain glaciers and snow cover have declined worldwide. Satellite data since 1978 show that the extent of Arctic sea ice during the summer has shrunk by more than 20 percent.
- Since 1961, the world's oceans have been absorbing more than 80 percent of the heat added to the climate, causing ocean water to expand and contributing to rising sea levels.
- If no action is taken to reduce emissions, the IPCC concludes that there will be twice as much warming over the next two decades than if the GHGs had been stabilized at their 2000 levels.
- The full range of projected temperature increase has now been revised to 1.1–6.4°C by the end of the century because higher temperatures reduce the amount of CO_2 that the land and ocean can hold, keeping more stored in the atmosphere.
- Warming is expected to be greatest over land and at most high northern latitudes and least over the Southern Ocean and parts of the North Atlantic Ocean.
- High latitude precipitation will increase, and subtropical lands (e.g., Egypt) will face drought.
- Extreme heat, heat waves, and heavy precipitation will become more frequent.
- Sea ice is projected to shrink in both the Arctic and Antarctic under all model simulations. Some projections show that by the latter part of the century, late-summer Arctic sea ice will disappear almost entirely.
- Increasing atmospheric CO_2 concentrations will lead to increasing acidification of the oceans, destroying coral and other fragile marine ecosystems.

The IPCC also states that it is very likely that the Atlantic Ocean Conveyor Belt will be 25 percent slower on average by 2100 (with a range from 0 to 50 percent). Nevertheless, Atlantic regional temperatures are projected to rise overall due to more significant warming from increases in heat-trapping emissions. The models used by the IPCC project that by the end of this century, the global average sea level will rise between 17 and 58 centimeters above the 1980–1999 average. In addition, recent observations show that meltwater can run down cracks in the ice and lubricate the bottom of ice sheets, resulting in faster ice flow and increased movement of large ice chunks into the ocean, contributing to sea-level rise (IPCC, 2007).

Working Group II Report: Impacts, Adaptation, and Vulnerability

WGII describes climate change's effects on society and the natural environment and some of the options available for adapting to these effects. The IPCC has determined that anthropogenic warming over recent decades is already affecting many physical and biological processes on every continent. Of the 29,000 observational pieces of data reviewed, almost 90 percent showed changes that were consistent with the response expected of climate change. In addition, the observed physical and biological responses have been the greatest in the regions that warmed the most.

The major conclusions stated in this report include the following:

- Hundreds of millions of people face water shortages that will worsen as temperatures rise. The most at risk are regions currently affected by drought, areas with heavily used water resources, and areas that get their water from glaciers and snowpack such as the western United States, Europe, and Asia.

- The land area affected by drought is expected to increase, and water resources in affected areas could decline as much as 30 percent by mid-century. U.S. crops that are already near the upper end of their temperature tolerance range or depend on strained water resources could suffer with further warming.

- More than one-sixth of the world's population currently lives near rivers that derive their water from glaciers and snow cover; these communities can expect to see their water resources decline over this century.

- Melting glaciers in areas like the Himalayas will increase flooding and rockslides, while flash floods could increase in northern, central, and eastern Europe.

- The IPCC expects food production to decline in low-latitude regions (near the equator), particularly in the seasonally dry tropics, as even small temperature increases decrease crop yields in these areas.

- The IPCC projections show drought-prone areas of Africa to be particularly vulnerable to food shortages because of a reduction in the land area suitable for agriculture; some rain-fed crop yields could decline as much as 50 percent by 2020.

- Under local average-temperature increases, regions such as northern Europe, North America, New Zealand, and parts of Latin America could benefit from increased growing season length, more precipitation, and/or less frost, depending on the crop. However, these regions may also expect more flooding. In addition, depending on existing soil types, agriculture may or may not even be feasible.

- Up to 30 percent of plant and animal species could face extinction if the global average temperature rises more than 1.5–2.5°C relative to the 1980–1999 period. Many say the low range could be reached by mid-century.

- Spring has been arriving earlier, influencing the timing of bird and fish migration, egg laying, leaf unfolding, and spring planting for agriculture and forestry. It can threaten and endanger species by altering the timing of migration, nesting, and food availability, causing them to be out of sync.

- Many species and ecosystems may not be able to adapt to the effects of climate change and its associated disturbances (including floods, drought, wildfire, and insects), causing mass extinctions.

- Experts expect coral reefs and mangroves in Africa to be degraded to the point that fisheries and tourism suffer. Some areas, such as the national parks of Australia and New Zealand and many parts of tropical Latin America, are likely to experience a significant loss of biodiversity.

- Flooding caused by sea-level rise is expected to affect millions of additional people every year by the end of this century, with small islands and the crowded delta regions around large Asian rivers (such as the Ganges-Brahmaputra) facing the highest risk.

- Regions especially at risk are low-lying areas of North America, Latin America, Africa, the popular coastal cities of Europe, crowded delta regions of Asia that face flood risks from both large rivers and ocean storms, and many small islands (such as those in the Caribbean and South Pacific) whose very existence is threatened by rising seas.

- Scientists expect heat waves, droughts, wildfires, floods, severe storms, and dust transported between continents to cause locally severe economic damage and substantial social and cultural disruption. The IPCC projects an extended fire season for North America as well as increased threats from pests and disease.

- In cities that experience severe heat waves, scientists project an increase in the incidence of cardiorespiratory diseases caused by the higher concentrations of ground-level ozone (smog) that may accompany higher air temperatures. Some infectious diseases, such as those carried by insects and rodents, may also become more common in regions where those diseases are not currently prevalent (such as dengue fever, malaria, yellow fever, encephalitis, Lyme disease, and visceral leishmaniasis).

- Many of the unavoidable short-term consequences of climate change can be addressed through adaptation strategies such as building levees and restoring wetlands to protect coasts, altering farm practices to grow crops that can survive higher temperatures, building infrastructure that can withstand extreme weather, and implementing public health programs to help people in cities survive brutal heat waves. This is a more serious problem, however, for developing countries that lack the economic wherewithal to build appropriate infrastructures.

Working Group III: Mitigation of Climate Change

There are several strategies available today that the IPCC believes could slow climate change and prevent the worst environmental consequences if they were implemented immediately. Although there has been some criticism that implementing proper measures to halt climate change would be too expensive, the IPCC has determined that the economic impact on the world economy would only be a fraction of a percent reduction in the annual average growth rate of global gross domestic product.

The IPCC also warns that the policies that have been put into place so far have not been robust enough to stop the growth of global emissions caused by the increased use of fossil fuels, deforestation, overpopulation, and wildfires. It is critical that clean technologies are developed in order to reduce emissions and stop climate change. Although there has been much talk about reducing emissions, there has been an increase in heat-trapping gases of 70 percent from 1970 to 2004. Of these, CO_2 emissions account for 75 percent of the total anthropogenic emissions. The emission growth rate is expected to continue if serious changes are not made immediately.

In 2004, developed countries (such as the United States) had 20 percent of the world population and contributed nearly three-quarters of the global emissions. Developing countries generated only one-quarter of the emissions. The IPCC has projected that CO_2 emissions from energy use are projected to increase 45–110 percent if fossil fuels continue to dominate energy production through 2030, with up to three-fourths of future emission increases coming from developing countries (such as China and India).

The IPCC analyzed several mitigation options—some of them efficient enough to bring about a 50–85 percent reduction in emissions of green-house gases by 2050 (compared with 2000 levels). Predictions with these models put GHG concentrations at the end of the century at 445–490 ppm. As a comparison, the IPCC says if mitigation of this nature does not take place and GHG levels continue to increase, concentration levels could reach 855–1,130 ppm. The IPCC believes there will be more mitigation technologies available before 2030 that could lead to even greater emissions reductions. They believe that the search for energy efficiency will play a key role in the future and support larger investments in research and development to stimulate deployment of new technological advances. They also stress the importance of increasing government funding for research, development, and demonstration of carbon-free energy sources.

The Bali Conference (2007)

The 13th Conference of the Parties to the United Nations Framework Convention on Climate Change was held in Bali, Indonesia, on December 3–15, 2007. The outcome and purpose of the Bali Conference was to put in place and adopt the Bali Road Map as a 2-year plan to finalize a binding agreement at the conference to be held in Copenhagen in 2009. With the understanding that the Kyoto Protocol would expire in 2012, the participating countries knew there would be a significant amount of planning necessary in order to be able to reach some kind of a unified working agreement at that point.

The world's leaders acknowledged that the Kyoto Protocol was not the final step in the process—it was merely a tool to get the world working together to solve one of the biggest problems it had ever collectively faced. They also acknowledged the importance of putting new targets in place so that all the prior work participating countries had accomplished with negotiated reductions and trading in carbon markets was not lost. Equally important were the new business ventures that had been forged toward renewable energy and technology. It was imperative that the momentum be maintained and the successes that had been won remain in the forefront of not only policy makers but the public eye as well.

Another significant advancement that has occurred since the inception of the Kyoto Protocol has been further advancement of the scientific understanding of climate change, the true nature of greenhouse gases, and the climate's long-term sensitivity to them. This advancement can be seen in the IPCC's Fourth Assessment, released in 2007. With its refinement of scientific principles, it also shows where the Third Assessment was incorrect—the projections it made were far too conservative. Serving as a wake-up call, the Fourth Assessment shows that scientists had discovered that the impacts of climate change were being felt even more rapidly than they would have even thought possible. For example, one thing that shocked the scientific community was the rapid rate of ice melt in the polar areas and the fact that the Northwest Passage was open to shipping for the first time in recent history. Ten years ago, no one would have believed that was possible.

Again, with the scientific evidence overwhelming, although the majority of the nations represented at the conference were eager to move forward with negotiations on a new treaty, the United States balked once again, opposed to including binding emissions reductions. Interestingly, at the 11th Conference of the Parties, held in Montreal in December of 2005, the senior U.S. negotiator opposed any consideration of binding targets in new talks and walked out of the negotiations, which nearly ruined the decision to hold the Bali conference in the first place. It was only after it became obvious that the United States would be publicly blamed for the failure of the 11th Conference talks did the U.S. negotiator return to the table.

Then, early in 2007, the United States again put a dark cloud over proposed negotiations for the 2007 G8 Summit. German Chancellor Angela Merkel proposed adopting a new international treaty, which would:

- Hold global temperature increases below 2°C (beyond that scientists felt a catastrophic tipping point would be reached)
- Create a 20 percent improvement in the energy efficiency of each G8 economy by 2020
- Create a 30 percent increase in the energy efficiency of each nation's utility sector by 2020

Following this proposal, Japan's Prime Minister Abe added his own proposal that the G8 leaders adopt a 50 percent reduction in greenhouse gas emissions by 2050. U.S. opposition, however, ruined those negotiations as well, and the end result was a weak endorsement of achieving a new international agreement under the UN process by 2009, and a willingness "to seriously consider" a 50 percent emissions reduction by 2050 as a long-term goal (Clapp, 2007).

Unfortunately for the Bali Conference, most of the attending nations were not pleased nor supportive of the United States because of their established history of breaking treaty commitments on global warming, of refusing or watering down strong emissions reduction goals, and especially of failing to act to reduce their own emissions while pointing the finger at everyone else and telling them that they must reduce theirs. Basically, the United States was, and still is, at odds with the rest of the world. Other nations long ago concluded that voluntary international agreements had failed and that binding agreements were the only measures that had any chance of success. The United States refuses to enter any binding measurable agreements, instead opting for the "scout's honor" approach.

For the Bali talks, the United States was planning on a set of talks focused on individual national emissions reduction plans all under an umbrella of a long-term emissions reduction goal, but without any international enforcement mechanism or shorter term targets or carbon trading systems. The European Union, on the other hand, envisioned a negotiation that included binding targets. In the end, a weak compromise was agreed to: negotiations could proceed, the United States could discuss its voluntary approach, and under the auspices of the Kyoto Protocol, the United States could not block discussion of binding targets. The consensus was that under the Bush administration, there was no other solution. On a more positive note, it was at the Bali Conference that developing countries came to the table willing to discuss emissions reductions of their own for the very first time.

The outcome of the Bali Conference was the Bali Road Map, which was a 2-year process to finalizing a binding agreement in 2009 at the scheduled Copenhagen Climate Change Conference. The Bali Action Plan, which is part of the Road Map, opens discussion on "measurable, verifiable, and reportable" emissions reduction actions that may be undertaken by developing countries. This is an important point, especially for China. China has never before been willing to consider emissions reduction measures reportable to an international body. Also important for China, verifiable emissions reduction measures are a threshold requirement for China to receive credit for emissions reduction measures in world carbon markets.

In addition, rain-forest nations were also willing to discuss quantifiable emissions reduction measures. They proposed negotiations on how emissions reductions achieved through policies to fight deforestation might qualify for market credits. Another black eye for the United States on this one: for the country that invented the cap-and-trade market-based approach to emissions reductions and expected the rest of the world to embrace it at Kyoto, U.S. representatives blocked the rain-forest nations' proposal. By the end of the conference, the United States had opposed and rejected everything that was offered at the talks and offered no proposals as to what the United States would commit to in reducing its own emissions. The position of the United States was not viewed favorably by any other country.

The final outcome of the Bali Road Map included support for the mitigation commitment by all developed countries, provided enhanced cooperation to "support urgent implementation" of measures to protect poorer countries against climate change, provided improved access to predictable and sustainable financial resources, and launched formal negotiations on a new international global warming agreement and set a firm deadline of December 2009, when the 15th Conference of the Parties would meet in Copenhagen for its adoption.

Copenhagen Climate Change Conference (2009)

According to the Bali Road Map, the new direction that all the nations would agree to take once the Kyoto Protocol expired in 2012 was supposed to be decided. On the last day of the conference, however, the international media released that the "climate talks were in disarray." Reports were issued of a summit collapse; a weak political statement was expected to be issued at the conclusion of the conference.

The Copenhagen Accord was drawn up as the document from the conference, written by the United States, China, India, Brazil, and South Africa. Referred to as a "meaningful document" by the United States government, it was not adopted but merely "taken note of." It was also not passed unanimously by all the participating countries. Basically, the document recognized that climate change stands as one of the greatest challenges of the present day and that actions need to be taken to keep temperature increases below the 2°C limit warned against by climate scientists. The document is not legally binding and does not contain any legally binding commitments for reducing CO_2 emissions. Many of the countries in attendance were outright opposed to the document.

Analyses of the conference have been mixed and varied. Many perceive it a complete failure, meaningless and weak. Others consider the fact that it may prove useful if it pushes people to begin analyzing some of the underlying misconceptions and then work toward a new view of the situation. Perhaps clarifying goals and standards could help to gain the support and understanding of some of the developing countries. The United States agreed to aid the developing world with $30 billion over the next 3 years, rising to $100 billion per year by 2020, in order to help poor countries adapt to climate change. The Accord also has a provision for developed countries' paying developing countries to reduce emissions from deforestation and degradation, a program known as REDD. Earlier proposals that would have tried to limit temperature rises to 1.5°C and cut CO_2 emissions by 80 percent by 2050 were dropped. Although the agreement made was nonbinding, U.S. President Obama said that countries could show the world their achievements. He said that if they had waited for a binding agreement, no progress would have been made. He made the comment that the agreement would need to be built on in the future and that "We've come a long way but we have much further to go" (BBC, 2009).

Based on the agreements made in Copenhagen, if the Accord is followed it allows countries to set their own greenhouse gas emissions reduction goals for 2020. The probability of meeting the 2°C target, however, is only given a 50 percent chance for a success rate. In its favor, the agreement will, at the very least, cut greenhouse gases, set up an emissions verification system, and reduce deforestation, which is still a step forward, and this time, the United States is included. Copenhagen also stands for a fundamental statement that there is an acceptance of the science and an acceptance that there will have to finally be verification and accountability by all participants—which is in itself a major step forward.

Although most people hoped for more, perhaps Rajendra Pachauri, chairman of the IPCC, summed it up best. He saw three major achievements come out of Copenhagen:

- The acceptance of a 2°C limit for temperature increase, along with reference to the scientific basis for doing so. This, to Pachauri, indicated that science had finally had an influence on negotiators defining what would represent dangerous anthropogenic interference with the climate system.
- The BASIC countries (Brazil, South Africa, China, and India) became part of the final agreement.
- There was an agreement for funding developing countries' actions from 2010 to 2012.

What Pachauri hoped for next was for the major countries to get busy before the next major Conference in Cancun to create an inclusive agreement involving all the countries of the world.

Cancun Convention (2010)

The UNFCCC Cancun Climate Change Conference was held in December 2010, and the resultant product from those talks was a document called the Cancun Agreements. These talks—more productive overall than the previous ones in Bali—are viewed as a collection of "significant decisions by the international community to address the long-term challenge of climate change collectively and comprehensively over time and to take concrete action now to speed up the global response." The talks determined key steps toward concrete plans to reduce greenhouse gas emissions, as well as to help developing nations protect themselves from climate impacts and also build their own sustainable futures (United Nations Framework Convention on Climate Change, 2011).

The main objectives of the Cancun Agreements include the following:

- Establish clear objectives for reducing human-generated greenhouse gas emissions over time in order to keep the global average temperature rise below 2°C.
- Ensure the international transparency of the actions that are taken by countries and ensure that global progress toward the long-term goal is reviewed in a timely manner.
- Encourage the participation of all countries in reducing these emissions, in accordance with each country's different responsibilities and capabilities to do so.
- Mobilize and provide scaled-up funds in both the short and long term to enable developing countries to take greater and effective action.
- Mobilize the development and transfer of clean technology to boost efforts to address climate change, getting it to the right place at the right time and for the best effect.

- Assist the particularly vulnerable people in the world to adapt to the inevitable impacts of climate change.
- Protect the world's forests, which are a major repository of carbon.

This marks a milestone in negotiations because it succinctly represents the most critical components of climate change management. Also importantly, it addresses the very issues the United States has been balking at for the past two decades—committing to an action plan toward solving the problem. Also of note, it recognizes that there will be inevitable impacts that will need to be adjusted to and that no longer can the world's countries adopt a "let's wait and see" attitude. The UNFCCC has published what they feel are the significant ramifications of the key agreements reached at Cancun:

- They form the basis for the largest collective effort the world has ever seen to reduce emissions, in a mutually accountable way, with national plans captured formally at an international level under the banner of the United Nations Framework Convention on Climate Change.
- They include the most comprehensive package ever agreed upon by governments to help developing nations deal with climate change. This encompasses finance, technology, and capacity-building support to help them meet urgent needs to adapt to climate change and to speed up their plans to adopt sustainable paths to low-emission economies, which can also resist the negative impacts of climate change.
- They include a timely schedule for nations under the Climate Change Convention to review the progress they make towards their expressed objective of keeping the average global temperature rise below 2°C. This includes an agreement to review whether the objective needs to be strengthened in the future, on the basis of the best scientific knowledge available (United Nations Framework Convention on Climate Change, 2011).

The governments of every country now have a tall order in front of them. They now need to turn their pledges into real action immediately. For 2011, they are tasked with putting together workable plans on just how they are going to implement plans and programs to achieve the goals their pledges indicate. According to the UNFCCC, the following new institutions will be developed as a result of the agreements:

- A green climate fund to house the international management, deployment, and accountability of long-term funds for developing country support.
- A technology mechanism to get clean technologies to the right place at the right time.

- An adaptation framework to boost international cooperation to help developing countries protect themselves from the impacts of climate change.
- A registry where developing countries will detail their voluntary plans to limit greenhouse gas emissions and the support they need to achieve them.

The UN also acknowledges that although the emissions reduction targets and actions that were announced in Cancun are the most comprehensive and ambitious global effort to date, they are still inadequate in the long term to keep the agreed maximum global temperature rise of 2°C (United Nations Framework Convention on Climate Change, 2011). Countries would meet again in Bangkok in April 2011 to further clarify their pledges and refine specific methodologies on just how they would accomplish their goals. One new guideline that came out of the Cancun meetings was that industrialized countries will boost the regular reporting of progress toward their targets by submitting detailed annual inventories of greenhouse gas emissions and by reporting on progress in emissions reductions every 2 years. In addition, industrialized countries agreed to develop low-carbon strategies or plans, which should ensure firm foundations are built that will be able to be upheld.

According to the analysis of an independent science-based assessment called Climate Action Tracker, which tracks the emission commitments and actions of countries, with current pledges the projected temperature rise is within the 2.6–4.0°C range. Therefore, individual countries must still refine and tighten their plans to bring temperature rise to within the agreed upon 2°C limit. Table 4.3 lists where various countries fall short according to their current plan's effectiveness, according to Climate Action Tracker's (2011) analysis of the proposals provided from the Cancun meetings. Therefore, the next step to refine the process and commitment was at the next scheduled meeting in April 2011 in Bangkok.

Bangkok (2011)

The Bangkok climate talks held on April 3 and 4, 2011, unfortunately did not close the gap between the emissions reduction pledges agreed to previously in Cancun and what is needed to get the world on track in order to limit global warming to 2°C above preindustrial levels, as was hoped. At its conclusion, there were still no developed countries that volunteered to increase their reduction target as was requested in the Cancun Agreements and in the mandate for the workshop held in Bangkok. There was little clarification for positive action by countries to the extent that it would raise ambition levels; rather the effect of most of the countries' clarifications still remained unclear.

Major uncertainty issues still remain a key issue, such as:

1. Accounting for forests
2. Carryover of unused allowances

TABLE 4.3

Individual Country Assessment from Cancun Agreement Proposals

Role Model	Sufficient	Medium	Inadequate
Maldives	Bhutan	Chile	Argentina
	Brazil	Iceland	Australia
	Costa Rica	India	Belarus
	Japan	Indonesia	Canada
	Norway	Israel	China
	Papua New Guinea	Mexico	Croatia
	South Korea	Singapore	EU27
		South Africa	Kazakhstan
		Switzerland	Moldova
			New Zealand
			Russian Federation
			Ukraine
			United States

Role model, emission targets are more ambitious than the 2°C range; sufficient, pledges in this area are in the more stringent part of the 2°C range; medium, pledges in this area are in the least stringent part of the 2°C range; inadequate, emission targets in this area are less ambitious than the 2°C range.

3. "Business as usual" emissions in developing countries
4. What the exact conditions are for implementing more ambitious pledges

Major issues that still need to be clarified and addressed are the details of the comprehensive international agreement for some of the developed countries, as well as the scale of the financial support needed for some of the developing countries. More work still needs to be done to come to a workable arrangement that will realistically cut GHG emissions to where they need to be in order to avoid the dangerous climate change threshold. The world awaits positive action while the clock continues to tick.

Conclusions

International climate change negotiation and effective management has taken a long and painful journey to reach any level of commonality among nations, with a definite dividing line between the United States and the remainder of the world's countries. Whether based on existing extensive infrastructure and an economy for the most part based solely on fossil fuels or not, getting the United States to cooperatively make significant, honest changes has been very difficult. While good progress has been made and measured by other countries around the world, the sad reality is that without U.S. cooperation and funding, the program is hurting, and it is doubtful any real progress will be made until they are fully engaged in the effort as one of the world's major players.

Whether it be through public criticism, the momentum of the green movement, a change in administration, or a combination of all three, we can be cautiously optimistic that the past political and environmental stagnation may be coming to an end. For the sake of the environment and the developing countries of the world, it is to be hoped that this is the case. What will be needed in the future is strong leadership, focused objectives, dedication to the cause, and a clear understanding of not only the science of climate change, but also the politics, economics, psychology, and sociology behind it. It is also apparent from recent data released by scientific think tanks that these things cannot come a moment too soon—they are desperately needed now. Whether through better public leadership, political agendas, or other compelling reasons, it is time the United States comes on board with some concrete solutions and a willingness to work. It is also important that every other country—including China and India—do the same. Timing is now critical, and there is no room left for anyone to merely slide by anymore expecting a free ride—everyone is now behind the steering wheel.

Discussion

1. With the smooth implementation of the Montreal Protocol for the control of ozone-depleting chemicals, why do you think the Kyoto Protocol did not follow suit? Why was it not as successful right from its initial implementation?

2. What do you think should be put into effect: an emissions reduction goal without any international enforcement mechanism or a binding agreement with enforceable mechanisms? Why do you feel your choice would be more effective?

3. Do you think the United States is just procrastinating and really has no intention of ever cutting back on greenhouse gas emissions? Do you think the federal government has too many close business relations with oil companies? Automobile companies? Why or why not?

4. Do you think the United States is stalling and just aiming for a "business as usual" approach with no real intent of changing? If so, to what do you attribute the lack of real commitment?

5. If you were a U.S. negotiator, how would you suggest the United States fix its relationship with the other negotiating countries? Do you think it is even possible at this point?

6. Do you think the outcome of the Cancun meeting from 2010 will finally make the United States accountable for its greenhouse gas emissions? Do you believe the United States is finally willing to conform to the same rules as the rest of the developed nations? Why or why not? On what do you base your opinion? What are your thoughts on Congress's split support on the issue?

7. Do you believe the ambitious nature of the Cancun Agreement is realistic given today's economy and business drivers? Justify the continuing popularity of the sport-utility vehicle and recreational vehicle market and suggest ways to handle that.

8. If you were a negotiator for a country, how would you propose to meet your country's goals, both short and long term?

References

BBC. 2009. Copenhagen deal reaction in quotes. *BBC News*, December 19. http://news.bbc.co.uk/2/hi/8421910.stm (accessed March 2, 2011). Discusses the opinions from the Copenhagen conference.

Broder, J. M. 2010. Support grows for expansion of ozone treaty. *The New York Times*, November 12. http://green.blogs.nytimes.com/2010/11/12/support-grows-for-expansion-of-ozone-treaty (accessed December 10, 2010). Discusses using the Montreal Protocol for climate change negotiations.

Brown, P. 2005. U.S. cities snub bush and sign up to Kyoto. *The Guardian*, May 17. http://www.guardian.co.uk/environment/2005/may/17/usnews.climate change.

Clapp, P. 2007. Statement on the decisions of the 13th Conference of the Parties to the UNFCCC. Pew Environmental Group. December 19.

Climate Action Tracker. 2011. Bangkok climate talks hoped to clarify pledges. http://www.climateactiontracker.org (accessed February 2, 2011). Provides analysis of countries' climate-mitigation pledges.

CNN. 2002. Cool response to global warming plan. *CNN World*, February 15. http://articles.cnn.com/2002-02-15/world/japan.climate_1_kyoto-treaty-warming-kyoto-protocol?_s=PM:asiapcf (accessed February 2, 2011). Discusses the international response to the U.S. reaction to the Kyoto Protocol.

Farman, J. C., B. G. Gardiner, and J. D. Shanklin. 1985. Large losses of total ozone in Antarctica reveal seasonal ClO_x/NO_x interaction. *Nature*. 315: 207–210.

Intergovernmental Panel on Climate Change. 2001. Climate change 2001: Synthesis Report: Summary for policymakers. http://www.ipcc.ch/pdf/climate-changes-2001/synthesis-spm/synthesis-spm-en.pdf (accessed December 9, 2010).

Intergovernmental Panel on Climate Change. 2007. *WG1: The Physical Science Basis.* Cambridge, UK: Cambridge University Press.

Molina, M. J., and F. S. Rowland. 1974. Stratospheric sink for chlorofluoromethanes: Chlorine atomic-atalysed destruction of ozone. *Nature.* 249: 810–812.

United Nations Environment Programme. 2007. Climate proofing North American cities and communities key challenge for 21st century. http://www.unep.org/Documents.Multilingual/Default.asp?ArticleID=5553&DocumentID=504&l=en (accessed February 21, 2011). Discuses what the U.S. and Canada will face unless adaptation measures to climate change are taken, incorporated, and practiced economically.

United Nations Framework Convention on Climate Change. 2011. The Cancun Agreements: An assessment by the Executive Secretary of the United Nations Framework Convention on Climate Change. http://cancun.unfccc.int (accessed February 25, 2011). Provides information on the outcome of the climate conference meeting at Cancun in December 2010.

U.S. Embassy. 2001. United States policy on the Kyoto Protocol. http://www.usembassy.at/en/download/pdf/kyoto.pdf (accessed January 15, 2011). Provides a summary of the Bush administration's stance on the Kyoto Protocol.

West, L. 2001. Bush announces U.S. plan to tackle global warming. About.com Environmental Issues. http://environment.about.com/od/governments corporations/a/bush_plan.htm (accessed February 19, 2011). Discusses the Bush administration's view on the Kyoto Protocol.

World Meteorological Organization. 2006. Scientific assessment of ozone depletion. http://www.wmo.int/pages/prog/arep/gaw/ozone_2006/ozone_asst_report.html (accessed January 3, 2011). Provides results for the ozone monitoring of the stratosphere under the Montreal Protocol.

Suggested Reading

Hansen, L. J., and J. R. Hoffman. 2010. *Climate Savvy: Adapting Conservation and Resource Management to a Changing World.* Washington, DC: Island Press.

Intergovernmental Panel on Climate Change. 2007. *WG1: The Physical Science Basis.* Cambridge, UK: Cambridge University Press.

5

Sociological Connections to Climate Change

Overview

This chapter takes a look at the sociological issues of climate change and explores several of the current key aspects, such as the public reactions to environmental issues in general and the reasons for those reactions— why some people seem to care so deeply about the environment around them and its future, yet others seem to give it no thought at all. It also looks at the public's slow shift in recent years toward understanding the basic concepts of climate change. In fact, the shift in emphasis from the name global warming to climate change was partly for sociological reasons: the label global warming implied to the public that rising temperatures were the only characteristic of the phenomenon. Because of that, for areas that were experiencing cooler temperatures or flooding, the public was not making the connection that it was part of the same phenomenon. Therefore, scientists began to transition to the term climate change to make it more understandable that there was more involved to the issue than rising temperatures.

This chapter touches on some of those sociological issues. It discusses how sociology is tied to the environment and environmental issues and why social systems and cultural values are tied in with biodiversity. It examines the organization of environmental movements and the interesting chain of events that occurred in order for the first Earth Day to come about over 40 years ago. Next it takes a close look at the effects that climate change has on society and who and where feels the brunt of the change more than others. Following that, this chapter touches on immigration issues that are expected to arise as the climate changes and how that could present a problem for some countries. In conclusion, it discusses some interesting theories about what may get people to listen to messages about the environment based on recent research, as well as specific thoughts from individual sociologists on the subject.

Introduction

An important component of the whole climate change picture is the socio-logical aspect, yet it is one of the least looked at and understood. Even though societies around the earth are currently being affected and scientists have projected future scenarios for several locations worldwide and issued warn-ings to begin preparing by investing in mitigation efforts, planning defense strategies, becoming proactive in changing lifestyle choices, and planning for various future scenarios, the general public still displays tendencies that are reluctant and lukewarm.

What seems puzzling in this case is the general feeling of having to push people to act and respond to climate change when it is a situation that affects their very lives and futures. This raises questions as to where the commu-nication process is failing. Is it an issue of education, politics, culture, eco-nomics, lifestyles, mindsets, or some other unidentified factor? There has even been concern expressed in the sociological field that sociologists are not paying enough attention to the issue, yet the sociological aspect of the big picture is one of the most crucial components. With such a diversity of cultures worldwide, there are several sociological issues to consider.

Development

One of the most overlooked aspects of both the study and management of climate change is the sociological aspect. Although sociologists have repeat-edly brought this issue to public and scientific attention, it still remains one area significantly lacking in the attention it deserves, which is ironic because both the short- and long-term effects of climate change affect the environ-ment in multiple ways that impact human lives every day.

The Environment and Sociology

Climate change and its effect on society fall under the sociological study of societal-environmental interactions or environmental sociology. The principal focus of the field is the relationship between society and the environment—specifically social factors that lead to environmental problems, the resultant societal impacts of those problems, and the subsequent efforts to adequately solve those problems. A fairly new specialized emphasis in sociology, it did not gain real distinction until the environmental movement of the 1960s and early 1970s, particularly in the United States and Europe.

This time period in the United States became characterized as the "Environmental Decade" and is marked by the time interval when the

United States Environmental Protection Agency was created and such acts as the Endangered Species Act, the Clean Air Act, and the Clean Water Act were implemented. This was also the time frame when Earth Day was instituted, which represented a significant paradigm shift in public thought and is discussed in greater detail in a later section.

Over the past few decades, there have also been several notable anthropogenic environmental disasters that have played a significant role in making the public more aware of damage that can be caused to the environment—damage that is sometimes permanent. As society becomes more aware of the environment and how altering it can have a negative impact, it has spurred the development in recent years of many environmental awareness groups committed to helping preserve our surroundings. Fortunately, these groups have had a significant effect in educating political representatives, policy makers, educators, and the general public.

One of the worst human-caused disasters to occur was the Russian nuclear power plant explosion in Chernobyl in 1986, when a reactor exploded during a failed cooling system test and ignited a massive fire that burned steadily for 10 days. The accident released radioactivity four hundred times more intense than that of the Hiroshima bomb during World War II. The accident affected a huge area—parts of the Ukraine, Belarus, and Russia (the former Soviet Union), covering an area of approximately 233,000 square kilometers. The ground was contaminated with unhealthy levels of radioactive elements. The town of Pripyat, which was built specially to house the employees of Chernobyl, was evacuated after the accident, displacing 50,000 people. Even today, more than 20 years later, the town has never been reoccupied. The people who lived in the area at the time of the accident who were not killed outright in the explosion suffered many serious health problems, including radiation sickness, thyroid cancer, leukemia, and birth defects causing cancer and heart disease. Researchers have estimated that roughly seven million people were affected by this accident.

Another major human-caused environmental disaster was the nuclear accident on the Susquehanna River in Pennsylvania at Three Mile Island in March 1979 (Figure 5.1). This was the disaster that began the controversy over the safety of using nuclear energy in the United States. In this incident, the water pumps in the cooling system failed, causing cooling water to drain away from the reactor, which partially melted the reactor core. The accident released about one thousandth the amount of radiation as the Chernobyl disaster did. The reactor core barely escaped meltdown because of the implementation of safety measures. Scientists do not know for certain how much radiation was released during the accident. There was an evacuation of an 8-kilometer radius as a safety precaution. Experts do believe that several elderly people died from being exposed to the radiation. Dairy farmers reported the deaths of many of their livestock. Some local residents developed cancer. Some studies also indicate that premature death and birth

FIGURE 5.1
Three Mile Island was the site of one of the nation's worst anthropogenic environmental disasters. (Courtesy of EPA, http://www.epa.gov/history/images/1980b.jpg.)

defects resulted as well. The cleanup for the accident began in August 1979 and finally ended in 1993, with a price tag of $975 million. Nearly 91 metric tons of radioactive fuel was removed from the area.

Times Beach, Missouri, was the site of another well-known environmental scare that got the public's attention when high dioxin levels were found in the soil. Dioxin is a hazardous chemical used in Agent Orange, a highly toxic chemical warfare agent. Levels in the soil were determined to be one hundred times higher than the threshold considered to be toxic to humans. The dioxin had been mistakenly added to an oil mixture that was used to spray the roads in the 1970s to keep the dust problem under control. Many illnesses, miscarriages, and animal deaths at the time were blamed on the levels of dioxin in the area. This episode was one of the environmental disasters in the United States that spurred the enactment of federal action and the implementation of the Comprehensive Environmental Response Compensation and Liability Act, a piece of legislation commonly referred to as Superfund because of the fund established within the act to help the cleanup of locations like Times Beach.

In the Baia Mare gold mine in Romania in January of 2000, cyanide, which is used to purify gold from rocks, overflowed into the Tisza River. The wastewater that the cyanide was in also contained lead and other hazardous materials. By February, it had impacted the Danube River, a major river that also flows through Serbia and Hungary. This incident poisoned many of the fish in the river, and many people along the river had to be treated for eating contaminated fish.

According to the U.S. Environmental Protection Agency, Love Canal is one of the most horrifying environmental tragedies in U.S. history. Love Canal was originally planned to be a "perfect" community on the eastern edge of Niagara Falls in New York. In order to generate power for the community, a canal was to be dug between the upper and lower Niagara Rivers so that power could be generated inexpensively for the residents. The canal was never finished for generating energy, however. In the 1920s, all that was left was a remnant of the original digging, and it was turned into a municipal and industrial chemical dumpsite. Then, in 1953, the Hooker Chemical Company, who then owned the property and the canal, covered all the waste with dirt and sold it to the city for the sum of one dollar. Unfortunately, the city used the ground for a new development and constructed about 100 homes and a school on it. During an extremely wet period, where the area received ample rainfall, the leaching process began. The waste disposal drums began to corrode and started breaking up through the residents' back yards. The vegetation in the area—trees, shrubs, gardens, and grass—began to die and turn black. Chemicals began to pool in people's backyards and basements. Children got burns on their hands and faces when they played outside. An increase in birth defects began to occur, as well.

At the time, according to a report issued by the Environmental Protection Agency (EPA), one father whose child was born with birth defects as a result of the tragic situation remarked: "I heard someone from the press saying that there were *only* five cases of birth defects here. When you go back to your people at EPA, please don't use the phrase '*only* five cases.' People must realize that this is a tiny community. Five birth defect cases here are terrifying" (Beck, 1979).

On August 7, 1978, New York Governor Hugh Carey told the residents of Love Canal that the New York State Government would purchase the homes that were affected by the chemicals. The same day, President Jimmy Carter approved emergency financial aid for the area. These were the first emergency funds ever to be approved for a human-caused disaster rather than a natural disaster. In addition, the United States Senate approved a "sense of Congress" amendment saying that federal aid should be forthcoming to relieve the serious environmental disaster that had occurred. President Carter remarked of the situation: "The presence of various types of toxic substances in our environment has become increasingly widespread—one of the grimmest discoveries of the modern era" (Beck, 1979).

A total of 221 families had to be relocated as a result of this disaster. Today, agencies such as the EPA, working under governing laws such as the Clean Air and Water Acts, the Pesticide Act, the Resource Conservation and Recovery Act, and the Toxic Substances Control Act, strive to protect the American public and the environment.

The Valley of the Drums in Kentucky gained national attention in 1979 and quickly became known as one of the United State's worst abandoned hazardous waste sites. Over a period of 10 years in Bullitt County, people had disposed of thousands of barrels of hazardous wastes. They had been haphazardly thrown in pits and trenches or just strewn about (Figure 5.2).

FIGURE 5.2
Valley of the Drums. This site was used as a disposal area for hazardous chemical waste, causing serious environmental pollution. Today it is still remembered as an example of environmental irresponsibility. (From Publitek, Inc., Waukesha, Wisconsin.)

The drums sat so long exposed to the outdoor elements that they began to deteriorate and leak. When it rained, the barrels would fill with water, overflow, and wash the chemicals that were inside into nearby Wilson Creek, which led to the Ohio River. Chemicals were found in them such as toluene (which is associated with liver and kidney damage and respiratory illness, is harmful to developing fetuses, and can cause death) and benzene (which can cause leukemia and neurological problems and weaken immune systems). This incident gained much national attention; it helped spur the creation of the Superfund.

In March of 1989, the *Exxon Valdez* struck Bligh Reef in Prince William Sound, Alaska, creating the largest oil spill in U.S. history. The oil slick spread more than 7,770 square kilometers and onto more than 563 kilometers of beaches in Prince William Sound, at that time known as one of the most pristine and beautiful natural areas in the world (Figure 5.3). The spill polluted about 1,900 kilometers of shoreline and was devastating to the wildlife in the fragile ecosystems. It killed approximately 250,000 sea birds, 2,800 sea otters, 250 bald eagles, and roughly two dozen killer whales.

The spill released more than 11 million gallons and cost more than $3.5 billion to clean up. This was one of the environmental disasters that received a great deal of media attention and brought significant environmental awareness to the public. Many people were outraged at the damage done to the

FIGURE 5.3
The tanker *Exxon Valdez* accident in Prince William Sound, Alaska, in March 1989. NOAA responders survey the oil-soaked beaches of Prince William Sound. (Courtesy of NOAA, http://www.photolib.noaa.gov/.)

wildlife habitat and the fragile ecosystems, and this single incident in particular served as a strong reminder that human behavior can have far-reaching impact on the environment.

One topic that has been in the news quite a bit the past few years and has subsequently made the public more environmentally aware is the melting of the polar ice as climate change becomes more of a problem. *Time* magazine has run several special editions covering the melting of glaciers, rising seas, and diminishing icepack (such as in the April 9, 2001; April 3, 2006; April 9, 2007; and October 1, 2007, issues). Likewise, *National Geographic* featured an article on melting icecaps and rising sea levels in their June 2007 issue. Environmental scientists have said that because the Arctic ecosystems are so sensitive and fragile, when they are stressed, they respond quickly. Because of this, they provide an early warning of the effects of climate change. Scientists at the National Oceanic and Atmospheric Administration (NOAA) and National Aeronautics and Space Administration (NASA) believe the Arctic will be one of the first areas to react to the effects of global warming and show subsequent signs of impact. They are warning the public today that the average temperatures in the Arctic region are rising twice as fast as they are anywhere else in the world. NASA has taken satellite images of the polar ice cap and has determined that it is shrinking at a rate of 9 percent each decade. They predict that if this trend continues, there may not be any ice left in the Arctic during the summer season by the end of this century. This would upset the food chain and negatively impact the balance of the ecosystem, affecting the native people, wildlife, and vegetation.

Scientists are already seeing changes in the feeding and migration patterns of walrus, seals, polar bears, and whales. This makes it difficult for native inhabitants of the area (such as the Eskimo and Inuit) to hunt and obtain the food supply on which they rely for survival. Melting ice also contributes to rising sea levels. Today, many of the native villages that have existed for centuries along the coastlines are being flooded and swamped as the sea levels rise and put the coasts under water. This is a very serious example of how climate change is threatening the identity, culture, way of life, and the very existence of certain cultures.

Climate change scientists have recently announced that rising temperatures are already affecting Alaska. According to Deborah Williams, executive director of the Alaska Conservation Foundation, annual temperatures have increased 6.7–8.3°C, and winter temperatures have warmed 13–17°C—more than any other place on earth and more than four times the global average. Permafrost is melting, causing more than 600 families so far to lose their homes.

According to Steven Amstrup, a polar bear specialist with the U.S. Geological Survey, "As the sea ice goes, that will direct to a very great extent what happens to polar bears" (Carlton, 2005). Polar bears could become extinct within the next century because they have adapted to hunting on the ice. If they try to swim, they are more likely to tire and drown.

There is no lack of scientists warning the public of the ill consequences to society if climate change is allowed to continue. In fact, they have delivered ample warnings over the past two decades, which have been well publicized (Figure 5.4). Why, then, do certain populations display a general trend toward ignoring the significance of the warnings? In particular, why does the United States not heed the warnings when they have the means at their disposal to immediately implement the changes that will make a real difference? While studies are being published, newscasts are being aired of the negative effects currently in progress, and projections are being announced that will impact every human life in the near future, the business as usual attitude seems to prevail in the United States. Other countries, such as those of the European Union, Canada, and Australia, for example, seem to be much more aware of the situation—to the point where they have already incorporated meaningful changes into their lives to help mitigate, prepare, and accommodate those changes as they occur. For example, Europe is much more oriented toward public transportation rather than driving personal vehicles.

In a study conducted by Bassett et al. (2008), participants from Europe, North America, and Australia were studied to determine whether there was a relationship between active transportation (defined as the percentage of trips

SPECIAL REPORT GLOBAL WARMING

TIME

BE WORRIED. BE VERY WORRIED.

Climate change isn't some vague future problem—it's already damaging the planet at an alarming pace. Here's how it affects you, your kids and their kids as well

EARTH AT THE TIPPING POINT
HOW IT THREATENS YOUR HEALTH
HOW CHINA & INDIA CAN HELP
SAVE THE WORLD—OR DESTROY IT
THE CLIMATE CRUSADERS

FIGURE 5.4
Time magazine. (From Wasatch Images, Salt Lake City, Utah.)

taken by walking, bicycling, and public transit) to obesity rates. What was interesting about this study from a climate change perspective were the percentages of active transportation participants between the three geographic areas. Europe dominated the active transportation—they walked more than residents of the United States: 382 vs. 140 kilometers per person per year. They also bicycled more: 188 vs. 40 kilometers per person per year. The conclusion of the study was that Europeans walked and bicycled more per year than did those from the United States, Australia, and Canada (Bassett et al., 2008). From a climate change perspective, this is a more desirable behavior because they are using less fossil fuel. The study cited the following reasons as to why Europeans did not drive as much:

- They have densely packed cities that generate shorter trips.
- Europe has more restrictions on car use (car-free zones, "no through" zones, etc.).
- They have extensive safe, convenient facilities for cycling and walking.
- They provide bikeways, pedestrian walkways, and bicycle parking lots in conjunction with ample public transportation areas.
- Traffic regulations and enforcement policies favor pedestrians and cyclists over motorists.
- Owning and operating a personal car is extremely expensive.

There are probably many reasons why people choose to utilize public transportation, walk, bicycle, carpool, or drive a personal vehicle, and those reasons are diverse, depending on cultural values, economic opportunities, educational opportunities, geographic factors, and other key reasons. Each of those diverse reasons, however, has an impact on climate change, because the choices people make—especially those that involve the burning of fossil fuels somewhere along the production line—are going to come into play as temperatures rise.

In a study conducted by Constance Lever-Tracy in 2008, she stated that while the public had become aware over the years of human-caused climate change, most sociologists, as a general rule (excluding environmental sociologists), had surprisingly little to say about the possible future social avenues that may occur as a result of it. She also stated that in general, because sociologists were usually not in a position to either validate or not validate the claims of natural scientists, there was an established tendency to "look the other way." In the case of climate change, however, not paying attention and becoming directly involved could spell disaster for the future of sociology because the consequences of climate change could spell disaster for society. Therefore, it would be in the best interests of sociologists to pay close attention, because as she said, "these developments can affect the very core of our discipline's concerns. We need a cooperative multidisciplinarity of social and natural scientists working together."

This observation is very astute, because the consequences do directly affect society. Perhaps if more sociologists analyzed the potential effects and what that could mean to the future of society as a whole, then more people would be likely to listen. Lever-Tracy points to the multitude of events that occurred in 2005 as the crucial "tipping point" that should have rallied the professional attention of sociologists. In particular, she discusses evidence, such as the 26 tropical storms and three category 5 hurricanes that occurred that year (including Hurricane Katrina) and the fact that that social responses to these disasters were mixed. The announcement by NASA that the oceans were getting warmer, the escalating greenhouse effect, the drastic melting of the Siberian permafrost releasing trapped methane, and the unexpected shrinking glaciers on the Antarctic ice sheet shocked the scientific community, yet received no real reaction from sociologists.

Lever-Tracy's explanation for the nonalarmist response does make sense. She explains it as a flaw in the science of sociology and attributes it to two factors. The first factor is the mirroring of an indifference sociologists find in contemporary society towards to the future. The second factor is sociologists' continuing foundational suspicion of naturalistic explanations for social facts, which generally leads sociologists to question or ignore the authority of natural scientists.

Regarding the first reason, she explains that the rate of change of natural processes has been shrinking toward the time scales of human society. For human society, there is a strong emphasis on immediate gratification and a decline in long-term direction or plans. This "desensitizing," if you will, has made it so that even threats just decades away now barely register. It is as if people today live in a "permanent present" reality. In other words, people today care much less about future scenarios: They do not think about the future as much because they live in the now. This is the here and now, the disposable society, the time when people focus on immediate consumption. It extends further: Politicians are more interested in the next election than the climate 50 years from now. Business leaders are more focused on the next annual report than what sea level will be in 20 years and what islands will no longer exist and which coastal areas will be long submerged.

Lever-Tracy's suggestion to solve this disparity is for both social and natural scientists to work together toward a common end, and she explains that it is just as necessary to focus into the future. As she summarizes: "The alternative is irrelevance or worse—an effective complicity with the vested interests of fossil fuel corporations" (Lever-Tracy, 2008).

In response to Lever-Tracy's thoughts on climate change, Steven R. Brechin (2008) pursued the topic further. He believes that climate change presents the most significant challenge facing the world today, and that in itself should be more than sufficient to mobilize the world to act and respond rapidly in a meaningful way. He also believes that the human response will be made, in part, to conditions not yet fully realized, which presents society with a unique challenge: it is not a matter of simply reacting, but in thoroughly thinking things through and being prepared. He also believes that it is most likely that only after global societies are restructured by human-altered natural processes will sociology have a new focal point. One of the biggest obstacles he points out is that for the affluent, influential northern countries, change is going to be harder to see come about because they will not be affected as much as other countries. Because of this, they will not feel like they are "living on the edge." Therefore, they will be harder to motivate toward a new mindset or paradigm. To them, Brechin states: "environmental concerns have become and continue to be seen as no more than background noise." He also states: "social change is difficult, especially when response must be made to conditions not yet fully realized."

As frustrating as it might be, the human factor is a significant part of the big picture, and motivating people to get prepared for disaster or to be proactive about an environmental concern that may be a generation away in their eyes can sometimes feel like an impossible task.

The Need for Biodiversity

There is a strong link between biodiversity and human well-being (Figure 5.5). Biodiversity contributes both directly and indirectly to many factors of human well-being, such as security, basic materials for a good life, good social relations, good health, and freedom of choice and action (Royal Society, 2007). Over the past century many societies have benefited from converting natural ecosystems to human-dominated ecosystems and exploiting biodiversity, but not everyone has benefited. Others have experienced a decreased well-being, and some social groups have experienced an increase

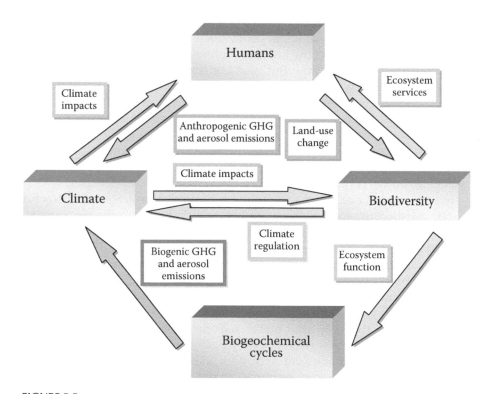

FIGURE 5.5
The relationships between climate change, biodiversity, and human well-being. (From Wasatch Images, Salt Lake City, Utah.)

in poverty. For example, with the issue of deforestation, the societies doing the harvesting have benefited, but the societies living in the forest have experienced a lowering of biodiversity and resulting poverty.

The way humans treat the biodiversity of an area is critically important, because if the biodiversity is lowered, then the environment is not as resilient to abrupt or unexpected change. Then, if a significant change does occur, the area may not be able to support enough diversity to remain strong enough, which can cause the ecosystem to be compromised and die. Following that logic, if the ecosystem becomes unhealthy, it cannot support the human and animal populations that depend on it, which then compromises the existence of the life within it.

Over the past few hundred years, human activity has significantly changed the surface of the earth. As a result, the earth's climate is also being changed. Species becoming extinct is not new news; in fact, it is happening at rates faster than it ever has before. The disregard being shown toward the earth's ecosystems today is taking its toll on their health. In particular, according to the Millennium Ecosystem Assessment (2005), there are four categories of services provided by ecosystems to society, as shown in Figure 5.6, which are supporting, regulating, provisioning, and cultural services. Supporting services are what all other ecosystem services depend upon, and include carbon cycling and water and nutrient cycling. The regulating services provide the mechanisms that moderate the impact of stresses and shocks on ecosystems and include climate and disease regulation. This also determines the distribution of provisioning services, such as food, fuel, and fiber, and cultural services, such as spiritual and aesthetic clues (Kinzig and Perrings, 2007).

An assessment of ecosystem transformation is largely determined by who is doing the assessment and exploitation. If it is the individual who has benefited, then the transformation will be valued; if it is an individual who did not benefit, then he was short-changed.

Environmental Movements: The Classic Case of Earth Day

One of the most notable days in support of the environmental cause via a socialistic voice was the inception of Earth Day in 1970—a truly landmark event. Founded by U.S. Senator Gaylord Nelson, the idea for it actually began in 1962. An environmentalist himself, Nelson was openly concerned that the environment was not an issue that the politicians in the United States seemed to care anything about. He met with President Kennedy and persuaded him to give visibility to the issue by going on a national conservation tour. The outcome was a 5-day, eleven-state conservation tour in September 1963. To Nelson's surprise, the tour was not a success: it did not succeed in putting the issue onto the national political agenda.

Following his first failure, Nelson did not give up. He continued to speak on environmental issues to a variety of audiences in approximately twenty-five

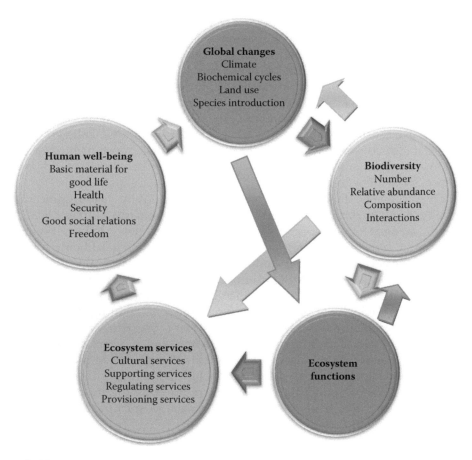

FIGURE 5.6
The relationships between ecosystem services, biodiversity, and global change. (From Wasatch Images, Salt Lake City, Utah.)

states. Everywhere he went, he saw evidence of environmental degradation. He also found something else: public interest and support. For the next 6 years, however, Nelson was unable to get the concept of environmentalism onto the political agenda. Then, in the summer of 1969, Nelson was out West on a speaking tour and he took note of the Vietnam War demonstrations called "teach-ins." They had spread to college campuses all across the nation and had the ability to rally large groups together.

They gave Nelson the idea he needed. He decided to organize a huge grassroots protest over what was happening to the environment. He was convinced that if he could tap into the environmental concerns of the general public and generate that same student energy in anti-war protests he had seen on campuses into the environmental cause, he could generate a demonstration that would force this issue onto the political agenda. It was a gamble, but he was game for a try.

At a conference in Seattle in September 1969, he announced that in the spring of 1970 there would be a nationwide grassroots demonstration on behalf of the environment and invited everyone to attend and participate. The news services carried the story from coast to coast. The response was electric. It took off with enthusiasm. Crowds of people chipped in on the effort and sent letters, telegrams, and made phone calls. Finally, the American people felt that they had a forum to express their concern about what was happening to the land, rivers, lakes, and air—and they all joined in. The 4 months following that, Nelson ran an Earth Day affairs office out of his office in Washington, DC, coordinating all the details. Then, 5 months before the scheduled day for the rally, the *New York Times* ran an article, and in it was quoted: "Rising concern about the environmental crisis is sweeping the nation's campuses with an intensity that may be on its way to eclipsing student discontent over the war in Vietnam…a national day of observance of environmental problems…is being planned for next spring… when a nationwide environmental 'teach-in'…coordinated from the office of Senator Gaylord Nelson is planned" (Hill, 1969).

When the day arrived—April 22, 1970—it became immediately obvious that it was a spectacular success—a credit to the spontaneous response at the grassroots level. There were more than twenty million demonstrators and thousands of schools and local communities that participated. According to Gaylord Nelson: "That was the remarkable thing about Earth Day. It organized itself" (Envirolink, 2011).

The Ramifications of Climate Change on Society

When discussing sociology and the effects of climate change, it is also important to look at environmental sustainability. Along with the concept of the "here and now" mentality, it would seem in order to be thinking and planning ahead for methods to sustain society. There are several major concerns with society: the inequitable use of resources and hence the inequitable impact on the environment; the extreme variance of mindsets ranging from those who practice strictly green lifestyles and pull more than their weight being environmentally responsible and those who live each day like there were four earths of resources at their disposal; the uneven natural distribution of resources; the uneven distribution of wealth; the variable control of trade and commerce; and the disparate distribution of power—the list goes on and on. All of these social factors play a role in how climate change will affect an individual's outlook. For example, the cartogram in Figure 5.7 is a graphic statistical representation of carbon emissions. As expected, the United States is bloated, as is much of Europe and China. Then if you look at Africa, it is hardly visible, along with South America. That evidence is graphic enough, but the even harder piece of information to swallow is what the map would look like if we were illustrating those negatively impacted by the effects of climate change. That map would appear much different. In that version, the

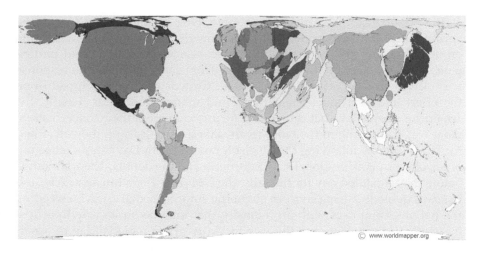

FIGURE 5.7
(See color insert.) A cartogram depicting world global warming. This illustrates the world in terms of carbon emissions. The United States, Europe, and China are disproportionately large, whereas Africa is barely visible. (Courtesy of SASI Group, University of Sheffield, and Mark Newman, University of Michigan, http://www.worldmapper.org/display.php?selected= 295, 2006.)

United States and Europe would be very small, whereas Africa would be blown up like a balloon, as would Asia. Unfortunately, there is nothing fair about this picture: with climate change, the wealthiest nations are the biggest offenders (for the most part) but will be impacted the least; and the world's poorest nations, who emit relatively few greenhouse gases, are destined to receive the brunt of the detrimental impacts from climate change.

Climate Justice

In an article by Randy Poplock (2007), the issue of "climate justice" is introduced. Poplock discusses the way in which low-income and minority populations generally bear the brunt of human health consequences during emergency situations, such as Hurricane Katrina in 2005, because they tend to have limited access to health insurance. In a recent study conducted in Seattle, residents reported that white populations had an 89 percent health insurance rate compared with 75 percent for black and 61 percent for Latino populations.

One dangerous ramification from this is that heat waves—one of the effects common to climate change—are particularly devastating to low-income and minority residents because they generally lack air-conditioning, live in higher crime areas (where doors and windows remain closed for security reasons, cutting air circulation), and congregate in highly urbanized neighborhoods (where temperatures are usually higher because of the urban heat effect).

Heat waves are not the only danger putting minorities at higher risk, either. Floods and wildfires are just as dangerous and devastating—and they are also associated with climate change. In this case, low-income populations are less mobile, often lack access to warning systems (such as the Internet), or do not understand English warnings, which makes them more susceptible to catastrophe. It is also this population that tends to lack adequate property or homeowner's insurance, making wildfire and flood damage even more devastating.

Another serious issue to consider is raising prices for commodities. As items, such as water, become more in demand, they will become more expensive. Electricity rates could start climbing, making it difficult for some people to have access to power. As farms begin suffering from projected water shortages, food may start becoming scarce, causing grocery prices to escalate, furthering impacting low-income families. Poplock points out, with some irony, that low-income families are the ones who get penalized, but are generally not the ones who contribute the most to the emission of greenhouse gases. He also points out that although the news commonly portrays these types of scenarios about other countries, it happens in the United States, as well. For example, when Hurricane Katrina hit the South in 2005, those who were hit the hardest were the low-income minorities. Poplock says, "Climate justice must be considered in policy decisions and public dialogue. Every environment issue poses a social justice concern, and climate change is no exception" (Poplock, 2007).

Immigration Issues

Another very real issue that needs to be considered today is the ramification of immigration. One need only pick up a couple of newspapers to read headlines about mass extinction of species, global warming, deforestation, melting of polar ice caps, the rapid depletion of fossil fuels, sea-level rise, drought, failed crops, dust storms, and polar bears stranded on icebergs to get a taste of the consequences of climate change. However, rarely, in comparison, does one read headlines detailing environmental nonsustainability, political instability, economics, and poverty, yet they are all relevant social issues that bear extreme weight and need to be dealt with in some way.

Another issue also not much discussed is that of climate models and their analytical results. Most climate models agree that by the close of this century, the polar and subpolar areas will receive more precipitation, and the subtropics, the area between the tropical and temperate zones, will receive substantially less. These correspond with the areas that currently depend heavily on rainfall for crop irrigation—areas that have been suffering from drought for years. This will have negative ramifications on food production, prohibiting those regions from being able to produce enough food to adequately support their populations, increasing the incidence of starvation. For example, in

Africa, where conservative estimates show that only 6 percent of its cropland receives irrigation, there may be greater starvation and political unrest than currently exist. Elizabeth Kolbert's article, "Changing Rains," in the April 2009 issue of *National Geographic*, states that changes in rainfall patterns could be contributing to the conflict in Darfur, for nomadic herders and farmers constantly fight over cropland. She also points out that there are already several tensions in the Middle East over allocation of water in places with severe shortages. In addition, several ecologists have warned that future wars will be waged all over the world not over oil, but over water (Kolbert, 2009).

In addition to wars over water, the lack of water will also trigger another social side effect of climate change with international refugees: Water shortages will result in food shortages. When this occurs, it will leave people with only one feasible option: to immigrate to countries with modest water supplies and available food. Climate scientists have predicted that rising sea levels could displace billions of people—mostly low-income families—who live along the coastlines of Africa, Asia, and Latin America. Every country that could be a potential destination will suddenly be faced with how to handle a mass international immigration. This, in turn, could lead to extreme political instability (see Chapter 9). These are considerations that climate scientists are warning about now and have been for years, yet the general public seems too complacent to respond. It is a problem, however, that is critical to address.

One Explanation to Encourage Going Green

So how do you get someone's attention and convince them to go green? According to author Steve Stillwater (2010), there are several motivations that could be effective: the green economy is expected to produce new jobs, alternative energy technology is being developed to replace fossil-fuel-based energy sources, you do less harm to the environment, you can live more efficiently and be less wasteful, or you can save money. However, according to the *Wall Street Journal*, none of those is the correct answer. The correct answer is: The strongest motivating factor to cause someone to go green is good old-fashioned peer pressure (Simon, 2010).

A recent experiment illustrates the *Wall Street Journal's* point superbly: Two different placards were placed in hotel bathrooms to encourage guests to reuse their towels. Some of the rooms' placards stated: "Show your respect for nature." Other rooms' placards stated: "Join fellow guests in helping to save the environment." After the experiment was run and the results were tallied, it was found that only 25% of the guests that had the first placard reused their towels, yet 75% of the guest that had the second placard reused their towels. Based on these results, the hotel tried a second round of experiments and made the placard specific to the guest in the particular room. The placards then read similar to this: "75% of the guests who stayed in Room 295 reused their towels." This sign achieved an even higher compliance percentage, illustrating that simple old-fashioned peer pressure works and is more effective than using an abstract rationale about the benefits to the environment.

Based on these results, Steve Stillwater (2010) made a reference to companies marketing green products and the obvious implications of "using peer pressure and creating a guilt complex." He noted that most people would rather not be singled out as not being willing to go along with most other people in protecting the environment and that it was a much more powerful motivator than just quietly going green because you think it is the right step to take for the environment. So the logical next question: how do sociologists deal with that? If that is a legitimate mindset along with only living in the here and now and wanting to plan for the future, it sounds like sociologists have their work cut out for them, at the very least, on this one. What is also interesting is that maybe it is not really necessary to spend millions on studies to figure out how the human mind works and how to get the public to respond and adapt in a positive manner to climate change—perhaps the solution is much simpler than sociologists realize.

Other Social Factors to Consider

As climate change progresses, whether countries be large or small, wealthy or impoverished, there will be some social issues we all will have to deal with. One is the social impact of drought and desertification. Today, it has become an issue that affects the world. No longer confined to isolated areas, its effects will reach across the globe as temperatures climb. It also seems that even the wealthier nations are not any better prepared to cope with the devastation it can wreak. According to Raymond Anselmo (2009) of the Model United Nations Far West, even highly developed nations are experiencing the same difficulties, as evidenced by the significant droughts in portions of the United States recently. He points out that although there have been moves toward solving the problems associated with drought and desertification and addressing the accompanying socioeconomic complications, the problems are as great—if not greater—than they have ever been. He states in a 1990 estimate that worldwide 15 million acres are lost to desertification each year, along with 26 billion tons of topsoil lost because of erosion, even though experts are warning that decisive, comprehensive action is needed right now in order to rectify the situation. The problem still continues, and each acre lost to either process is an acre lost to the future of crop production. With growing populations and decreasing viable agricultural areas, it is as if society is willingly heading for its own destruction.

Anselmo also points out that although there are several root causes of drought and desertification, with a few exceptions, most are human-caused; most causes are because humans overtax ecosystems. Because of this, Anselmo believes the social impact of drought and desertification is fourfold in nature. First, a lack of food means starvation and ill health, causing many to either die or become incapacitated. Second, as people begin to starve, they flee to another location, causing internal and external refugee problems, which often lead to cultural clashes. Third, the lack of supplies, massive floods, and not being able to find work in the new location causes a breakdown of the traditional family structure. These issues can also cause people to rebel against their own government if they do not move elsewhere. Fourth, these problems can harm political advancement, such as when a government has to limit freedoms or tighten its economic belt in order to deal with the new conditions.

Anselmo also explains that solutions have been proposed from all areas and viewpoints, and a few nations have already put plans in place to solve their problems. Several nations currently have proposals and plans for drought and desertification problems such as encouraging the private sector to develop new technology, dams and storage reservoirs in monsoon areas and other areas prone to regular-flooding, canals, windbreaks, better irrigation, water resource and forest management, tree and grass planting programs, higher reserves for economic protection, and reducing fossil fuel burning. The drawback, however, is that actually applying these ideas costs money, which is often in short supply.

Solutions have been offered by the United Nations to solve these problems, which are usually global or regional in nature. In 1974, the United Nations established its Plan of Action to Combat Desertification. In 1978, an account was established in order to finance projects related to desertification. The UN also established a program on agricultural methods to educate people living in drought-stricken nations.

In London, in February 2009, a lecture series was held on climate change and its impact on the developing world during the Ismaili Centre Lecture Series. Rather than look at climate change from the standard perspective in the natural environment, they looked at it as regards its social implications (Sachedina, 2009). Focusing on southern Africa, they discussed climate-induced changed rainfall patterns. They pointed out that southern Africa is currently receiving less precipitation, rendering the region susceptible to drought, while other areas, such as the northern and eastern parts of Africa have had an increase in rainfall. They believed that changes like these would create winners and losers, which could create conflict. One of the panel members, Dr. Salim Sumar, Executive Officer of Focus Humanitarian Assistance Europe, noted that effective responses to disasters are less simple than they were in the past. He also said that more nongovernmental organizations are now necessary in order to help regions affected by floods. In addition, Smith pointed out that the impact of disasters is difficult to confine within borders (*The Ismaili*, 2009).

In reference to the United Kingdom's leadership role, Lord Adair Turner, Chairman of the United Kingdom's Government Committee on Climate Change and the country's Financial Services Authority, commented that "even if global targets are met, global temperatures may still increase by 2°C by 2020" (Committee on Climate Change, 2011).

Camilla Toulmin of the International Institute for Environment and Development, an independent think tank, believed there is a need for governments and civil society to "start measuring what we value, rather than just economic factors" (*The Ismaili*, 2009). She then predicted that this would lead to a shift toward cooperation and away from conflict. She emphasized that economic growth should not be the only driving force in decision making. Lord Turner also stated that it should be possible to solve the climate change problem and still maintain a healthy growth in the gross domestic product, as long as the economic component was not the principal focus (*The Ismaili*, 2009).

The panel also noted that over recent years, because climate change has been established as a fact, the real challenge is to slow its progression down and focus attention on low-income societies and mitigating its impact on their lives, jobs, and lifestyles. Camilla Toulmin summed up the consensus of the panel: "Real security comes from building a climate of trust, rather than building a climate of fear" (*The Ismaili*, 2009).

Dr. Craig A. Anderson from the Department of Psychology at Iowa State University completed a study concerning the effects climate change has on violence. He initially established that uncomfortably warm temperatures can increase the likelihood of physical aggression and violence. Three types of data support this: In the first, physical studies have shown that studies conducted in controlled laboratory settings have shown that uncomfortably hot temperatures increase physical aggression. Second, when looking at different geographical areas, hotter areas have higher violence incidence records. The third type looks at one region over different times. In a controlled study conducted by Anderson (2010) it was found in the U.S. crime rates over 55 years, that a 1°C temperature rise led to more than 7.5 more assaults and homicides per 100,000 population, which is believed to be a function of the thermoregulation and emotion regulation areas of the human brain. Likewise, they found that hotter summers led to larger summer increases in violent crime than cool summers. The implication for climate change is that areas that experienced increased temperature rise will also experience an elevation in interpersonal conflict and violent crime.

According to research, the heat-aggression link operates immediately and directly on an individual. The most high-risk groups include those living in extreme poverty, those with poor childhood nutrition, broken family units, and low IQ, and those growing up in violent or unstable neighborhoods. Violent tendencies do not have to be instantaneous, either; they can also take years to manifest, which also has implications for climate change. It is possible that climate change will increase the violent risk behavior of many individuals in a slower and more indirect way, as well. According to Anderson, it has been shown that victims of flooding, prolonged drought, and civil unrest are exposed to many of the known risk factors for the development of violence-prone adolescents and adults. Violent tendencies can manifest themselves, even though it may take many years. Examples of this are evident from incidences such as the Dust Bowl of the 1930s and Hurricane Katrina in the United States.

Anderson does suggest some solutions. His first suggestion—and the most desirable—is for the world's countries to decrease their greenhouse-gas emissions immediately. For the damage that has already been done or is too late to completely avoid, he also suggests:

1. Education: Aggression can be reduced by making people aware of the likelihood and reasons for their hostility, and people can be taught coping mechanisms in order to help them deal with their anger in a more productive way.

2. Government coordination and action: If governments began preparing ways to relieve the stress of heat, such as providing food, shelter, education, and relocation of high-risk individuals, this could help defuse future tensions. This would enable those at risk to still maintain their cultural ties.

Although the second option would be extremely expensive and would require cooperation among countries, if it were possible, it could avoid a lot of trauma later on.

Conclusions

When dealing with the issues of climate change, it is always easy to get caught up in the cause/effect relationships of the physical environment because many of them are so visually obvious: polar bears drowning in the seas, barren landscapes cracked and dried that have not seen water for months, wilted crops that have failed their growing season, angry floods ravaging everything in their path, rising sea levels consuming coastlines. However, what is critical—and unfortunately gets overlooked—is the human factor, the impact climate change is having on social systems and cultures around the world. Even more horrifying is the damage done to societies in undeveloped countries that never contributed significantly to the problem in the first place and that are feeling the most severe of the impacts. To add insult to injury is the sad realization that many who live in the industrialized countries that emitted the greenhouse gases in the first place, causing the problem to escalate to the point it is at today, will never know of their suffering and discontent. What a shame, indeed, but it does not need to be that way, and it is not too late to change it.

One of the critical factors that is absolutely necessary for the solution of the problem is education. These are the news stories that should be reaching prime time. Education is important in the entire scheme of things with climate change. Knowledge is power, and through knowledge this problem can be researched, understood, and mitigated, yet even in educated, developed countries, it is surprising the number of the general population who still do not truly understand the causes and ramifications of climate change. This is a topic ready for sociologists to study, explore, and embrace. There are people right now feeling the harsh effects of climate change, and there will be more to come. The time to begin making a difference is as soon as it is understood that there is a problem.

Discussion

1. Why do you think some countries are more receptive than others to making adaptations to climate change?

2. If more countries modeled Europe's infrastructure to support pedestrian and bicycle traffic, do you think that would encourage more people to participate in those activities rather than choose to use a personal vehicle?

3. What are the principle reasons why people do not take public transportation when it is available? If you were involved in public transportation, what measures would you take to improve it so that more people relied on it? Why do you think your suggestions would make a difference?

4. Do you think we live in a disposable society that is concerned principally with the here and now? Does this put one at a disadvantage for focusing on a solution for climate change? If so, what are some effective methods that could alter that mindset and make a feasible solution easier to obtain?

5. If you were in a public leadership position and needed to get people motivated, how could you encourage society to prepare beforehand for a future event that scientists have warned about rather than just wait until it happens and then put an emergency plan together? Can you think of any public programs or campaigns that could get residents involved without making it mandatory through law?

6. Why do you think the public so readily embraced the celebration of Earth Day, but it is so difficult today to get participation with climate change issues?

7. What reforms could you suggest to assist low-income and minority families so that they do not take the brunt of natural disasters? Is there a feasible, cost-effective solution?

8. What are your thoughts on peer pressure? Do you think it is necessary to use manipulation to convince people to go green, or can you think of a better way?

9. Do you think a "climate of fear" has been created in society around the topic of climate change? If so, what evidence can you cite to back your claims? If not, do you see that as a possibility in the future? For either yes or no, what is your response to Camilla Toulmin's comment at the lecture series in London where she said: "Real security comes from building a climate of trust, rather than building a climate of fear"?

10. What types of impact do you think heat stress will have on people and their tendency for violence? Do you foresee violence between nations because of climate change? If so, what would you suggest as a solution to this?

References

Anderson, C. A. 2010. Climate change and violence. *The Encyclopedia of Peace Psychology.* New York: Wiley-Blackwell. Outlines some of the unexpected social ramifications of climate change.

Anselmo, R. 2009. Social and economic impact of drought and desertification. *Economic and Social Council, Model United Nations Far West.* http://www.munfw. org/archive/41st/ecosoc1.htm (accessed January 13, 2011). Discusses the ramifications of what to expect with the escalation of climate change.

Bassett, D. R., Jr., J. Pucher, R. Buehler, D. L. Thompson, and S. E. Crouter. 2008. Walking, cycling, and obesity rates in Europe, North America, and Australia. *Journal of Physical Activity and Health.* 5: 795–814. Examines the rates of activity of people for each of the three geographic areas and how that corresponds to the rates of obesity in those countries.

Beck, E. C. 1979. The Love Canal tragedy. Environmental Protection Agency. http:// www.epa.gov/history/topics/lovecanal/01.htm (accessed January 6, 2011). Provides a historical account of the Love Canal tragedy.

Brechin, S. R. 2008. Ostriches and change: A response to "Global Warming and Sociology." *Current Sociology,* May. 56(3): 467–474. Discusses the reasons why the general public is loathe to react to scientists' warnings of climate change.

Carlton, J. 2005. Scientists say drowned polar bears ominous; population down 22% from 1987. *Underwater Times.com,* December 18. http://www.underwatertimes. com/news.php?article_id=80101367459 (accessed August 10, 2011).

Committee on Climate Change. 2011. Climate targets. http://www.theccc.org.uk/ topics/science-and-environment/climate-targets (accessed April 1, 2011). Discusses physical impacts of climate change.

Envirolink. 2011. How the first Earth Day came about. http://earthday.envirolink. org/history.html (accessed February 2, 2011). Provides information on how Earth Day was organized and the intent behind it.

Hill, G. 1969. The environment may eclipse Vietnam as college issue. *New York Times*, November 30.

The Ismaili. 2009. A changing climate: Exploring the social impact of global warming, March 27. http://www.theismaili.org/cms/684/A-changing-climate-Exploring-the-social-impact-of-global-warming (accessed February 27, 2011). Discusses the impact of climate change on developing countries.

Kinzig, A. P., and C. Perrings. 2007. Biodiversity and human wellbeing. Center for Climate Change and Environmental Studies, Nigeria. http://www.center4 climatechange.com/biodiversity.php (accessed August 10, 2011).

Kolbert, E. 2009. Changing rains. *National Geographic*, April. http://ngm.national geographic.com/2009/04/changing-rains/kolbert-text (accessed March 22, 2011). Discusses the issues of flooding and drought during climate change.

Lever-Tracy, C. 2008. Global warming and sociology. *Current Sociology*, May. 56(3): 445–466. http://csi.sagepub.com/cgi/content/abstsract/56/3/445 (accessed October 5, 2010). Discusses the need for a cooperative multidisciplinary network of social and natural scientists working together to solve the problems of climate change.

Millennium Ecosystem Assessment. 2005. *Ecosystems and Human Well-being: Biodiversity Synthesis*. Washington, DC: World Resources Institute. Discusses the effects of climate change on the world's ecosystems.

Poplock, R. 2007. The poor are hit hardest by climate change, but contribute the least to it. *SeattlePi.com*, August 19. http://www.seattlepi.com/default/article/The-poor-are-hit-hardest-by-climate-change-but-1246942.php (accessed March 1, 2011). Discusses low-income and minority populations and what they endure during disaster situations.

The Royal Society. 2007. *Biodiversity-Climate Interactions: Adaptation, Mitigation, and Human Livelihoods*, June. http://www.royalsociety.org/WorkArea/DownloadAsset.aspx?id=5538 (accessed July 8, 2011). Provides information on biodiversity of ecosystems and the effects of climate change.

Sachedina, S. 2009. A changing climate: Exploring the social impact of global warming. *The Ismaili*, March 27. http://www.theismaili.org/cms/684/A-changing-climate-Exploring-the-social-impact-of-global-warming (accessed March 2, 2011). Explores the positive benefits of preparing societies for climate change well ahead of time and teaching responsibility.

Simon, S. 2010. The secret to turning consumers green. *The Wall Street Journal*, October 18. http://online.wsj.com/article/SB10001424052748704575304575296243891721972.html (accessed August 10, 2011).

Stillwater, S. 2010. Green living motivations: What makes someone go green? *EzineArticles.com*. http://EzineArticles.com/?expert=Steve_Stillwater (accessed February 2, 2011). Discusses the realities of convincing others to adopt the green lifestyle.

Suggested Reading

Calthorpe, P. 2010. *Urbanism in the Age of Climate Change*. Washington, DC: Island Press. Discusses how modern cities can be planned and built with climate change mitigation in mind.

Crate, S. A. and M. Nuttall (editors). 2009. *Anthropology and Climate Change: From Encounters to Actions*. Walnut Creek, CA: Left Coast Press. Provides a well-rounded understanding of how societies around the globe perceive and adapt to climate change from the perspective of their own unique sociocultural framework.

DeLisi, M. 2005. *Career Criminals in Society*. Thousand Oaks, CA: Sage.

United Nations Human Settlement Program. 2011. *Cities and Climate Change: Global Report on Human Settlements 2011*. London: Earthscan Publications. Discusses the effects of climate change on the world's cities.

6

Human Psychology and the Media

Overview

One critical element to the future of climate change is perception. The way each individual views climate change, whether they see it as an important issue that must be positively dealt with or not, will determine the future of every person on earth. Many people's perceptions today are shaped by what they obtain from the media—the programs shown and the stories reported—and how they are presented—on the TV, news, radio, Internet, or other source. This chapter takes a look at those very issues and the effect they can have on the progression and development of a scientific issue. The chapter begins by looking at how the messages we receive are constantly being shaped by several factors going on around and within us. They are influenced by a unique combination of public input, peer pressure, cultural values, and our own mix of personal perception based on our own experiences and life events. Next, this chapter explores the power of the media, the responsibility the media has of reporting accurately, and what the ramifications can be when stories are not presented responsibly. It also delves into journalistic balance and why it is important overall, but also why it can serve as a roadblock in controversial scientific issues such as climate change and how the reader can recognize the difference—whether undue significance is being placed on an issue that does not hold enough scientific merit—and keep the real issues in perspective. This chapter then discusses the issue of the advancement of scientific theories and the subsequent evolution of thought and how the media often uses that to discredit controversial scientific issues. Finally, it covers the occasional data flaw and the interesting results that can happen with media interaction in relation to human psychology.

Introduction

Because climate change is such a controversial issue, it is important for people to take upon themselves the personal responsibility of becoming thoroughly educated about the topic and being aware of what climatologists and other specialists know, what they suspect, and what the controversies are about. Fortunately, today there are several organizations whose purpose is to educate others about the most critical environmental subjects—such as climate change—and teach environmental responsibility through conservation. From both the governmental and private sectors, there are many organizations that offer opportunities to get involved in fighting climate change, becoming educated, and educating others about the latest discoveries and developments, such as the Pew Environmental Group, National Aeronautics and Space Administration (NASA), NASA Goddard Institute for Space Studies (GISS), the Environmental Defense Fund, National Geographic, and the U.S. Environmental Protection Agency, to name just a few. These organizations are beneficial because they are generally well connected with political and administrative information. They also keep up with the latest research techniques and present a good source of unbiased scientific information. Many also attend the formal political negotiations and conferences and offer reports that are generally more neutral and objective, often giving the reader better, more direct information.

Again, this is another example of the value of education, as discussed in the previous chapter. Becoming knowledgeable about the issues is critical. Although this may sound obvious, on many occasions misinformation has been released to the public and has muddled and damaged the progression of solving critical issues—and climate change is an excellent case in point. The damage caused by the release of wrong or antagonistic information often slows the progress of research and mitigation efforts—sometimes even significant political headway—making the acquisition of a sound understanding of the psychological aspects of media attention important to have.

Development

Climate Change, Human Psychology, Cultural Values, and the Media

The media has an enormous influence on what the public hears. It is the media that disseminates information through newscasts, magazines, newspapers, and the Internet, providing an unparalleled opportunity not only to inform the public of the latest issues but also to play a role in how that information is perceived. Another component that contributes to how information is received is different for each person and is based on preferences, perceptions, and beliefs that are influenced by psychology and value systems. These are the sometimes-subtle forces at work shaping people's opinions about highly controversial subjects, such as climate change.

According to Dr. H. Steven Moffic, a professor of psychiatry and behavioral medicine at the Medical College of Wisconsin, "Global warming is a concept that everyone hears about, but many are slow to respond to. The problems and risks of climate change seem to be far in the future—they might be 25 or 50 years away—so why would people pay attention to those issues when there are so many day-to-day problems to deal with?" (Bloom, 2008).

Dr. Moffic believes the ability to ignore climate change is very human. "Our brains in many ways have not evolved much from when humans started to develop thousands of years ago. We are hardwired to respond to immediate danger—we call this the 'fight or flight response'—but there is no similar mechanism that alerts us to long-term dangers" (Bloom, 2008). He believes that these reactions are just part of human nature: "People are so preoccupied with immediate problems like jobs and health and the economy that it's hard to pay attention to climate change, and to willingly take on another challenge" (Bloom, 2008). Moffic also explains that people might distance themselves from the issue so that they do not feel responsible: "The issue of how much humans contribute to the cause of global warming may also contribute to why we tend to ignore its impact. Who wants to believe they might be guilty for contributing to a problem that could destroy the Earth?" (Bloom, 2008).

In order to put the issue into perspective, Dr. Moffic suggests everyone identify and do simple things that do not require big changes. He believes that each individual can have a large effect on others and, through example, influence others to take action. He also suggests that everyone "try to make global warming a more immediate issue—whether it is thinking about your kids, grandkids, the future of the whole Earth, or your health. Try to think about ways in which this issue is important to you right now" (Bloom, 2008).

In work done by Elke U. Weber at the Center for Research on Environmental Decisions at Columbia University on why the subject of climate change has not scared more people yet, she attributes it to universal characteristics of human nature. According to Weber, behavioral decision research over the last 30 years has given psychologists a good understanding about the way humans respond to risk; specifically in the decisions they make to take action to reduce or manage those risks. One of the biggest motivators to respond to risk is worry. When people are not alarmed about a risk or hazard, their tendency is not to take precautions.

Weber points out that with the issue of climate change, personal experiences with notable and serious consequences are still rare in many regions of the world. In addition, when people base their decisions on statistical descriptions about a hazard provided by others, it is not a big enough motivator for action (Weber, 2006).

An example of this can be seen in a scenario such as the rapid rise in the price of gasoline in 2008. When prices skyrocketed at the pumps, it caught the public's attention and raised an immediate interest in hybrid cars, alternative fuels, and using public transportation, because the consumer was hit hard financially. Then, when gasoline prices dropped again, consumers thought less about energy conservation and alternative fuels because they were no longer immediately suffering the direct consequences. Human nature dictates that if something negative happens elsewhere in the world, the mindset of an individual is "it only happens to others."

The stark reality about climate change is the inertia it engenders. Other locations may be suffering through droughts (such as Africa) or sea-level rise (such as the Pacific or Caribbean Islands), but people think it will not happen in the United States or wherever else they may live. Sadly, when it eventually does happen, it will already be too late, and just as it is human nature to procrastinate when an immediate threat is not looming, eventually the public will be caught in the mindset: "I wish I had done something about it sooner."

Weber also believes that the reason people tend to avoid taking action against long-term risks is related to two psychological factors: the finite pool of worry hypothesis and the single action bias. The finite pool of worry hypothesis posits that people can only worry about so many issues at one time, and of the issues they worry about they are prioritized from greatest to least. Generally, the greatest worries are those most directly affecting their lives at the moment. As an example, Weber pointed out that the finite pool of worry theory was demonstrated by the fact that in the United States there was a rapid increase in concern about terrorism after the attacks on 9/11. Because of the intense focus on terrorism, other important issues—such as environmental degradation or restrictions on civil liberties—took an immediate backseat (Weber, 2006).

The single action bias is described by Weber as follows: Decision makers are very likely to take one action to reduce a risk that they encounter and worry about but are much less likely to take additional steps that would provide incremental protection or risk reduction. The single action taken is not necessarily the most effective one, nor is it the same for different decision makers. However, regardless of which single action is taken first, decision makers have a tendency to not take any further action, presumably because the first action suffices in reducing the feeling of worry or vulnerability. Weber concludes that based on behavioral research over the past 30 years attention-grabbing and emotionally engaging information interventions may be required to ignite the public concern for action in response to climate change (Weber, 2006).

A country's cultural values also play a significant role in public perception—and reaction—to climate change. An individual's values promote public action on issues such as civil rights, feminism, the jobs and social justice movements, the peace movement, the organic food and alternative health care movements, and the environmental movement. According to the State of the World Forum, these movements have gained strength over the past 50 years. In the United States alone, they estimate that more than fifty million people support some sort of groups based on personal values, such as those seeking to protect the environment. The numbers continue to grow and in Europe are even more numerous. These movements have power over political decision makers. Organizations with influence include The Pew Charitable Trusts, Defenders of Wildlife, World Wildlife Fund, and Union of Concerned Scientists.

The Power of the Media

Reporting about climate change by the media has run the gambit in recent years. Because there are many points of view, the question is: Where does the truth lie? Reports and stories concerning climate change have ridiculed scientists and environmental groups. Reports have shown big businesses and countries (such as the United States) openly challenging the facts of climate change. Industries, such as oil companies, have accused the media of misinforming the public about the ill effects of burning fossil fuels. Other news stories have accused the Bush administration of silencing critics, including leading government climate scientists who have warned the public openly of the consequences of climate change. As further reports about climate change continue to reveal a bleaker future, some are concerned that it will encourage fear tactics from environmentalists, whitewashing by some business interests, and a show by governments to illustrate reductions in emissions.

A few media reports claim climate change to be a fraud; still others claim it is simply a cause designed to harm the U.S. economy and make the United Nations more powerful. Others say it is driven by academia and the simple desire of climate scientists to make a lot of money by using fear as a tool to earn more research grants.

All of this misinformation presents a challenge. A trend that has emerged is that the mainstream media in recent years has turned toward reporting actions and solutions, but there does seem to be a fine line on what the public expects. Some climate change researchers have expressed concern that too much reporting will lead to climate fatigue whereby the public will become desensitized to the issue. Others feel that the media should be used as an educational tool, that there is so much potential to educate the general public in ways that are not fatiguing. As an example, consider Figures 6.1 and 6.2. The billboard advertises a pickup truck dependent on fossil fuels and not rated with high fuel efficiency, but the advertising has emotional appeal by suggesting the luxury, comfort, and status that will be bestowed upon the buyer of the vehicle. When the public looks at this type of advertisement, they are not reminded of climate change issues or the health of the environment and future generations.

The movie poster sends an entirely different message. Focused on the earth and those who live on it, it communicates very well the connections between life on land, in the oceans, and the overall connection to everything on earth. This type of media representation serves not only as entertainment but also as a strong positive approach to public education applicable to people of

FIGURE 6.1
Commercial media is geared to appeal to a consumer's ego, desires, comfort level, and status. By presenting a product in this way, it is much easier to generate personal interest, because making a sale is the goal. If climate change scientists were to recreate this advertisement, it would read much differently and carry a much different message. (From Wasatch Images, Salt Lake City, Utah.)

all ages. Instead of causing environmental fatigue, it sparks environmental interest through its creative storyline and breathtaking photography, giving the viewer a glimpse of the diversity and fragility of life on earth that they probably would never see otherwise.

Keeping a Journalistic Balance

Journalistic balance—giving each side of an issue a voice—is an important concept. The organization Fairness and Accuracy in Reporting states that a new study found that in U.S. media coverage of climate change, superficial balance—telling both sides—can actually be a form of informational bias. For example, the Intergovernmental Panel on Climate Change has reiterated that human activities have had a discernible influence on the global climate and that climate change is a serious problem that must be addressed immediately. The media, in the name of balance, has given disproportionate air play to the small group of climate change skeptics and allowed them to have their views greatly amplified.

FIGURE 6.2
The media has the power to contribute to public education about the environment in a positive manner. One highly successful way is through entertainment. With good narrative and photography, a strong impression can be made to individuals, strengthening public involvement around issues such as doing their part in solving the climate change problem. (From Wasatch Images, Salt Lake City, Utah.)

When reading reports from the media, it is important to clearly note who is being interviewed and whether or not the source is reputable and noteworthy. For example, the Intergovernmental Panel on Climate Change is a reputable organization. It consists of top scientists from around the globe who employ a decision-by-consensus approach. To back their reputability, D. James Baker, administrator of the U.S. National Oceanic and Atmospheric Administration (NOAA) and undersecretary for oceans and atmosphere at the Department of Commerce under the Clinton administration, has said,

"There's no better scientific consensus on this or any issue I know—except maybe Newton's second law of dynamics" (Boykoff, 2004).

In 1996, the Society of Professional Journalists removed the term *objectivity* from its ethics code. Today, the trend seems to lean more toward fairness, balance, and accuracy. Journalists currently are taught to identify the most dominant, widespread position and then tell both sides of the story. Robert Entman, a media scholar, says, "Balance aims for neutrality. It requires that reporters present the views of legitimate spokespersons of the conflicting sides in any significant dispute and provide both sides with roughly equal attention" (Boykoff, 2004).

In an attempt to make a coverage balanced, it is important to understand that this does not mean the coverage is accurate. In terms of global warming, efforts to keep a "balance" may allow skeptics—many of which are funded by the various carbon-based industries—to be frequently and inappropriately consulted and quoted in news reports on climate change. It is important here to make a distinction: When the issue is of a political or social nature, fairness—presenting arguments on both sides with equal weight—helps ensure that the reporting is not biased. This method can cause problems when it is applied to scientific issues, however, because it results in journalists presenting competing points of view on a scientific question as though they had equal scientific weight, when they most certainly do not. This is what has happened with climate change. There have been incidences where the media has let skeptics have too much voice, and it has enabled them to confuse the public and distort the seriousness of the problem.

Unfortunately, in giving equal time to opposing views, the major mainstream media has occasionally seriously trivialized scientific understanding of the important role that humans play in the climate change process. While there is a value to presenting multiple points of view, it does the reader an extreme disservice when scientific findings about which the world's top scientists and experts have come to a well-represented consensus (representing enormous investments of research and time) are presented next to, and with equal weight, the opinions of a few skeptical scientists. This results in confusing the readers, leaving them with the frustration that they no longer know what to believe because there are so many conflicting reports. Situations like this slow down the constructive progression of climate change research, and it is a travesty to true science.

An example of this can be seen in a *New York Times* article from April 10, 2009. Marc Morano sponsors a Web site (ClimateDepot.com) dedicated solely to the downplay of climate change. His chief goal is to debunk climate change as a serious issue. Kert Davies, the research director for Greenpeace, commented that he would "like to dismiss Mr. Morano as irrelevant, but could not. He is relentless in pushing out misinformation. In denying the urgency of the problem, he definitely slows things down on the regulatory front. Eventually, he will be held accountable, but it may be too late" (Kaufman, 2009).

As scientists who are actively involved in climate change research look into Morano's claims, they say "he may be best known for compiling a report listing hundreds of scientists whose work he says undermines the consensus on climate change. Environmental advocates, however, say that many of the experts listed as scientists on Morano's Web site have no scientific credentials and that their work persuaded no one not already ideologically committed" (Kaufman, 2009).

One of Morano's recent reports entitled "More than 700 International Scientists Dissent over Man-Made Global Warming Claims" was far from balanced. Kevin Grandia, who manages Desmogblog.com, which describes itself as dedicated to combating misinformation on climate change, said the report is filled with so-called experts who are really weather broadcasters and others without advanced degrees (Kaufman, 2009). Grandia also said Morano's report misrepresented the work of legitimate scientists and pointed to Steve Rayner, a professor at Oxford, who was mentioned for articles criticizing the Kyoto Protocol. Dr. Rayner, however, in no way disputed the existence of climate change or that human activity contributes to it, as Morano's report implied. In e-mail messages, Rayner had asked to be removed from the Morano report, but it was published with his name included. When asked about it, Morano replied that he had no record of Rayner's asking to be removed from the list and that the doctor must "not be remembering this clearly" (Morano, 2009). In cases like this, it is imperative that any information obtained about climate change—or any scientific issue, for that matter—be looked at critically and its validity assessed as to its scientific soundness and quality (Figure 6.3).

Scientists' Mindsets and Data Change

One way the media has negatively impacted the advancement of climate change research is to attack scientists when they have changed their theories or their positions on a scientific viewpoint. For example, the media brought up a theory postulated back in the 1970s that did not pan out and allowed outspoken critics to use it in an attempt to diminish the reputations of scientists today. Several mainstream media sources republished the stories from the 1970s about a coming age of global cooling and the climate disaster it would trigger. Because this nearly 40-year-old theory never panned out, some skeptics have said climate change will not pan out either, but scientists say that is an unfair comparison. Dr. William Connolley, a climate modeler for the British Antarctic Survey, says that "Although the theory got hype from the news media in the '70s, it never got much traction within the scientific community. New data and research over the decades have convinced the vast majority of scientists that global warming is real and under way" (Azios, 2007).

The issue in the 1970s centered around the possibility that nearly three decades of cooling experienced in the Northern Hemisphere since World

FIGURE 6.3
It is imperative to be selective on which literature is reviewed and believed. Any item in print from a source that is not reputable should not be assumed to carry the same weight and impact as the sources shown here, which are reputable and only publish quality material. (From Wasatch Images, Salt Lake City, Utah.)

War II might be the beginning of a new ice age. Data suggested that per-haps the huge increase in dust and aerosols from pollution and develop-ment might be stepping up the cooling process. The investigation did not last long, however, because temperatures began to rise again, and the issue was abandoned. Today, improved climate methodologies have revealed that although aerosols did have a cooling effect, CO_2 and other greenhouse gases were more potent in bringing about atmospheric change on a global scale. Improvements in technology over recent years have greatly aided the advancement and accuracy of scientific research, which continues to evolve and improve.

To go back to the issue of climatologists changing their minds, however, R. Stephen Schneider, a professor in the Department of Biological Sciences at Stanford University and a senior fellow at the Center for Environment Science and Policy of the Institute for International Studies, says, "Scientists are criticized by global-warming skeptics for making new claims and revis-ing theories, as if we are required to stay politically consistent. But that goes against science. We must allow for new evidence to influence us." Schneider also explains: "For some, the original speculation was that dust and aerosols would increase at a rate far beyond CO_2 and lead to global cooling. We didn't know yet that such effects were so regionally located. By the mid-1970s, it was realized that greenhouse gases were perhaps more likely to be shifting climate on a global scale" (Azios, 2007). Connolley agreed, stating, "Climate science was far less advanced in the 1970s, only beginning in a way, and ideas were explored in a tentative way that has later been abandoned" (Azios, 2007).

This represents an inherent issue of science in general. As additional knowledge is gained about a subject, processes and outcomes of phenomena may change. Scientists need to remain open-minded and objective. If they do not remain open-minded, they will miss critical pieces of scientific information and possibly risk the outcome of a scientific breakthroughs (Azios, 2007). One thing remains clear, however: The media, if used correctly, has enormous potential to guide the public and can play a significant role in helping people understand the science, the relevant issues, and the options for a better future.

The Occasional Data Flaw

Another thing the public seems to forget and the media seems to be completely unforgiving, if not antagonistic, about is the occasional data flaw. Unfortunately, there seems to be a public perception that scientists are up on some imaginary pedestal and everything they release is cast in stone, as just discussed above, with no room for change. That also applies to the occasional data flaw. Unfortunately, just like everyone else experiences, the occasional error does occur. Although this may seem obvious and to not warrant mentioning here, it is important, because when it happens, first the media and then the public often seem quick to jump on it, openly criticizing the source, often in an attempt to discredit not only the scientist and agency from which it originated, but also the issue in general.

An example of this is an incident that happened to NASA's GISS. GISS is one of the foremost research units on climate change, led by Dr. James E. Hansen, one of the most world-renowned climate change experts. Hansen has been involved for more than 30 years in global temperature data analysis, and the GISS, under his leadership, has provided the scientific community with exceptional data analysis that has led to the remarkable advancement seen in the climate change field in recent years. However, even an organization as distinguished as this one is not exempt from public criticism; perhaps it is even more vulnerable to criticism because of its status.

In 2001 the GISS published an updated analysis of a temperature record that had had additional data added to it, with the objective of improving the long-term temperature record. The change consisted of using the newest U.S. Historical Climatology Network analysis for several United States stations that were part of this network. The improvement, developed by NOAA researchers, adjusted station records that included station moves or other discontinuities. Unfortunately, GISS failed to recognize initially that the station records obtained electronically from NOAA did not contain the necessary adjustments. Therefore, there was a discontinuity in 2000 in the records of those stations (1999 and prior already contained the adjustments; 2000 had not been adjusted yet). Once GISS caught the error, it was immediately corrected. The resultant error that was originally reported (the data without the correction applied to the year 2000 data) averaged 0.15°C over the contiguous forty-eight states. This area covers approximately 1.5 percent of the globe, making the global error negligible, however. Therefore, one may think that catching the negligible error, correcting it, and subsequently rerunning and rereleasing the analysis would be sufficient, but that is not how the scenario played out once the media got hold of what had happened. The printed story was embellished and distributed to news outlets throughout the country with headlines such as "NASA Cooks the Data." The following headlines and excerpts are just a few of what was actually released:

1. NASA, NOAA Cooking the Data: "The cooks—er, 'scientists'—at NASA's Goddard Institute of Space Studies (GISS) have released their latest sky-is-falling temperature findings..." (http://www .iceagenow.com/NASA_NOAA_cooking_the_data.htm)

2. Global Warming Hysteria: Did NASA Cook the Books on Warming Statistics?: "I have no idea if this is true, but if it is—there should be a Congressional inquiry..." (http://www.firstthings.com/blogs/ secondhandsmoke/2011/01/30/global-warming-hysteria-did-nasa- cook-the-books-on-warming-statistics)

3. James Hansen: Cooking the NASA Books for Climate Change: "[M]uch of the world uses NASA's continually-revised data and graphs to determine weather history and policy. Which is unfortunate. Particularly, because NASA simply cannot be trusted to provide scientifically unbiased information on this subject. When not demonstrating incomprehensible incompetence, NASA cooks the books..." (http://deathby1000papercuts.com/2008/11/ james-hansen-cooking-the-nasa-books-for-climate-change)

4. NASA, NOAA Create Global Warming Trend with Cooked Data: "There is a major problem with the NASA and NOAA numbers, according to skeptical researchers who have dissected the data: They are inaccurate, the result of cherry-picking, computer manipulation and 'best guess' interpretation..." (http://www .examiner.com/seminole-county-environmental-news-in-orlando/ nasa-noaa-create-global-warming-trend-with-cooked-data)

Is this responsible journalism? Hardly. However, one has to wonder how many people read this and believed it. The bigger question is: How many people are led astray and kept from looking further into the truth and doing something constructive about fighting climate change because of media stories like these? This is a noteworthy illustration as to why a reader must be very wary of the sources of information and be educated in the interpretation of that source's story.

The next logical question that must be asked is: Why? What is the media's purpose in printing these types of stories? One only has to look at the tabloids obtainable at the grocery store, perhaps. There seems to be an insatiable need for drama, gossip, and sensationalism in certain cultures, and perhaps this is simply an extension of it, but when it is applied to a critical issue—such as climate change—and can do more harm than good, should that not be where the proverbial line is drawn? With regard to the incident discussed above, media stories also went on to say that "NASA has now silently released corrected figures, and the changes are truly astounding. The warmest year on record is now 1934" (Malkin, 2007). This report was also incorrect, however, because when both the before and after graphs prepared by NASA were compared, 1998 and 1934 were tied for warmest year (prior to 2000) in BOTH graphs, meaning the media had publically misinterpreted the graphs. Therefore, the obvious misinformation of the media reports, along with the subsequent absence of any effort to correct the stories after NASA pointed out the misinformation, may appear as if the media's aim may instead have been to create distrust or confusion in the minds of the public rather than to transmit accurate information.

This brings to mind a serious question: How are scientists supposed to handle a situation where a data error—no matter how quickly it is corrected—provides ammunition for people who may be more interested in launching a public-relations campaign rather than reporting science? Is it possible to eliminate these occasional data flaws or an unintentional occurrence of disinformation? Of course not. Scientists are human beings, too, and occasionally mistakes will occur, but when errors are recognized, they are also fixed, and data are rerun. It is when intentional attacks on integrity somehow enter the scientific picture that the true damage can be done. As in this example, it can serve to delay, or even weaken, the forward movement of advancements that need to happen for the benefit of the public.

Conclusions

As evidenced from the examples presented throughout this chapter, the media has enormous power and control over the general public. Along with that power needs to come responsibility and responsible journalism, which is especially critical when it comes to controversial and critical scientific issues such as climate change. The general public can be very susceptible to what is said in the news, and unfortunately, through a misrepresented story, much damage can be done to a sound scientific concept, causing setbacks in progress being made toward mitigation and public understanding. Each of us has a responsibility as well. It is important that as readers and viewers, we research and process what we are exposed to, thereby enabling us to define the difference between responsible accounting and hearsay. It is also imperative that we realize that our perceptions of issues are naturally processed through our own personal biases and cultural backgrounds, and it is up to us not to cloud issues by letting them cross personal boundaries, thereby reducing the inherent value of discoveries being made or slowing the direction that society needs to take to ensure its protection and well-being.

Discussion

1. Why would independent research organizations often provide a good source of unbiased scientific information? Why would they be a benefit to have in attendance at formal political negotiations and conferences?

2. List some public interest topics that you know of where the media has misinformed or sensationalized. How did this impact the issue?

3. List five viable reasons the general public has an appetite for sensationalism. What does this say about the corresponding cultural value system? How do you feel this impacts the scientific advancement toward understanding and mitigating climate change?

4. Define journalistic responsibility. How does this apply to the climate change issue? What positive and negative examples can you think of?

5. Elke Weber states that when people base their decisions on statistical descriptions about a hazard provided by others, it is not a big enough motivator for action. What do you think would be a big enough motivator to spur the general public to action in the case of climate change? Is there a big enough motivator?

6. Weber concludes that based on behavioral research over the past 30 years, attention-grabbing and emotionally engaging information interventions may be required to ignite the public concern for action in response to climate change. Do you agree? Why or why not? If you agree, what types of information interventions do you believe would suffice?

7. What reasons do you think explain the fact that independent research and environmental organizations have significant power over political decision makers?

8. Some climate change researchers have expressed concern that too much reporting will lead to climate fatigue, whereby the public will become desensitized to the issue. Others feel that the media should be used as an education tool, that there is so much potential to educate the general public in ways that are not fatiguing. What is your perception? How would you devise a media program that could be used as a successful, engaging educational tool?

9. If you were a media executive, what policies would you enact to deal with unfair media manipulation of the occasional scientific data flaw?

References

Azios, T. 2007. Global-warming skeptics: Is it only the news media who need to chill? *The Christian Science Monitor*, October 11. http://www.csmonitor.com/2007/1011/p13s03-sten.html (accessed October 5, 2010). Discusses the scientific processes involved in evolving theories over time as additional data becomes available.

Bloom, D. 2008. American psychiatrist explores the psychology of global warming. *RushprNews*, December 3. http://www.rushprnews.com/2008/12/03/american-psychiatrist-explores-the-psychology-of-global-warming (accessed January 5, 2011). Discusses the reasons why people are slow to react to significant issues such as climate change.

Boykoff, J., and M. Boykoff. 2004. Journalistic balance as global warming bias. Fairness and Accuracy in Reporting, Nov/Dec. http://www.fair.org/index.php?page=1978 (accessed November 22, 2010). Discusses the problem of the media creating controversy where science finds consensus.

Kaufman, L. 2009. Dissenter on warming expands his campaign. *The New York Times*, April 9. http://www.nytimes.com/2009/04/10/us/politics/10morano.html (accessed November 6, 2010). Discusses the problem of skeptics spreading their opinions about climate change and slowing down the progress of scientific advancement.

Malkin, M. 2007. Hot news: NASA quietly fixes flawed temperature data: 1998 was NOT the warmest year in the millenium. MichelleMalkin.com, August 9. http://michellemalkin.com/2007/08/09/hot-news-nasa-fixes-flawed-temperature-data-1998-was-not-the-warmest-year-in-the-millenium/ (accessed August 10, 2011).

Morano, M. 2009. More than 700 international scientists dissent over man-made global warming claims. Free Republic, March 16. http://www.freerepublic.com/focus/f-news/2207835/posts (accessed September 21, 2010). A post of information that presents climate change from a skeptical viewpoint.

Weber, E. U. 2006. Experience-based and description-based perceptions of long-term risk: Why global warming does not scare us (yet). *Climate Change*. 77(1–2): 103–120. Discusses why people are not more concerned about climate change.

Suggested Reading

Hansen, J. E. 2009. *Storms of My Grandchildren: The Truth about the Coming Climate Catastrophe and Our Last Chance to Save Humanity*. New York: Bloomsbury.

Norgaard, K. M. 2011. *Living in Denial: Climate Change, Emotions, and Everyday Life*. Cambridge, MA: The MIT Press.

Pooley, E. 2010. *The Climate War: True Believers, Power Brokers, and the Fight to Save the Earth*. New York: Hyperion.

Washington, H., and J. Cook. 2011. *Climate Change Denial: Heads in the Sand*. London: Earthscan Publications Ltd.

7

The Role of International Organizations

Overview

The existence of professional organizations that operate on an individual basis for the purpose of research, data gathering, and consulting officials on key issues both politically and economically is paramount. It is the international organizations—these objective, nonbiased independent groups—that often can make a difference of whether data or a study is objective and meaningful or not. This chapter discusses the key role of international organizations. It begins by taking a look at how international cooperation began and the role that international organizations subsequently took because of it. The chapter then focuses on some individual, notable groups to provide an overview of the roles they play in the international spotlight. Following that it will look at individual organizations and programs for an overview into the insight, function, and accomplishments of these groups. Finally, the chapter will look at the progress of individual countries and just how various nations of the world can work together toward the sustainable use of energy resources.

Introduction

Because of their objective and informative input, their involvement, and then subsequent reports to government decision makers, international organizations can often make a meaningful difference in international policy. Oftentimes, climate-policy experts work with government officials, explaining highly technical information and ramifications, to offer assistance in the interpretation and scope of a particular issue so that meaningful action toward the mitigation of climate change can take place. International organizations bridge an important gap between government officials, the scientific community, and the general public. They assist all three areas academically and practically and play a critical role in climate change research and negotiations. When there is international cooperation along with their involvement, meaningful results often occur.

Development

Because climate change is a global problem, it will take a global solution. It does not matter whether greenhouse gases are released in Los Angeles, London, Tokyo, or Paris; they obviously have the same impact on the atmosphere. Thus, if only a few countries make an effort to slow emissions, it will not solve the climate change problem. All countries must be involved in the solution in order to obtain successful results.

The Evolution of International Cooperation

The space exploration era not only gave scientists a new view of the earth and global science, but it also allowed for data to be recorded in new ways. Computers and modeling software led to new studies and discoveries. Some of the most interesting findings were the changing levels of CO_2 in the atmosphere (the Keeling Curve in 1958), climate cycles, paleoclimatology through interpretation of ice cores, and ocean/atmospheric circulation patterns.

In the late 1960s, an environmental movement was gaining momentum worldwide, and climate change became one of the most-discussed topics. The first significant conference where scientists discussed climate change was the Global Effects of Environmental Pollution Symposium held in Dallas, Texas, in 1968. Then, in 1970, a month-long Study of Critical Environmental Problems at the Massachusetts Institute of Technology was held. At this symposium, nearly all of the attendees were from the United States, and they felt the need for better international representation (Matthews et al., 1971). This led to a second gathering in which fourteen nations met in Stockholm in 1971, where a key topic discussed was climate change—a Study of Man's Impact on Climate. Each attendee returned home with a dire message to their nation: Rapidly melting ice and rapid climate change could occur in the next 100 years because of human activity. The recommendation of the scientists was to create a major international program to monitor the environment. It was from this recommendation that the United Nations Environment Programme (UNEP) was formed.

At this point, researching climate and gathering data had officially become one of the United Nations's environmental responsibilities. One of the milestones at the time was that the scientists involved pointed out that "the rate and degree of future warming could be profoundly affected by government policies" (UNEP, 1985). They called on governments to consider positive actions to prevent future warming. This was the tipping point where climate science shifted from a merely scientific issue to a political issue. As a result, in 1986, a small committee of experts, the Advisory Group on Greenhouse Gases, was formed.

This spurred international, national, and regional conferences, which further promoted research and scientific collaboration. The result in the 1980s was interesting. Studies, research, and conferences conducted by organizations such as the U.S. National Academy of Sciences gained momentum among climate scientists. According to the science writer Jonathan Weiner, "By the second half of the 1980s, many experts were frantic to persuade the world of what was about to happen. Yet they could not afford to sound frantic, or they would lose credibility" (Weart, 2004).

One of the big fears was that any push for policy changes would set the scientists against potent economic and political forces and also against some colleagues who vehemently denied the likelihood of climate change. The scientific arguments became entangled with emotions.

What was called for was more proof—more concrete data. So the scientists went back to work. New research concepts were developed. Scientists began looking at the issue as a climate system, using the input of all related scientific fields (geophysics, chemistry, biology, etc.). By looking at everything together, computer models could be developed to begin understanding how climate change worked and therefore how it could be prevented.

In 1982, through scientific work conducted by the UNEP, the Vienna Convention for the Protection of the Ozone Layer was held, and twenty nations signed the document created at the convention. When the ozone hole was discovered over Antarctica and shocked the world, it led to the 1987 Montreal Protocol of the Vienna Convention, where governments formally pledged to restrict emissions of specific ozone-damaging chemicals. The Montreal Protocol has had great success in reducing emissions of chlorofluorocarbons and thereby preventing further damage to the ozone layer. It has not, however, had a significant impact toward reducing climate change.

The success at Montreal was followed up the next year by a World Conference on the Changing Atmosphere: Implications for Global Security, also called the Toronto Conference. The conclusions drawn at this conference were that "the changes in the atmosphere due to human pollution represent a major threat to international security and are already having harmful consequences over many parts of the globe" (Weart, 2004).

For the first time, a group of prestigious scientists called on the world's governments to set strict, specific targets for reducing greenhouse-gas (GHG) emissions. They advised that by 2005, the world should push its emissions 20 percent below the 1988 level. Observers saw this goal as a major accomplishment, if only as a marker to judge how governments responded.

The Toronto Conference caught the attention of many politicians. Officials were impressed by the warnings of prestigious climate experts. Prime Minister Margaret Thatcher, herself a chemist, gave climate change an official endorsement when she described it as "a key issue" in a speech she delivered to the Royal Society in September 1988 (Weart, 2004). At that time, she also increased funding for climate research. She was the first major world leader to take a positive, strong position to do something to fight climate change.

In 1988, the World Meteorological Organization and the UNEP collaborated in creating the Intergovernmental Panel on Climate Change (IPCC). Unlike earlier conferences, national academic panels, and advisory committees, the IPCC was composed mainly of people who participated not only as science experts but also as official representatives of their governments—people who had strong links to national offices, laboratories, meteorological offices, and scientific research agencies like NASA. Today, most of the world's climate scientists are involved in the IPCC, and it has become a pivotal player in policy debates. Since 1988, climate change has been accepted as an international issue, both scientifically and politically.

The Role of International Organizations

An evolution of events led to the productive international cooperation that could effectively deal with climate change. Once international cooperation had been put in place, the creation of international organizations naturally followed. This section discusses some of those organizations.

Renewable Energy and Energy Efficiency Partnership

The Renewable Energy and Energy Efficiency Partnership (REEEP) is a worldwide public-private partnership that was originated by the United Kingdom, other business interests, and governments at the Johannesburg World Summit on Sustainable Development in August 2002. Its goals are to reduce GHG emissions, help developing countries by improving their access to reliable, clean energy, make renewable energy and energy efficiency systems more affordable, and help nations financially who engage in energy efficiency and use renewable resources.

The United Kingdom's rationale for developing REEEP was an effort to correct the fact that there was nothing else in place—either policy-wise or in a regulatory capacity—to promote renewable energy or energy efficiency. In addition, it was felt that current limits in a country's finances stood in the way of being able to make the transition, and economic assistance was needed. By removing these market barriers, it was hoped that more progress would be made toward achieving the long-term transformation of the energy sector.

REEEP relies on a bottom-up approach, where partners work together at regional, national, and then international levels to create policy, regulatory, and financing programs to promote energy efficiency. Currently, REEEP is funded by many governments, including Australia, Austria, Canada, Germany, Ireland, Italy, Spain, the Netherlands, the United Kingdom, the United States, and the European Commission. The European Commission is the executive branch of the European Union of which twenty-seven countries are members (Austria, Belgium, Bulgaria, Cyprus, Czech Republic, Denmark, Estonia, Finland, France, Germany, Greece, Hungary, Ireland, Italy, Latvia, Lithuania, Luxembourg, Malta, Netherlands, Poland, Portugal, Romania, Slovakia, Slovenia, Spain, Sweden, and the United Kingdom).

REEEP currently has nearly fifty ongoing projects covering roughly forty countries including China, India, Brazil, and South Africa. They work with over two hundred partners, thirty-four of whom are governments (including all the G8 countries, except Russia), countries from emerging markets and the developing world, businesses, nongovernmental organizations, and civilian volunteers. REEEP relies on partners' voluntary financial contributions, experience, and knowledge.

European Climate Change Programme

The European Climate Change Programme (ECCP) was begun in June 2000 by the European Union's European Commission. Their goal was to identify, develop, and implement all the necessary elements of a European Union (EU) strategy to implement the Kyoto Protocol. All EU countries' ratifications of the Kyoto Protocol were deposited on May 31, 2002.

The EU decided to work as a unit to meet its Kyoto emissions targets. The ECCP approaches this by using an emissions scheme known as the EU Emissions Trading System (ETS). In order to achieve their legally binding commitments under Kyoto, countries have the option of either making these savings within their own country or buying emissions reductions from other countries. The other countries still need to meet their Kyoto target reductions, but the use of a free market system enables the reductions to be made for the least possible cost. Most reductions are made where they can be made in the least-expensive manner, and excess reductions can be sold to other countries whose cuts are prohibitively expensive.

EU ETS is the largest GHG emissions trading scheme in the world. In 1996, the EU identified as their target a maximum of 2°C rise in average global temperature. In order to achieve this, on February 7, 2007, the EU announced their plans for new legislation that required the average CO_2 emissions of vehicles produced in 2012 to exceed no more than 130 g/km. Looking ahead to the time when the Kyoto Protocol expires in 2012, the ECCP has identified the need to review their progress and begin creating a plan of action to implement once the Kyoto Protocol expires. To launch their post-2012 climate policy, the EU held a conference on October 24, 2005, in Brussels. From this, the Second European Climate Change Programme was launched. The ECCP II consists of several working groups:

- The ECCP I review group (comprised of five subgroups: transport, energy supply, energy demand, non-CO_2 gases, and agriculture)

- Aviation
- CO_2 and cars
- Carbon capture and storage technology
- Adaptation
- EU emissions trading schemes

Some of the highlights of their work follow. In their assessment of aviation, the EU determined that it contributes to global climate change and that its contribution is increasing. Even though the EU's total GHG emissions fell by 3 percent from 1990 to 2002, emissions from international aviation increased nearly 70 percent. In spite of significant improvements in aircraft technology and operational efficiency, it has not been enough to neutralize the overall effect of aviation emissions, and they are likely to continue. Therefore, the EU issued a directive to include aviation in the EU ETS, which was published January 13, 2009. The intention is for the EU ETS to serve as a model for other countries considering similar national or regional schemes and to link these to the EU scheme over time. This way, the EU ETS can form the basis of wider global action.

There is also a new proposal to reduce the CO_2 emissions from passenger cars. On December 19, 2007, the European Commission adopted legislation to reduce the average CO_2 emissions of new passenger cars, which account for about 12 percent of the European Union's carbon emissions. The proposed legislation is to improve the fuel economy of cars and ensure that average emissions from the new cars do not exceed 120 g/km of CO_2 through an integrated approach.

The Commission's proposal will reduce the average emissions of CO_2 in the EU from 160 to 130 g/km in 2012, a 19 percent reduction of CO_2 emissions. This will make the EU a world leader in the production of fuel-efficient cars. Customers will benefit from fuel savings. From 2012, manufacturers will have to ensure that the cars they produce are meeting emissions standards. In addition, the curve is set so that heavier cars will have to improve more than lighter cars. Manufacturers' progress will be measured each year.

The EU also warns of the effects of climate change and the various adaptations that must take place to prepare for them. The EU stresses the importance of putting adaptation plans in place to soften impacts on society and the economy, including on water, agriculture, forestry, industry, biodiversity, and urban life. They also acknowledge that the impacts of climate change will hit locally and regionally in different ways and that adaptation measures will have to be planned out at local, regional, and national levels. To solve these issues and answer appropriate questions, there is currently an ECCP working group putting together an impact and adaptation plan, dealing with water resources, marine resources, coastal zones, tourism, human health, agriculture, forestry, biodiversity, energy infrastructure, and urban planning issues.

The *International Herald Tribune* reported on March 9, 2007, that the EU drafted an agreement that would make Europe a world leader in fighting climate change but that also compromised by allowing some of Europe's most polluting countries to limit their environmental goals (Bilefsky, 2007). The draft agreement committed the EU to reduce GHG emissions by 20 percent by 2020 and required the EU to obtain one-fifth of its energy from renewable energy resources such as wind and solar energy, as well as to fuel 10 percent of its cars and trucks with biofuels made from plants.

Under pressure from several of the former Soviet bloc countries, however, which currently rely heavily on cheap coal and oil for their energy and fought changing to more costly environmentally friendly alternatives, the EU agreed that individual targets would be allowed for each of the twenty-seven EU members to meet the renewable energy goal. Unfortunately, that means eastern Europe's worst polluters in the fastest-growing economies will most likely face the least stringent targets compared with their Western counterparts. Many of the eight former communist nations that joined the EU in May 2004 are significantly behind the rest of the Union in developing renewable energy. Poland, for example, currently derives more than 90 percent of its energy for heating from coal.

In response to the agreement in general, however, the European Commission president, José Manuel Barroso, called the measures "the most ambitious package ever agreed by any institution on energy security and climate change" and expressed hope that they would spur the world's biggest polluters, including the United States, China, and India, to take similar action (Bilefsky, 2007).

International Carbon Action Partnership

The International Carbon Action Partnership (ICAP), formed in October 2007, is a coalition of European countries, U.S. states, Canadian provinces, Australia, New Zealand, Tokyo metropolitan government, and Norway formed to fight climate change. The international and interregional agreement was signed in Lisbon, Portugal, on October 29, 2007, by U.S. and Canadian members of the Western Climate Initiative, northeastern U.S. members of the Regional Greenhouse Gas Initiative, members of the European Union and the European Commission, Australia, Tokyo metropolitan government, Norway, and New Zealand.

ICAP is designed to open lines of communication for sharing valuable information, such as research, effective policy initiatives, lessons learned, and new developments. By working together to establish similar design principles, ICAP partners are ensuring that future market systems, in conjunction with regulation in the form of enforceable caps, will boost worldwide demand for low-carbon products and services, provide a larger market for innovators, and achieve global emissions reductions at the fastest rate and lowest cost possible. The partnership supports the current ongoing efforts undertaken under the United Nations Framework Convention on Climate Change. ICAP is working toward finding global solutions by:

1. Monitoring, reporting, and verifying emissions and working to determine reliable sources appropriate for inclusion in a globally linked program

2. Encouraging common approaches and pushing partners to expand the global carbon market

3. Creating a clear price incentive to innovate, develop, and use clean technologies

4. Encouraging private investors to choose low-carbon projects and technologies

5. Providing flexible compliance mechanisms that ensure reliable reductions at the fastest pace and lowest cost

According to UK former prime minister Gordon Brown, "The launch of the International Carbon Market Partnership is a truly significant step forward in the global effort to combat climate change. Building a global carbon market is fundamental to reducing greenhouse gas emissions while allowing economies to grow and prosper. Trading emissions between nations allows us all to reach our greenhouse gas targets more cost effectively. And it therefore allows us to reduce emissions more than we could by acting alone" (Revkin, 2007).

Governor Jon Corzine of New Jersey commented, "My background as the former head of Goldman Sachs has given me a unique perspective on many market-based solutions to important public problems, such as environmental degradation. But it is my life in public service that has helped me understand that it will take the courage and commitment of a core set of leaders, like those of us gathered today, to drive implementation of smart, feasible, and measurable policies needed to address an issue as urgent as global warming" (*EnvironBusiness News*, 2011).

Former governor Eliot Spitzer of New York said, "Global warming is the most significant environmental problem of our generation, and by establishing an international partnership, we are taking the vital steps to address this growing concern. In the absence of federal leadership, New York is implementing a greenhouse gas emissions trading program that will achieve a 16 percent reduction in power plant emissions by 2019. Today, we continue that work by joining the ICAP where we can begin working with our global partners, share experiences, and address issues of program design and compatibility, thereby strengthening our markets" (*EnviroBusiness News*, 2011)

The Progress of Individual Countries

Several of the world's countries have already made significant progress toward reducing their GHG emissions. In order to keep the earth in a reasonable facsimile of what we know today, it will take the concerted effort of every nation on earth. The noteworthy progress accomplished so far is discussed below.

Iceland

For the past 50 years, Iceland has been decreasing its dependence on fossil fuels by tapping the natural power found within its natural resources. Its waterfalls, volcanoes, geysers, and hot springs have long provided its inhabitants with abundant electricity and hot water. Today, virtually 100 percent of the country's electricity and heating comes from domestic renewable energy sources: hydroelectric power and geothermal springs. The country is still dependent, however, on imported oil to operate their vehicles and fishing fleets. It is so expensive to import that the cost is roughly two dollars a liter (eight dollars a gallon) for gasoline (Figure 7.1).

As of September 2007, Iceland ranks fifty-third in the world in GHG emissions per capita, according to the U.S. Department of Energy's Carbon Dioxide Information Analysis Center. Professor Bragi Árnason of the University of Iceland has suggested using hydrogen to power the nation's transportation.

FIGURE 7.1
Sitting strategically on tectonic plate boundaries, Iceland has an abundance of geothermal energy that it can tap as a major energy source. (Courtesy of Ásgeir Eggertsson, http://commons.wikimedia.org/wiki/File:Krafla_geothermal_power_station_wiki.jpg.)

Hydrogen is a product of water and electricity, and as he points out, "Iceland has lots of both." He further comments, "Iceland is the ideal country to create the world's first hydrogen economy" (Mihelich, 2007). His suggestion caught the attention of car manufacturers, who are now using Iceland as a test market for their hydrogen fuel cell prototypes. One car that is receiving attention is the Mercedes Benz A-class F-cell, an electric car powered by a Daimler AG fuel cell.

Ásdis Kristinsdóttir, project manager for Reykjavik Energy, says, "It's just like a normal car, except the only pollution coming out of the exhaust pipe is water vapor. It can go about 161 kilometers on a full tank. When it runs out of fuel the electric battery kicks in, giving the driver another 29 kilometers—hopefully enough time to get to a refueling station. Filling the tank is similar to today's cars—attach a hose to the car's fueling port, hit 'start' on the pump, and stand back. The process takes about five to six minutes" (Mihelich, 2007).

In 2003, Reykjavik opened a hydrogen fueling station to test three hydrogen fuel cell buses. The station was integrated into an existing gasoline/diesel fueling station. The hydrogen gas is produced by electrolysis—sending a current through water to split it into hydrogen and oxygen. The public buses could run all day before needing refueling. They calculated that Reykjavik would need five additional refueling stations; the entire nation will need just fifteen refueling stations.

At that time, fuel cell cars were anticipated to go on sale to the public by 2010. The involved carmakers promised they would keep costs down, and the Icelandic government would offer its citizens tax breaks for driving them. Árnason figures it will take an additional 4 percent of power to produce the hydrogen. Once Iceland's vehicles are converted over to hydrogen, the fishing fleets will follow. He predicts Iceland will be completely fossil fuel–free by 2050. He said, "We are a very small country but we have all the same infrastructure of big nations. We will be the prototype for the rest of the world" (Mihelich, 2007).

Then, through an unexpected series of events with the downturn in the economy, in September 2008, Iceland announced it was thinking of halting its future hydrogen economy in favor of battery electric vehicles. In an article in the *New York Times*, it was reported that Iceland was having a difficult time acquiring hydrogen vehicles. At a Reykjavik conference focusing on alternate transportation, Iceland announced plans to team up with Mitsubishi Motors to supply the country with a fleet of tiny i-MiEV electric cars (which have a range per charge of about 100 miles with lithium-ion batteries). If these plans were to go ahead, it would make Iceland the first European country to have i-MiEVs, which were scheduled to go on sale in Japan in the summer of 2009 (Motavalli, 2008).

"Hydrogen cars are not mass produced anywhere," said Teitur Torkelsson, managing partner of FTO Sustainable Solutions. "But a majority of car makers are announcing electric cars to be produced in the next four or five years, so it becomes a big part of our energy solutions" (Romm, 2008). The plan at that point was to bring in a few electric cars to test, and if they performed well and residents of Iceland were willing to pay whatever price was asked (as yet unknown), Iceland would start seeing electric cars on their highways by 2010 instead of the hydrogen, as initially expected.

Then by February 2009, only 5 months later, Iceland abruptly changed gears again and found itself involved in one of its most ambitious projects ever: building a hydrogen infrastructure of filling stations for hydrogen cars, putting them back on track to their initial hydrogen transportation goal (Woodard, 2009). As of February 2009, Reykjavik had fourteen hydrogen-fueled cars along with hydrogen filling stations. They now lay claim to the world's first commercial hydrogen fueling station. The beauty of it is that it works in a similar fashion to the standard fueling pump for gasoline: pull up to the pump, swipe your credit card, attach the pump fixture, and in 5 minutes you'll be back on the road with a tank full of emissions-free fuel produced right at the filling station from water and sustainable generated electricity. In a second report issued later that year, Iceland reportedly now has twenty-two hydrogen vehicles, which makes them second only to Germany in the EU for a hydrogen-powered fleet (Figure 7.2).

FIGURE 7.2
Iceland is aiming for a hydrogen transportation infrastructure. This is one of the Shell operational stations. (Courtesy of Jóhann Heidar Arnason, http://en.wikipedia.org/wiki/File:Hydrogen_filling_station.JPG.)

The only exhaust a hydrogen-fueled vehicle produces is water out of the tailpipe, according to Jon Bjorn Skulason, general manager of Icelandic New Energy: "If we complete our plans, we will be a zero-emissions society. We would not have to import fuel from foreign sources, and we would be 100 percent sustainable, which must be the true future of the world" (Woodard, 2009).

Iceland's initial projection was to convert the country to hydrogen by 2040, but because of delays in automobile manufacturers' production, the global recession, and Iceland's own delay in building additional fueling stations because of its own financial concerns, it is now currently several years behind schedule. The good news is that there is still a schedule, and the country plans to convert to a hydrogen economy.

According to Bragi Árnason, the University of Iceland chemist who first promoted the hydrogen economy for Iceland, "You will use electricity wherever you can, but batteries do not have a sufficient range—maybe 200 or 300 kilometers. Most experts agree that hydrogen is candidate fuel No. 1, because it's the cheapest and easiest to make" (Woodard, 2009).

Notably, Iceland is committed to completely separating itself from fossil fuels by midcentury. Instead of importing oil to power its cars and fishing fleet, it plans to power them with electricity from hydroelectric and geothermal plants. Completely dedicated to renewable energy, Iceland has harnessed meltwater from massive ice sheets and the steam that pours from its volcano-dotted landscape, which together generate virtually all the island's heat and electricity.

Iceland is also actively involved in carbon-sequestration research. Icelandic, U.S., and French scientists have been studying chemical weathering and water/rock interactions for decades. They are interested in using Iceland as a location for carbon sequestration because the country's geologic formations are ideal for it, and Icelanders' extensive knowledge of geothermal energy makes them good candidates for understanding the chemical reactions between gases at the earth's depths.

Sigurdur Reynir Gislason, a research professor of geology at the University of Iceland, said, "We hope to show the world in this pilot study that a natural process can be used to transform CO_2 emissions into a solid state and to safely store them underground for thousands, if not millions, of years. We also believe this process could not only be possible in Iceland, but in other countries that also have basaltic rocks" (Wagner, 2006).

Eileen Claussen, president of the Pew Center on Global Climate Change in Arlington, Virginia, said she is encouraged by such projects. "The Pew Center, along with many others, believe that carbon capture and storage underground in geological formations can be a significant part of the solution to climate change. Investment in these technologies illustrates the magnitude of the challenge and the lengths people are willing to go in order to change the dangerous path we're on" (Wagner, 2006).

Norway

Norway is another country involved in a pioneering effort to store CO_2 through carbon capture and storage. They have designated four separate sites: Sleipner, Snøhvit, Mongstad, and Kårstø. Since 1996, one million metric tons of CO_2 from the Sleipner Vest oil field in the North Sea has been separated from the gas production and stored in Utsira (a geological formation), 1,000 meters beneath the sea floor. Because of environmental concerns of leakage, the CO_2 storage facility is closely monitored. Several nations, supported by the European Union, have been involved in direct research and monitoring of this storage project, and they have developed prediction methods for the movement of the CO_2 spanning many years into the future. The resulting data are able to pinpoint the exact subsurface location of the CO_2 plume and confirm that the CO_2 is indeed confined securely within the designated storage reservoir.

The Snøhvit project began actively storing CO_2 on April 24, 2008, in an underground storage system. Natural gas, natural gas liquid, and condensate flow from the gas field in the Barents Sea. Up to 700,000 metric tons of CO_2 are separated annually from the natural gas and reinjected and stored in a formation 2,600 meters under the seabed.

Mongstad, Norway, has plans to host the largest crude oil terminal and refinery. The Norwegian government and the oil company Statoil-Hydro have signed an agreement to establish a full-time CO_2 carbon capture and storage operation to offset a new gas-fired plant at Mongstad (Norway's largest crude oil terminal and refinery). The project will be completed in two phases. The first phase will cover construction and operation of the Mongstad CO_2 capture testing facility, which will be operational in 2011. The test facility will be able to capture at least 100,000 metric tons per year. The second phase will be full-scale capture of approximately 1.3 million metric tons of CO_2 per year. This project is expected to be finished by the end of 2014.

In Kårstø, an area where carbon-storage technology is already in existence, storage capacity will increase tenfold through a retrofit in 2011/2012. It will then capture and store approximately one million metric tons of CO_2 each year.

Japan

According to a *USA Today* article of June 6, 2006, Japan hopes to cut back their GHG emissions and fight climate change with a plan to pump CO_2 into underground storage reservoirs rather than release it into the atmosphere. Fighting climate change is a top priority for Japan. They release 1.2 billion metric tons of CO_2 each year into the atmosphere, making them one of the world's top polluters (Greimel, 2006). According to Masahiro Nishio, an official at the Ministry of Economy, Trade and Industry, Japan is planning to bury 181 million metric tons of CO_2 a year by 2020, which will cut their emissions by one-sixth.

Carbon capture and storage is a process whereby CO_2 is captured from factory emissions, pressurized into liquid form, and then injected into underground aquifers, existing gas fields, or existing natural gaps between rock strata (see Chapter 12 for a more in-depth discussion). The process is still under scientific investigation, although there is an experiment being conducted in joint partnership between the U.S. Department of Energy, the Canadian government, and private industry. It began in 2005 and involved piping CO_2 from the Great Plains Synfuels plant in Beulah, North Dakota (a by-product of coal gasification), to the Weyburn oil field in Saskatchewan, Canada. In comparison, the proposed project in Japan is much larger.

According to Nishio, "Underground storage could begin as early as 2010, but there may still be hurdles to overcome. Capturing carbon dioxide and injecting it underground is prohibitively expensive, costing up to $52 a ton. Under the new initiative, the ministry aims to halve that cost by 2020. We have much to study in development" (Greimel, 2006).

Safety concerns also need to be addressed to ensure that earthquakes or rock fissures do not allow a sudden release of millions of tons of CO_2 into the atmosphere. The IPCC estimates that if CO_2 is stored properly and safely, it should remain stable for up to a thousand years. Japan will begin their program by capturing CO_2 from their natural gas fields. Then, as they get the technology and program running systematically, they will also include CO_2 from steel mills, power plants, and chemical factories (Greimel, 2006).

Japan's first carbon capture plant, a 47-megawatt plant at Mikawa, is nearing completion in 2011. The company that built it, Toshiba Corp., now has plans to take their carbon-storage technology worldwide. They already have plans to build a pilot project in India. According to Kenji Urai, managing director of Toshiba India Private Ltd., "Now that we are almost finished in Japan, we'd like to bring that technology to other parts of the world, like India." India is currently looking for ways to reduce their CO_2 emissions after having agreed to reduce the greenhouse gas in proportion to gross domestic product by 25 percent from 2005 levels by 2020 (Mehrotra, 2011).

Nations Working toward Sustainability

An organization called the International Council for Local Environmental Initiatives (ICLEI) was established in 1990 as an international association committed to helping governments achieve sustainable development and mitigate and adapt to climate change. The ICLEI provides technical consulting, training, and information services tailored to countries' needs. They have worked with countries worldwide, such as in the examples discussed below.

Through ICLEI, farmers in the agricultural areas around Blantyre, Malawi, are currently changing their agricultural practices to support crops that need less water and nurture the soil. At the national level, the government has begun to increase the nation's grain reserve, anticipating more droughts and flooding in the years to come. It is also constructing a new dam in view of predicted future drought. The government is taking a proactive role in identifying measures it will need to take within the next 3 years in order to prepare itself for, and adapt to, climate change.

In Sapporo, Japan, ICLEI is involved in a project called Warm-Biz. This is a national program geared toward energy conservation. Run by the Japanese Ministry of the Environment, its purpose is to encourage people to wear more clothing to work to compensate for temperature settings being reduced by 2°. In a pilot test program in Sapporo, 96.7 percent of the respondents supported the program overall, and the citizens there learned that energy-efficiency programs offer one of the best ways to reduce climate change pollutants.

Residents of Canada Bay, Australia, are building a water mining plant that will save drinking water. The plant will save up to 165 million liters of drinking water each year by providing recycled water for the city's fields, golf courses, and parks. The plant will work by purifying wastewater and using mechanical methods and minimal chemicals to produce high-quality treated water.

London is planning to cut GHG emissions by 60 percent within the next 20 years. Their plan aims to reduce emissions at the local government, industrial, and business levels. Individual elements of their plan include awarding green badges of merit for local businesses adopting reduction strategies, offering subsidies to homeowners to insulate their homes, and switching one-fourth of the city's power supply from the old and inefficient national grid to locally generated electricity using combined heat and power plants. According to former London mayor Ken Livingstone, speaking of London's Climate Change Action Plan, "Londoners don't have to reduce their quality of life to tackle climate change, but we do need to change the way we live" (*BBC News*, 2007).

On November 17, 2007, in Valencia, Spain, United Nations Secretary-General Ban Ki-moon described climate change as "the defining challenge of our age." He also challenged the world's two largest GHG emitters—China and the United States—to "play a more constructive role" (Rosenthal, 2007a). His challenge was delivered two weeks before the world's energy ministers met in Bali, Indonesia, to begin talks on creating a global climate treaty to replace the Kyoto Protocol when it expires in 2012.

The IPCC, which was awarded the Nobel Peace Prize (jointly with Al Gore) in October 2007, said the world would have to reverse the growth of GHG emissions by 2015 to prevent serious climate disruptions. According to Dr. Rajendra Pachauri, chair of the IPCC, "If there's no action before 2012, that's too late. What we do in the next two to three years will determine our future. This is the defining moment" (Rosenthal, 2007b). He also said that since the IPCC began its work 5 years ago, scientists have recorded "much stronger trends in climate change," like a recent melting of Arctic ice that had not been predicted. "That means you better start with intervention much earlier" (Rosenthal, 2007b).

One of the major differences with the IPCC's fourth assessment report (released in 2007) over previous ones was that the data had not been softened, diluted, and sifted through. It was direct and to the point. It was the first report to acknowledge that the melting of the Greenland ice sheet from rising temperatures could result in a substantive sea-level rise over centuries rather than millennia. It added a sense of critical urgency and importance never seen before in a report. "It's extremely clear and is very explicit that the cost of inaction will be huge compared to the cost of action," said Jeffrey D. Sachs, director of Columbia University's Earth Institute. "We can't afford to wait for some perfect accord to replace Kyoto, for some grand agreement. We can't afford to spend years bickering about it. We need to start acting now" (Rosenthal, 2007c).

"Stabilization of emissions can be achieved by deployment of a portfolio of technologies that exist or are already under development," said Achim Steiner, head of the UNEP. However, he noted that developed countries would have to help poorer ones adapt to climate shifts and adopt cleaner energy choices, which are often expensive. Steiner emphasized that the report sent a message to individuals as well as world leaders: "What we need is a new ethic in which every person changes lifestyle, attitude, and behavior" (Rosenthal, 2007).

Conclusions

International cooperation is paramount to the solution and management of climate change. Without nations working together toward a common goal, the kind of progress necessary to make a real difference will not happen. The role of many professional international organizations is to identify, develop, and help in the implementation of protocol and technology designed to combat climate change. To date, their participation and guidance has been critical in guiding nations and getting many programs off the ground and operational. Through their expertise, many nations are now working in practical and productive ways toward sustainability, such as Iceland, who is leading the world in innovative ideas with hydrogen technology and geothermal energy implementation. There are many other countries that are also actively engaged in practical progress to address the pertinent issues of climate change, such as Norway, Japan, Malawi, Australia, the United Kingdom, and Spain. Professional organizations are a welcome addition to the scientific study of climate change and are helping immensely in educating the world about the issues at hand so that political leaders, economists, analysts, and the scientific community can gain and benefit from their hard work and pioneering spirit.

Discussion

1. What do you think the most significant contribution of international organizations is?

2. What further role do you feel professional organizations could take in convincing countries like the United States, China, and India to become more serious about reducing GHG emissions?

3. What do you feel is the biggest value of international collaborative efforts and scientific conferences?

4. What do you think the international arena of scientists should focus on most in the short term concerning the climate change issue? Why? Defend your answer.

5. As far as a political focus for the future, what should scientists recommend?

6. Where would you recommend international organizations focus most on the climate change issue?

References

BBC News. 2007. Mayor unveils climate change plan, February 27. http://news.bbc.co.uk/2/hi/uk_news/england/london/6399639.stm (accessed November 21, 2010). Outlines London's plan to conserve energy.

Bilefsky, D. 2007. EU drafts compromise agreement on climate change. *International Herald Tribune Europe,* March 9. http://www.foeeurope.org/press/2007/coverage/IHT_energyflag090307.pdf (accessed February 6, 2011). Discusses the EU's plans to lower GHG emissions.

EnviroBusiness News. 2011. International carbon action partnership (ICAP) launched by leaders of over 15 countries. http://www.environbusiness.com/News/Current1/icap.html (accessed August 11, 2011).

Greimel, H. 2006. Japan to fight global warming by pumping carbon dioxide underground. *USA Today,* June 26. http://www.usatoday.com/tech/science/2006-06-26-japan-greenhouse-gas_x.htm. Presents Japan's plans for a long-term carbon storage project.

Matthews, W. H., W. H. Kellogg, and G. D. Robinson. 1971. *Man's Impact on the Climate.* Cambridge, MA: MIT Press. Offers a summary of the symposium held in 1970.

Mehrotra, K. 2011. Oshiba in talks with NTPC for carbon capture project. *Bloomberg,* February 11. Discusses Japan's carbon storage project and its effects worldwide.

Mihelich, P. 2007. Iceland phasing out fossil fuels for clean energy. *CNN Tech,* September 18. http://articles.cnn.com/2007-09-18/tech/driving.iceland_1_electricity-and-hot-water-fuel-cell-icelanders?_s=PM:TECH (accessed August 11, 2011).

Motavalli, J. 2008. Iceland's future could be electric. *The New York Times*, September 19. http://wheels.blogs.nytimes.com/2008/09/19/icelands-future-could-be-electric (accessed December 5, 2010). Discusses Iceland's debate on whether to develop hydrogen transportation as planned.

Revkin, A. 2007. Two new (and very different) roadmaps for climate progress. *The New York Times*, October 29. http://dotearth.blogs.nytimes.com/2007/10/29/two-new-and-very-different-roadmaps-for-climate-progress/ (Accessed August 11, 2011).

Romm, J. 2008. Electric vehicles crowd out hydrogen brethren at sustainable driving conference. *Grist*, September 28. http://www.grist.org/article/iceland-gives-hydrogen-the-cold-shoulder (accessed August 11, 2011).

Rosenthal, E. 2007a. Ban calls climate change "defining challenge of our age." *New York Times*, November 17. http://www.nytimes.com/2007/11/17/world/europe/17iht-climate.1.8372066.html (accessed August 11, 2011).

Rosenthal, E. 2007b. UN report describes risks of inaction on climate change. *The New York Times*, November 17. http://www.nytimes.com/2007/11/17/science/earth/17cnd-climate.html?pagewanted=print (accessed December 5, 2010). Outlines the UN's advice on climate change management.

Rosenthal, E. 2007c. U.N. chief seeks more leadership on climate change. *New York Times*, November 18. http://query.nytimes.com/gst/fullpage.html?res=9D00EFDE123AF93BA25752C1A9619C8B63. (accessed August 11, 2011).

United Nations Environmental Programme. 1985. Statement of the UNEP/WMO/ICSU International Conference on the assessment of the role of carbon dioxide and of other greenhouse gases in climate variations and associated impacts, October, Villach, Austria. *Scope 29: The Greenhouse Effect, Climate Change, and Ecosystems.* http://www.icsu-scope.org/downloadpubs/scope29/statement.html (accessed August 11, 201)

Wagner, T. 2006. Iceland set to capture carbon in its rocks. *MSNBC.com*, April 12. http://www.msnbc.msn.com/id/12034963/ns/world_news-world_environment/t/iceland-set-capture-carbon-its-rocks (accessed January 12, 2011). Discusses Iceland's carbon sequestration activities.

Weart, S. R. 2004. *The Discovery of Global Warming (New Histories of Science, Technology, and Medicine).* Cambridge, MA: Harvard University Press. Traces the history of the climate change concept through a long process of incremental research rather than a dramatic revelation.

Woodard, C. 2009. Iceland strides toward a hydrogen economy. *Christian Science Monitor*, February 12. http://www.csmonitor.com/Innovation/Energy/2009/0212/iceland-strides-toward-a-hydrogen-economy (accessed March 15, 2011). Discusses Iceland's hydrogen power economy and its use for transportation.

Suggested Reading

Dean, C. 2007. The problems in modeling nature, with its unruly natural tendencies. *The New York Times*, February 20. http://www.nytimes.com/2007/02/20/science/20book.html (accessed May 2, 2010). Discusses the inherent limits of mathematical models and appropriate assumptions concerning their usage.

Dow, K., and T. E. Downing. 2006. *The Atlas of Climate Change: Mapping the World's Greatest Challenge*. Los Angeles: University of California Press. Offers maps and geographic statistics and information on climate change, global warming, economics, and other related scientific topics worldwide.

Gelbspan, R. 1997. *The Heat Is On: The High Stakes Battle over Earth's Threatened Climate*. Reading, MA: Addison Wesley. Offers a look at the controversy environmentalists often face when they deal with fossil fuel companies.

Greinel, H. 2006. Japan to fight global warming by pumping carbon dioxide underground. *USA Today*, June 26. http://content.usatoday.com/topics/more+stories/Places,%20Geography/Countries/Norway/48 (accessed April 25, 2009). Explores the option of carbon sequestration as a viable way to counteract the effects of global warming.

Houghton, J. 2004. *Global Warming: The Complete Briefing*. New York: Cambridge University Press. Outlines the scientific basis of global warming, describes the impacts that climate change will have on society, and looks at solutions to the problem.

Reuters (London). 2008. Multinationals fight climate change. *The New York Times*, January 21. http://www.nytimes.com/2008/01/21/business/21green.html (accessed April 26, 2009). Looks at the joint efforts of eleven companies using renewable energy.

8

The Political Arena

Overview

This chapter discusses the current political climate both in the United States and internationally on the climate change issue. It treats the United States and its policy development separately because of its unique nature—over the past two decades its policy has developed more or less independently from the rest of the participating countries; the majority of the other developed countries of the world have gone through a process of cohesion, while the United States has historically stood on the sidelines watching but doing very little politically toward curbing the emission of greenhouse gases and controlling climate change. The chapter then takes a look at some of the current legislation being considered in the United States and the evolutionary track taken to get there. Next, this chapter focuses on the political positions and actions of other nations and their involvement in organized policy and looks specifically at the pledges from the Copenhagen Accord and what that means for the future.

Introduction

In order to get climate change effectively under control, it will take the efforts of every country worldwide. Because of the immensity of the issue, the backing of national governments is critical—both legislatively and economically. Even more important is whether or not individual countries will be able to work together in accomplishing both their short- and long-term goals. For many of the world's countries, new negotiations post-Kyoto Protocol are just a continuation and evolution in a long line of negotiations and agreements already forged, but for some nations—who are also major greenhouse gas emitters, such as the United States and China—these new negotiations resulting from the Copenhagen Conference are breaking new territory, and only time will tell how well these new players will adjust to the rigor of GHG emission commitments and control and whether or not they will be successful. For the world as a community, the post-Kyoto period will open up a new era of opportunity and possible achievement. Each country must determine its true commitment to the issue at hand, but one thing is certain: the political decisions and actions of each country weigh heavily on the future—a future in which we all hold stock.

Development

The Current Political Climate in the United States

Historically, the United States has not been a leader in stressing the importance of the climate change issue. In fact, they openly chose not to join with the rest of the countries in ratifying the Kyoto Protocol, stating that it was unfair China and India were not held accountable at that time as undeveloped nations for their GHGs when they were rapidly becoming developed and still would not be held to the same standards as developed countries. According to "The One Environmental Issue," a January 1, 2008, *New York Times* editorial, when Al Gore ran for president in 2000 he could have made the climate change issue a key point in his campaign, but his advisers persuaded him that it was too complicated and forbidding an issue to sell to ordinary voters. John Kerry's ideas for addressing climate change and broaching the idea of lessening the nation's dependence on foreign sources of oil made no headway either.

Although some politicians have tried to get involved in environmental issues, the overall trend has been one of inaction. However, times seem to be changing. Severe weather events are occurring, species are becoming endangered, glaciers are melting, and areas are suffering from drought. The media seems to have finally taken on the role of making the public aware of the effects of a warming world. The big question still remains to be answered, however: To what extent are Americans willing to accept responsibility for the threat, take action, and make the personal sacrifices necessary to control the problem? To be specific—are Americans finally willing to pay slightly more for alternate, renewable energy and significantly change their lifestyles in order to reduce their use of fossil fuels?

Even though Al Gore did not focus on climate change during his campaign, he has had considerable influence since and played a critical role in educating the public about the issue and why it has to be dealt with now. His film and book, *An Inconvenient Truth*, have made the public well aware of the issue. So much so, in fact, that survey polls show that the American population is becoming increasingly alarmed. In 2007, the Intergovernmental Panel on Climate Change and Al Gore shared the Nobel Peace Prize for their efforts to bring the issue to the world's attention (Gore, 2006).

One thing that has frustrated many Americans is that the U.S. government—typically a leader in global issues—has seemed to move so slowly to take action to halt the emissions of greenhouse gases (GHGs). Fortunately, state governments are not holding back and waiting any longer. Governors from half of the states have put into effect agreements to lower GHGs. Even federal courts have ordered the executive branch to start regulating GHGs.

This portion of the chapter will look at U.S. political policy beginning with the 2008 presidential elections as a dividing line of reference in the evolution of climate change policy and action. A look at the time period of the election and the events unfolding during that time span to set the stage is warranted, because that is when a noted turning point in policy was set in motion with the change in administration; then this chapter will look at what has transpired since, followed by a review of the political progress and evolution of important initial efforts that occurred prior to that which played a role and had some influence in setting the initial stage for everything that has happened to date. Key information will be presented concerning the current progress of climate legislation taking place in Congress, as well as executive branch action—including the Executive Office of the President, the Environmental Protection Agency, and the Department of Energy.

It was promising to see that during the 2008 presidential campaign and election, environmental issues became important points for discussion, and they need to be established as key issues from now on. John McCain—who had encouraged taking positive action to fight climate change all along—was serious about dealing with the issue. In 2003, with Joseph Lieberman, Senator McCain introduced the first Senate bill aimed at mandatory reductions in emissions of 65 percent by mid-century. In the Democratic race, all of the original candidates promised that major investments would be made in cleaner fuels and delivery systems, including underground carbon storage for coal-fired plants. They also promised efforts to work toward a new international agreement to replace the Kyoto Protocol when it expired in 2012.

In a *New York Times* article on April 1, 2009, entitled "Democrats Unveil Climate Bill," a new bill to stop heat-trapping gases and wean the United States off foreign sources of oil was also announced. The bill, now known as the Waxman–Markey bill, H.R. 2454: American Clean Energy and Security Act of 2009, written by Representatives Henry A. Waxman (D-CA) and Edward J. Markey (D-MA), set an ambitious goal for capping heat-trapping gases. The bill required that emissions be reduced 20 percent from 2005 levels by 2020. The proposal would reduce GHGs by about 80 percent by 2050 (Broder, 2009).

The bill would require the nation to produce one-fourth of its electricity from renewable energy sources, such as solar, wind, or geothermal by 2025. It also called for a modernization of the nation's electric grid, production of more electric vehicles, and major increases in energy efficiency in buildings, appliances, and the generation of electricity.

What the proposal did not address, however, was how pollution allowances would be distributed or what percentage would be auctioned off or given for free. It also did not address how the majority of the billions of dollars raised from pollution permits would be spent or whether the revenue would be returned to consumers to compensate for higher energy bills.

Waxman, who served as the chairman of the Energy and Commerce Committee, said that his measure would create jobs and provide a gradual transition to a more efficient economy. "Our goal is to strengthen our economy by making America the world leader in new clean-energy and energy-efficiency technologies" (Energy and Commerce Committee, 2009).

For coal-producing states, the bill offers $10 billion in new financing for the development of technology to capture and store emissions of CO_2 from the burning of coal. A coalition of business and environmental groups, United States Climate Action Partnership, said the measure is a "strong starting point" for addressing emissions of heat-trapping gases and that it had incorporated many of the partnership's recommendations (Broder, 2009).

President Obama's Outlook on Climate Change at Election

On January 20, 2009, when Barack Obama was sworn in as the forty-fourth president of the United States, he delivered a speech after taking the oath of office. In it, he stressed that "Each day brings further evidence that the ways we use energy strengthen our adversaries and threaten our planet." He also affirmed that the energy challenges the nation faces today are a very real crisis that must be dealt with, and he promised a waiting nation that "we will harness the sun and the winds and the soil to fuel our cars and run our factories...in an effort to roll back the specter of a warming planet." He also promised that the nation would no longer "consume the world's resources without regard to effect" (NPR, 2009).

Prior to his inauguration address, Obama had sent a video message to an international summit meeting on climate change organized by Governor Arnold Schwarzenegger of California, held in Beverly Hills, California, on November 18–19, 2008. Obama stressed that despite the continuing economic turmoil, reductions in GHG emissions would remain a central component of his energy, environmental, and economic policies. The message he sent was clear. The need to curb heat-trapping gases would be a priority for his administration. He also stressed that the energy revolution the nation could expect from his administration would overcome what he called America's "shock and trance" cycle as oil prices spike and collapse. The following transcript is his explanation of the that cycle (Clips & Comments, 2008):

> *Steve Kroft:* When the price of oil was at $147 a barrel, there were a lot of spirited and profitable discussions that were held on energy independence. Now you've got the price of oil under $60.

Senator Obama: Right.

Steve Kroft: Does doing something about energy, is it less important now than…?

Senator Obama: It's more important. It may be a little harder politically, but it's more important.

Steve Kroft: Why?

Senator Obama: Well, because this has been our pattern. We go from shock to trance. You know, oil prices go up, gas prices at the pump go up, and everybody goes into a flurry of activity. And then the prices go back down and suddenly we act like it's not important, and we start, you know, filling up our SUVs again. And, as a consequence, we never make any progress. It's part of the addiction, all right. That has to be broken. Now is the time to break it.

The following is a transcript of the video message Senator Obama sent to Schwarzenegger at the summit meeting on climate change (taken from Revkin, 2008):

Few challenges facing America—and the world—are more urgent than combating climate change. The science is beyond dispute and the facts are clear. Sea levels are rising. Coastlines are shrinking. We've seen record drought, spreading famine, and storms that are growing stronger with each passing hurricane season. Climate change and our dependence on foreign oil, if left unaddressed, will continue to weaken our economy and threaten our national security. I know many of you are working to confront this challenge. We've also seen a number of businesses doing their part by investing in clean energy technologies. Too often, Washington has failed to show the same kind of leadership. My presidency will mark a new chapter in America's leadership on climate change that will strengthen our security and create millions of new jobs in the process. That will start with a federal cap-and-trade system. We will establish strong annual targets that set us on a course to reduce emissions to their 1990 levels by 2020 and reduce them an additional 80 percent by 2050. We will invest in solar power, wind power, and next-generation biofuels. The United States cannot meet this challenge alone. Solving this problem will require all of us working together. I look forward to working with all nations to meet this challenge in the coming years. Now is the time to confront this challenge once and for all. Delay is no longer an option. Denial is no longer an acceptable response. The stakes are too high. The consequences, too serious. Stopping climate change won't be easy. It won't happen overnight. But I promise you this: When I am president, any governor who's willing to promote clean energy will have a partner in the White House. Any company that's willing to invest in clean energy will have an ally in Washington. And any nation that's willing to join the cause of combating climate change will have an ally in the United States of America.

Then, in a political presentation given on January 26, 2009, President Obama delivered a speech concerning jobs, energy, and climate change, during which he made the following points about his policy on climate change:

- "Year after year, decade after decade, we've chosen delay over decisive action. Rigid ideology has overruled sound science. Special interests have overshadowed common sense. Rhetoric has not led to the hard work needed to achieve results, and our leaders raise their voices each time there's a spike on gas prices, only to grow quiet when the price falls at the pump."

- "Now America has arrived at a crossroads. Embedded in American soil, in the wind and the sun, we have the resources to change. Our scientists, businesses, and workers have the capacity to move us forward."

- "It falls on us to choose whether to risk the peril that comes with our current course or to seize the promise of energy independence. And for the sake of our security, our economy, and our planet, we must have the courage and commitment to change."

- "It will be the policy of my administration to reverse our dependence on foreign oil while building a new energy economy that will create millions of jobs."

- "Today I'm announcing the first steps on our journey toward energy independence, as we develop new energy, set new fuel efficiency standards, and address greenhouse gas emissions."

- "We will make it clear to the world that America is ready to lead. To protect our climate and our collective security, we must call together a truly global coalition. I've made it clear that we will act, but so too must the world. That's how we will deny leverage to dictators and dollars to terrorists, and that's how we will ensure that nations like China and India are doing their part, just as we are now willing to do ours."

- "We have made our choice: America will not be held hostage to dwindling resources, hostile regimes, and a warming planet. We will not be put off from action because action is hard. Now is the time to make the tough choices. Now is the time to meet the challenge at this crossroad of history by choosing a future that is safer for our country, prosperous for our planet, and sustainable."

Obama stressed that the federal government must work with, not against, the individual states to control climate change. His plan also outlined the goal of requiring cars to meet a 35 mpg fuel-efficiency standard by 2020 and vowed to "help the American automakers prepare for the future, build the cars of tomorrow, and no longer ignore facts or science." Climate change is real, and Obama's energy policy will be dictated to deal with climate change and will free U.S. dependence on foreign oil for security purposes.

The Beginnings of Change in Legislation

The ultimate goal of political action on climate change is to limit and/or reduce the concentration of GHGs in the atmosphere. Political action is a critical component necessary to make any significant global change because without the implementation of the necessary laws and regulations—such as GHG-emissions limits, regulatory frameworks within which carbon trading markets can operate, reportable and trackable systems of accountability, and tax incentives or funding assistance—productive and long-term change is not feasible.

Although the United States had a slow start toward addressing the climate change issue, current legislation is now percolating, and progress is slowly being made. The climate change issue has also made it to the Supreme Court. On April 2, 2007, in one of its most important environmental decisions in years, the U.S. Supreme Court ruled that the Environmental Protection Agency (EPA) now has the authority to regulate heat-trapping gases in automobile emissions. The Court further stipulated that the EPA could in no manner "sidestep its authority to regulate the greenhouse gases that contribute to global climate change unless it could provide a scientific basis for its refusal" (EPA, 2009b). This gives the EPA the right to regulate CO_2 and other heat-trapping gases under the Clean Air Act. According to Justice John Paul Stevens, "The only way the agency could avoid taking further action now was if it determined that greenhouse gases do not contribute to climate change or provides a good explanation why it cannot or will not find out whether they do" (Greenhouse, 2007).

The Supreme Court also heard another case concerning the Clean Air Act, giving the EPA a broader authority over factories and power plants that want to expand or increase their emissions of air pollutants. Under this broader reading, they made a ruling of nine to zero against the Duke Energy Corporation of North Carolina in favor of the EPA, which made environmentalists ecstatic, marking a historic occurrence in the U.S. Supreme Court as a positive step toward the mitigation of climate change. Interestingly, since the ruling on the first case, there has been a growing interest among various industrial groups in working with environmental organizations on proposals for emissions limits.

According to a *New York Times* article on April 3, 2007, Dave McCurdy, president of the Alliance of Automobile Manufacturers, said in response to the decision that, "The Alliance looks forward to working constructively with both Congress and the administration in addressing this issue. This decision says that the EPA will be part of this process." Although many claimed victory with the Supreme Court's decision, not everyone was satisfied. Chief Justice John G. Roberts, Jr., believed the court should never have addressed the question of the agency's legal obligations in the first place (Greenhouse, 2007).

On April 17, 2009, the EPA formally declared CO_2 and five other GHGs to be pollutants that endanger public health and welfare. This landmark decision will now put in motion a process that will lead to the regulation of GHGs for the first time in U.S. history. According to the EPA, "The science supporting the proposed endangerment finding was compelling and overwhelming." The decision received diverse reactions. Many Republicans in Congress and industry spokesmen warned that regulation of CO_2 emissions would raise energy costs and kill jobs. Democrats and environmental advocates, however, said the decision was long overdue and would bring long-term social and economic benefits. Lisa P. Jackson, the EPA administrator, said, "This finding confirms that greenhouse gas pollution is a serious problem now and for future generations. Fortunately, it follows President Obama's call for a low-carbon economy and strong leadership in Congress on clean energy and climate legislation" (Greenhouse, 2007).

The ruling will be followed by a grace period for comments to be made and legislation to emerge from Congress. Once this has occurred, the EPA will determine specific targets for reductions of heat-trapping gases and new requirements for energy efficiency in vehicles, power plants, and industry. At that point, the EPA will begin the process of regulating the climate-altering substances under the Clean Air Act.

A *New York Times* article of December 18, 2007, stated that Congress plans to create a huge new industry with the purpose of converting agricultural wastes and other plant material into fuel, citing as its primary motive the reduction of the nation's dependence on foreign sources of oil and the cutting back of greenhouse gas generation. What Congress is proposing has far-reaching objectives—the fuel types proposed have not been produced commercially in the United States before and not everyone backs the idea. Some critics claim the technology is immature, the economics are uncertain, hundreds of new factories will be required, and a huge capital investment will be necessary. According to Mark Flannery, head of energy equity research at Credit Suisse, when asked about the plan's feasibility: "It's not clear that it is doable, but it wasn't clear you could send a man to the moon, either. You don't know until you try" (Krauss, 2007).

Historically, Washington's efforts to find new solutions to energy demand and efficiency were to develop more fuel-efficient cars, not alternative-fuel cars, making this new approach by Congress significant. Other portions of the bill are equally groundbreaking. The bill calls for a significant increase in the amount of ethanol used in the nation's fuel supply. Congress is proposing to double the nation's current level of production to 57 billion liters. It also foresees that by 2022, an additional 79 billion liters a year of ethanol or other biofuels will be produced by developing technology that can obtain useful energy from biomass such as straw, tree trimmings, corn stubble, and even common garbage.

Another reason why political involvement is crucial is that in order to accomplish these goals, the nation's key scientists and business leaders will need political and financial support to successfully deal with the technical, environmental, and logistical obstacles they will encounter. Martin Keller, the director of the Department of Energy (DOE) BioEnergy Science Center at the Oak Ridge National Laboratory in Tennessee, said, "We have the opportunity to revolutionize the way we create fuel for transportation. If we focus on this, we can replace between 30–50 percent of our gasoline consumption with new biofuels" (Krauss, 2007). Christopher G. Standlee, executive vice president of Abengoa Bioenergy, remarked, "It certainly is a challenge, but an achievable challenge."

Under the new legislation, corn ethanol use would reach 57 billion liters by 2015. Mandates for next-generation biofuel use would reach 34 billion liters in 2017 and 79 billion liters by 2022. The bill does contain an escape clause, allowing the government to modify the mandates if they do not prove feasible.

The measure is not without uncertainty or critics. Some have expressed concern at the short time line of only 5–15 years. According to Aaron Brady, an ethanol expert at Cambridge Energy Research Associates, "Congress is making the assumption that the technology will appear. To make billions of gallons of next-generation biofuels, a lot of things have to go right within the space of only a few years" (Krauss, 2007). Brady estimates that more than one hundred additional corn ethanol plants will be required, along with at least two hundred other biomass fuel plants, a number that could rise depending on how technology develops. He also figures that 635,000 metric tons of biomass would be needed each year for a distillery to produce 189 million liters of ethanol, which adds up in energy costs to transport it.

Some environmentalists remain uneasy because ethanol produced from corn still requires energy and fertilizer involving the use of natural gas, oil, and coal. Some food producers argue that the plan would require growing eight million hectares of corn—leaving fewer farming acres for fruits, vegetables, soybeans, alfalfa, and other crops and leading to higher food prices.

As with all important issues, there are always pros and cons that must be taken into account when making decisions. To date, there are a number of congressional acts, bills, and legislative proposals concerning the climate change issue.

Current Executive Branch Action

President Obama issued the Executive Order on Federal Sustainability (EO 13514) on October 5, 2009. This executive order was designed to make reducing greenhouse gases a priority of the federal government by requiring federal agencies to set a 2020 GHG-emissions reduction target within a 90-day period, devise a workable plan to reduce their auto-fleet fuel consumption, support sustainable communities, increase their energy efficiency, and leverage their federal purchasing power to promote environmentally responsible products and technologies.

The Council on Environmental Quality (CEQ) coordinates Federal environmental efforts and is designated to work with other federal agencies as well as the White House offices to develop environmental initiatives and polities, such as EO 13514. CEQ resides within the Executive Office of the President, established by Congress as part of the National Environmental Policy Act of 1969, and the Environmental Quality Improvement Act of 1970 provided additional responsibilities. Current chair of the CEQ is Nancy Sutley, who also acts as the principal environmental policy advisor to the president.

President Obama said on the subject: "As the largest consumer of energy in the U.S. economy, the federal government can and should lead by example when it comes to creating innovative ways to reduce greenhouse gas emissions, increase energy efficiency, conserve water, reduce waste, and use environmentally responsible products and technologies. This executive order builds on the momentum of the Recovery Act to help create a clean energy economy and demonstrates the federal government's commitment, over and above what is already being done, to reducing emissions and saving money."

The American Recovery and Reinvestment Act (ARRA) was also passed (February 13, 2009), which included $787 billion in new spending and tax incentives. Part of the spending involved $42 billion in energy-related investments, $21 billion in vehicles and transportation spending, such as for energy-efficient fleets and transit assistance, and approximately $570 million in climate science research spending. It also includes energy-related tax incentives, such as extending the renewable energy production tax credit and additional funding in Clean Renewable Energy Bonds.

The EPA, also part of the executive branch, is responsible for protecting human health and safeguarding the natural environment, including air, water, and land. They are responsible for, and have jurisdiction over, the following initiatives: Best Available Control Technology Guidance, Tailoring Rule, Endangerment Findings, Mandatory Greenhouse Gas Reporting Rules, Federal Vehicle Standards, and Renewable Fuel Standards. A summary of each follows.

BACT Guidance

On November 10, 2010, the EPA released its guidance for the Best Available Control Technology (BACT) requirements for GHG emissions from major new or modified stationary sources of air pollution. This falls under the requirement in the Clean Air Act that stipulates that any major new source or any major modifications to existing sources that emit greenhouse gases must employ technologies that now limit GHG emissions. For each facility, the BACT requirements must address the specific conditions of that particular facility and direct them to complete the maximum degree of emission reduction that has been demonstrated through available methods, systems, and the particular environmental considerations of that facility. What it does not provide for, however, is carbon capture and sequestration technology, unless the facility happens to be located next to an operating oil field whose operator agrees to purchase carbon dioxide for enhanced oil recovery.

The Tailoring Rule

Finalized on May 13, 2010, the tailoring rule is designed to focus new source review and permitting requirements on the largest stationary sources. Under the Clean Air Act, any pollutant subject to regulation under any provision of the act triggers additional requirements. For example, if an industry begins making a new product or makes modifications to an old process, they are subject to new regulation reviews by the EPA for compliancy. On April 1, 2010, the EPA issued its first standards for GHG emissions when it finalized the rules it was using. Beginning January 2, 2011, sources currently subjected to new source review permitting requirements for other pollutants would also have to meet the requirements for GHG emissions if they exceeded 75,000 tons per year. Beginning July 1, 2011, these requirements also apply to new sources with GHG emissions greater than 100,000 tons per year no matter the amount of emissions of other pollutants they emit.

Endangerment Finding

On April 17, 2009, the EPA issued its proposed endangerment finding, defining the basis for the proposed determination that six key GHGs—CO_2, CH_4, N_2O, hydrofluorocarbons (HFCs), perfluorocarbons, and SF_6—were responsible for contributing to climate change, which results in a threat to the public health and welfare of current and future generations. In addition, under the Clean Air Act, the EPA must also determine that particular sources "cause or contribute" to emissions, which threaten public health and welfare. To this, the EPA stated that the combined emissions of these GHGs from new motor vehicles and new motor-vehicle engines contribute to the GHG pollution, which threatens public health and welfare. Therefore, on December 7, 2009, the EPA issued its final endangerment finding and its final cause or contribute finding for light-duty vehicles.

Although this action in itself does not impose any restrictions, it is the required first step in the process leading to specific regulation of GHG emissions. The EPA has proposed to regulate GHG emissions from light-duty vehicles and engines in a joint proposal in September 2009 with the National Highway Traffic and Safety Administration. According to the EPA, the final rule went into effect on July 6, 2010. This represents the first-ever national GHG emissions standard under the Clean Air Act and Corporate Average Fuel Economy (CAFE) standards under the Energy Policy and Conservation Act.

The new standards apply to new passenger cars, light-duty trucks, and medium-duty passenger vehicles, covering model years 2012 through 2016. The EPA GHG standards require these vehicles to meet an estimated combined average emissions level of 250 grams of CO_2 per mile in model year 2016, equivalent to 35.5 miles per gallon if the automotive industry were to meet this CO_2 level all through fuel economy improvements.

In addition, on May 21, 2010, President Obama signed a Presidential Memorandum directing the EPA and Department of Transportation to create a first-ever national policy to increase fuel efficiency and decrease GHG pollution from medium- and heavy-duty trucks for model years 2014 to 2018. These trucks currently emit 20 percent of all GHG pollution related to transportation. In addition, Obama also called for an extension of the national program for cars and light-duty trucks to model year 2017 and beyond.

Mandatory Greenhouse Gas Reporting Rule

On September 22, 2009, the EPA announced that it is now requiring large emitters of GHGs to begin collecting data under a new reporting system. Fossil fuel and GHG suppliers, manufacturers of both motor vehicles and engines, and facilities that emit 25,000 metric tons or more of CO_2 equivalent per year will now be required to report their GHG emissions each year directly to the EPA. Data collection began in January 2010, to be compiled into the first annual report, due January 2011. This represents a significant development in climate change and its control in the United States. According to Lisa P. Jackson (the EPA's current administrator): "For the first time, we begin collecting data from the largest facilities in the country, ones that account for approximately 85 percent of the total U.S. emissions. The American public and industry itself, will finally gain critically important knowledge and with this information we can determine how best to reduce those emissions" (Greenway, 2009).

Federal Vehicle Standards

On May 19, 2009, President Obama made a historic announcement to set new national GHG and corporate average fuel-economy standards for light-duty vehicles. Then, on September 15, 2009, the EPA and the Department of Transportation issued a joint proposal to enhance these standards, which were finalized on April 1, 2010.

These new standards account for over 60 percent of the GHG emissions from the transportation sector. New federal standards accelerate fuel-economy improvements required under the Corporate Average Fuel Economy program, which is administered by the National Highway Traffic Safety Administration, under the Department of Transportation. This is a change to the Energy Independence and Security Act of 2007, which aimed for an average fuel economy of 35.7 mpg by 2020. This change speeds up the process by requiring an average fuel economy of 25.5 mpg by 2016. It also includes a GHG-emission limit per vehicle, which would be set by the EPA using its authority under the Clean Air Act. This is generally set up through a joint rule-making process between the EPA and National Highway Traffic Safety Administration. Tables 8.1 and 8.2 illustrate the projected emissions target under the CO_2 standards (g CO_2/mi) and the projected fuel economy standard (mpg), respectively.

TABLE 8.1

Projected Emissions Target under the CO_2 Standards (g CO_2/mi)

	2012	2013	2014	2015	2016
Passenger cars	261	253	246	235	224
Light trucks	352	341	332	317	302
Combined cars and trucks	295	286	276	263	250

The estimated benefits from this are significant—over the lifetime of the vehicles sold from 2012 to 2016, CO_2 emissions would be reduced by 950 million metric tons and save 1.8 billion barrels of oil. This also means a reduction on our reliance on foreign oil, which also equates to a cost savings. The EPA and the Oak Ridge National Laboratory estimated that a reduction of U.S. imported oil would result in a total energy security benefit of $12.38 per barrel of oil, partly through the reduction in defense spending (EPA, 2009a). The finalized standards would also reduce CO_2 emission from the light-duty vehicles by 21 percent in 2030 as compared to projection if the standard was not in place.

Renewable Fuel Standard

The renewable fuel standard (RFS) is a requirement that a certain percentage of petroleum transportation fuels be displaced by renewable fuels. RFS1 began with the Energy Policy Act of 2005. Congress then updated the standard in 2007, and it became the Energy Security and Independence Act of 2007. This new-renewable fuel standard is known as RFS2 and is a renewable-fuel standard for biofuels *only* that requires obligated parties to sell a certain amount of biofuels per year through 2022. This standard contains a four-part mandate for life-cycle GHG-emissions levels relative to a 2005 baseline of petroleum for: renewable fuel, advanced biofuel, biomass-based diesel, and cellulosic biofuel. Therefore, in order to be classified under one of these categories, a fuel must meet the percentage reduction in life-cycle GHG emissions. As shown in Figure 8.1, the RFS2 slowly ramps up advanced biofuels (cellulosic, biomass-based diesel, and noncellulosic advanced) until they overtake conventional biofuels in consumption levels by 2022. RFS2 was

TABLE 8.2

Projected Fuel Economy Standard (mpg)

	2012	2013	2014	2015	2016
Passenger cars	33.6	34.4	35.2	36.4	38.0
Light trucks	25.0	25.6	26.2	27.1	28.3
Combined cars and trucks	29.8	30.6	31.4	32.6	34.1

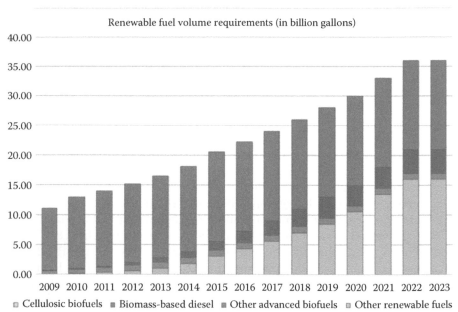

Renewable fuel volume requirements (in billion gallons)

□ Cellulosic biofuels   ■ Biomass-based diesel   ■ Other advanced biofuels   ▣ Other renewable fuels

FIGURE 8.1
(See color insert.) This schematic illustrates the working plan for the Renewable Fuel Standard 2 (RFS2), which contains a four-part mandate for life-cycle GHG-emissions levels relative to the 2005 baseline of petroleum for renewable fuel, advanced biofuel, biomass-based diesel, and cellulosic biofuel.

published by the EPA on March 26, 2010. The following are the identified impacts:

- RFS2 will displace about 13.6 billion gallons of petroleum-based gasoline and diesel fuel in 2022, which represents about 7 percent of expected annual gasoline and diesel consumption in 2022.
- RFS2 will decrease oil imports by $41.5 billion and will result in additional energy security benefits of $2.6 billion by 2022.
- By 2022, gasoline costs should decrease by 2.4 cents per gallon and diesel costs should decrease by 12.1 cents per gallon because of the increased use of renewable fuels.
- RFS2 will reduce GHG emissions by 138 million metric tons in 2022, which is equivalent to taking about 27 million vehicles off the road.
- RFS2 will increase net farm income by $13 billion dollars (equivalent to 36 percent) in 2022.
- RSF2 will increase the cost of food $10 per person in 2022.

The U.S. Department of Energy's Recovery Act spending is also part of the executive branch Action for Climate Change. The American Recovery and

Reinvestment Act of 2009 (Recovery Act or ARRA) is the economic-stimulus package that was passed by Congress on February 13, 2009, and signed by President Obama on February 17, 2009. About $63 billion from the Recovery Act is targeted for energy, transportation, and climate-research spending, with an additional $21 billion in climate energy tax incentives. The DOE received $36.7 billion of the climate- and energy-related funds, with nearly $33 billion for direct grants and the remainder for loan guarantees. The brief describes how DOE funds from ARRA have been appropriated, awarded, and spent. To date, only 3.9 percent has been spent and 55.7 percent has been awarded.

Current Congressional Action

The current U.S. House of Representatives in the 112th Congress (in effect from January 2011 through January 2013) has a Republican majority, making the atmosphere of this Congress much different from the 111th (January 2009 through January 2011). Unfortunately, the current Congress has focused more on the EPA and trying to enforce measures to prevent them from regulating GHG emissions now under its existing authority. In the 111th Congress, focus instead was aimed at measures to reduce GHG emissions.

During the process of passing continuing resolutions in trying to agree on a budget, continuing resolutions were passed in 2010–2011. During this time frame, H.R. 1 was passed, but not enacted, which was fortunate and gave sharp insight into the thinking by the current House regarding climate policy. H.R. 1 would have made the position of Assistant to the President for Energy and Climate Change and the position of Special Envoy for Climate Change illegal, blocked the EPA from regulating GHG emissions from stationary sources, and prohibited funds for the Intergovernmental Panel on Climate Change. H.R. 1 failed in the Senate by a vote of forty-four to fifty-six. Although this is indeed fortunate for those who care to be proactive about climate change, it is worrisome to think the House would not support efforts to mitigate the negative effects of climate change.

The Obama administration has opposed restraints on the EPA's GHG regulatory authority in the past, but currently members of both parties have introduced legislation to either delay or remove the EPA's authority to reduce GHGs under the Clean Air Act. Plans have also been introduced to allow the EPA to still keep their regulatory voice and delay action.

President Obama called for the adoption of a clean energy standard in his 2011 State of the Union address. He called for 80 percent of American energy to be obtained from clean sources by 2035. Action is not expected right away, because it may take some time for people to be able to fully sort out the issues.

The following list provides a summary of the introduced bills related to greenhouse gas emissions in the 112th Congress.

House of Representatives:

- H.R. 97: Introduced by Rep. Blackburn (R-TN). This bill amends the Clean Air Act to exclude carbon dioxide, water vapor, methane, nitrous oxide, hydrofluorocarbons, perfluorocarbons, and sulfur hexafluoride. It also prohibits the use of the Act for regulations related to climate change.
- H.R. 153: This bill, introduced by Rep. Poe (R-TX), prohibits the use of EPA funds to enact a cap-and-trade program or regulation of GHGs from stationary (point) sources (such as industrial facilities, smoke stacks, etc.).
- HR 199: Introduced by Rep. Capito (R-WV), this bill delays the regulation of CO_2 and methane from stationary sources for 2 years from enactment.
- HR 279: Proposed by Rep. Fortenberry (R-NE), this bill prohibits the regulation of methane from livestock using the Clean Air Act.
- HR 910: Proposed by Reps Upton (R-MI) and Whitfield (R-KY) and Sen. Inhofe (R-OK), this proposal questions human-caused climate change, amends the Clean Air Act to prohibit the EPA from issuing regulations concerning GHG for the purposes of addressing climate change, and excludes GHGs from the definition of "air pollutant." It also repeals 11 rules and actions issued since 2009, including the Endangerment Finding, the Tailoring Rule, and New Source Review rules. Exemptions from this include the joint rulemakings for emissions standards CAFE standards for light-duty vehicles (May 2010) and medium- and heavy-duty vehicles (November 2010) and statutorily authorized programs addressing climate change.

Senate:

- S. 228: Introduced by Sen. Barrasso (R-WY), this bill prohibits the president or any federal agency from promulgating regulations to control GHGs, considering climate effects of GHGs in any rule, or taking other actions unless controls of the gas are related to non-climate effects. It also repeals eleven rules and actions issued since 2009, including the Endangerment Finding, the Tailoring Rule, and New Source Review rules, as well as any actions related to GHGs in state or federal implementation plans. Exemption to this is made for the joint rulemakings for emissions standards and Corporate Average Fuel Economy Standards for light-duty vehicles (May 2010).
- S. 231: Introduced by Sen. Rockefeller (D-WV), this bill will delay EPA GHG regulations for stationary sources for 2 years and exempt

light-duty and medium/heavy-duty vehicle standards from that delay.

- S. 482: This is the Senate version of H.R. 910.

Congressional Input and Contributions toward Climate Change Legislation with Obama

The following takes a look at the 111th Congress record (2009–2011) for comparison with that of the 112th Congress. The 111th Congress opened its session with high hopes for the implementation of successful climate legislation. As shown earlier, shortly after being elected, President-elect Barack Obama released a statement placing enactment of a comprehensive climate and energy bill near the top of his list of legislative priorities. Then, following Obama's lead, the U.S. House of Representatives passed the American Clean Energy and Security Act of 2009 (H.R. 2454) by a vote of 219 to 212 in June 2009, only 5 months into the new Congress session. This act was an important one in the Congressional record of climate change legislation because its success represents the first time Congress approved a bill meant to curb the heat-trapping gases scientists have linked to climate change.

The American Clean Energy and Security Act of 2009

The act contained five distinct titles:

- I. Clean energy
- II. Energy efficiency
- III. Global warming pollution reduction
- IV. Transitioning to a clean energy economy
- V. Agriculture- and forestry-related offsets

Title I had provisions related to federal renewable electricity and efficiency standards, carbon-capture and -storage technology, standards for new power plants that used coal, research and development for electric vehicles, and support for the development of the electric smart-grid. Title II provided provisions related to building, appliance, lighting, and vehicle energy efficiency programs. Title IV hosted provisions to preserve domestic competitiveness and support workers, provided assistance to consumers, and provided assistance for domestic and international adaptation initiatives. Titles III and V dealt with a GHG cap-and-trade program.

The bill covered seven greenhouse gases: CO_2, methane, nitrous oxide, hydrofluorocarbons, perfluorocarbons, sulfur hexafluoride, and nitrogen trifluoride. Emitters that would be included under the regulation would include large stationary sources emitting more than 25,000 metric tons per year of GHGs; producers (i.e., refineries) and importers of all petroleum fuels; distributors of natural gas to residential, commercial, and small industrial users (i.e., local gas distribution companies); producers of "F-gases"; and other specified sources. The proposal also called for regulations to limit black-carbon emissions in the United States (black carbon is formed through the incomplete combustion of fossil fuels, biofuel, and biomass and is emitted in both anthropogenic and naturally occurring soot).

The bill had set up progressive targets over time. It established emission caps that would reduce aggregate GHG emissions for all involved facilities to 3 percent below their 2005 levels in 2012, 17 percent below 2005 levels in 2020, 42 percent below 2005 levels in 2030, and 83 percent below 2005 levels in 2050. Commercial production and imports of HFCs would be addressed under Title VI of the existing Clean Air Act and are covered under a separate cap.

The bill also used the value of emission allowances to offset the cost impact to consumers and workers, to aid businesses in transitioning to clean energy technologies, to support technology development and deployment, and to support activities aimed at building communities that are more stable against climate change. It was also designed to protect consumers from higher energy prices. Low- and moderate-income households would receive a refundable tax credit or rebate. In the first few years of the cap-and-trade program, about 20 percent of the allowances would be auctioned. This percentage would increase over time to about 70 percent by 2030. The bill still needed to be voted on and passed in the Senate and signed into law by the president.

Once the House passed the bill, the next step would have been for the Senate to have passed its own comprehensive climate and energy bill. Unfortunately, the Senate was unable to do so, despite much work by key committees and senators. The Senate Energy and Natural Resources Committee passed the American Clean Energy Leadership Act of 2009 (S. 1462) in June 2009 on a bipartisan vote of fifteen to eight, amending it by unanimous consent in May 2010. The bill would have established a renewable energy standard and addressed several other energy-related issues. In November 2009, the Senate Environmental and Public Works Committee passed the Clean Energy Jobs and American Power Act of 2009 (S. 1733), drawing heavily from the American Clean Energy and Security Act of 2009 bill in establishing a GHG cap-and-trade system. S. 1733 passed the committee by a vote of eleven to one, with all seven Republican members boycotting the final vote to protest the process by which the bill had been managed in committee.

Then Senators John Kerry (D-MA), Joseph Lieberman (I-CT), and, for a time, Lindsey Graham (R-SC), worked outside of the committee process in an attempt to broaden the base of support for the legislation within the Senate. Kerry and Lieberman released a draft discussion on their American Power Act in May 2010. The American Power Act would have established a GHG cap-and-trade system for utilities and industry while establishing a fee for transportation fuels.

Several other senators worked on and introduced the Carbon Limits and Energy for America's Renewal Act (S. 2877), which would have capped CO_2 emissions while allowing only very limited emissions trading and rebating the revenue from this system directly back to the public on a per capita basis. The Practical Energy and Climate Plan (S. 3464) was introduced by Senators Richard Lugar (R-IN) and Lisa Murkowski (R-AK), and this plan was intended to reduce oil imports, improve and create new efficiency standards, and establish a clean energy standard.

In conclusion, Senate Majority Leader Harry Reid (D-NV) was expected to combine the various elements of the climate and energy legislative proposals into a comprehensive climate bill that he would bring to the Senate floor. Citing a lack of bipartisan support in the Senate, however, Reid announced in July 2010 that upcoming energy legislation would not include a cap on GHG emissions. This effectively ended action on climate legislation for the 111th Congress.

KEY LEGISLATIVE ACTION OF THE 111TH CONGRESS
- The American Clean Energy and Security Act of 2009 (H.R. 2454)
  - Status: Passed out of the U.S. House of Representatives on June 2, 2009. Died in the Senate.
- The Clean Energy Jobs and American Power Act of 2009 (S. 1733)
  - Status: Passed out of the Senate Environment and Public Works Committee on November 5, 2010.
- The American Clean Energy Leadership Act of 2009 (S. 1462)
  - Status: Passed by the Senate Energy and Natural Resources Committee on June 17, 2009.
- The Clean Energy Partnerships Act of 2009 (S. 2729)
  - Status: Introduced by Senator Stabenow (D-MI) on November 4, 2009, and referred to the Senate Environment and Public Works Committee.
- The Clean Energy Act of 2009 (S. 2776)
  - Status: Introduced by Senators Alexander (R-TN) and Webb (D-VA) on November 1, 2008.
- The Carbon Limits and Energy for America's Renewal Act of 2009 (S. 2877)
  - Status: Introduced by Senators Cantwell (D-WA) and Collins (R-ME) on December 1, 2009.
- The American Power Act
  - Status: Discussion draft released by Senators Kerry (D-MA) and Lieberman (I-CT) on May 12, 2009.
- The Lugar Practical Energy and Climate Plan (S. 3464)
  - Status: Introduced by Senator Lugar (R-IN) on June 9, 2010.
- The Renewable Energy Promotion Act of 2010 (S. 3813)
  - Status: Introduced by Senators Bingaman (D-NM), Brownback (R-OK), Dorgan (D-ND), and Collins (R-ME) on September 21, 2010.

Past Congressional Input and Contributions toward Climate Change Legislation

Two other proposed acts that were introduced in 2007 and 2008, the Global Warming Pollution Reduction Act of 2007 and the Consolidated Appropriations Act of 2008, warrant some attention.

The Global Warming Pollution Reduction Act of 2007

The Global Warming Pollution Reduction Act of 2007 (S. 309), also known as the Sanders–Boxer bill, was proposed as a bill to amend the Clean Air Act to reduce emissions of CO_2. Introduced in the 110th Congress by Senators Bernie Sanders (I-VT) and Barbara Boxer (D-CA) on January 15, 2007, it was based on the increasing scientific evidence that "global warming is a serious threat to both the national security and economy of the United States, to public health and welfare, and to the global environment; and that action can and must be taken soon to begin the process of reducing emissions substantially over the next 50 years" (Sanders, 2007). The bill is considered the most aggressive bill on climate change and is backed by former Vice President Al Gore.

The bill listed several targets, incentives, and requirements that the EPA would employ to reduce emissions and help stabilize global concentrations of GHGs. The bill set the goal of reducing U.S. GHG emissions to a stable global concentration below 450 ppm—a level advised by leading climate-change scientists. It required the United States to reduce its emissions to 1990 levels by 2020 and make additional reductions between 2020 and 2050. Specifically, by 2030, the United States would have to reduce its emissions by one-third of 80 percent below 1990 levels; by 2040, emissions must be reduced by two-thirds of 80 percent below 1990 levels; and by 2050, emissions must be reduced to a level that is 80 percent below 1990 levels. The National Academy of Sciences would be the reporting agency to the EPA and Congress.

The bill also included a combination of economy-wide reduction targets, mandatory measures, and incentives for the development and diffusion of cleaner technologies to achieve the goals, as well as contained the following items:

- Vehicle greenhouse gas emissions standards
- Power plant greenhouse gas emissions standards
- Standards for geologic disposal of greenhouse gases
- Climate change research and development
- Energy-efficiency standards in electricity generation
- Reporting system for climate change pollutants
- Clean energy task force to support development and implementation of low-carbon technology programs

The bill was never passed into law, although it was proposed in sessions of Congress for the past 2 years. It can be reintroduced. Several environmental groups, such as the Sierra Club, Greenpeace, the National Audubon Society, and the Union of Concerned Scientists, supported the measure.

The Consolidated Appropriations Act of 2008

The Consolidated Appropriations Act of 2008, which became Public Law 110-161 on December 26, 2007, directed the EPA to develop a mandatory reporting rule for GHGs. The measure was included in a $500 billion omnibus budget that was signed into law by President Bush and will require U.S. companies to report their GHG emissions. The law did not specify, however, which industries must report or how often they must report.

Overall, the EPA would inventory approximately 85–90 percent of U.S. GHG emissions—from about 13,000 facilities across the nation. The GHGs included in the inventory include CO_2, CH_4, N_2O, HFCs, perfluorocarbons, SF_6, and other fluorinated gases, including nitrogen trifluoride and hydrofluorinated ethers. The collected data will include the total GHG emissions from all sources as well as each gas by category. Once a facility has met the requirements in one year, that facility will continue to report GHG emissions in future years. Companies must reevaluate each facility's emissions whenever there is a process change or other change that may increase the facility's emissions. Facilities that fail to satisfy the reporting requirements are subject to enforcement and penalties under the Clean Air Act. According to the EPA, data collected would be used in future policy decisions and serve as a benchmark to measure annual progress toward emissions reduction targets. This action is viewed as a first step toward a massive, comprehensive national climate change regulation.

The EPA recommends that as companies work to comply with the proposed rule, they should remain focused on the global issue of climate change and the necessity of preparing for possible further federal mandates to reduce GHG emissions. They stress that because of the importance of this issue, reducing emissions is not just a question of compliance; it is now the foundation of business performance. From now on, it should be viewed as part of the cost of doing business.

Because this act represents the first major step toward national comprehensive greenhouse gas emissions regulation, the EPA has proposed some guidelines in order to calculate an initial baseline emission measurement. Any owner or operator of a facility in the United States that directly emits GHG from specific source categories or emits 25,000 metric tons or more of CO_2 emissions annually from stationary combustion will be required to report emissions data under the regulation. The first report was due in 2011 for calendar year 2010. Exempt from this are motor vehicle and engine manufacturers, which would start their reporting for model year 2011. The Mandatory Reporting of Greenhouse Gases Rule was published in the *Federal Register* on October 30, 2009, and became effective on December 29, 2009. The final rule was changed slightly from its April 2009 version. For example, it now exempts research and development activities from reporting, adds additional monitoring options, and requires more data to be reported rather than kept as records so that the EPA can more easily verify reported emissions.

The EPA also foresees a future role for the individual states that are already ahead in reporting and controlling emissions. It views these states as an asset for education. States could take the role in educating the public and businesses and ensuring compliance. Progressive estimates place implementation of any U.S. legislation dealing with climate change to take effect no later than 2012 or 2013.

NOTABLE LEGISLATIVE ACTION FOR CLIMATE CHANGE*

111TH CONGRESS (2009–2011)

- The American Clean Energy and Security Act of 2009 (H.R. 2454)
- The American Clean Energy Leadership Act of 2009 (S. 1462)
- The American Power Act of 2009
- The Clean Energy Partnerships Act of 2009 (S. 2729)
- The Clean Energy Jobs and American Power Act of 2009 (S. 1733)
- The Clean Energy Act of 2009 (S. 2776)
- The Carbon Limits and Energy for America's Renewal Act of 2009 (S. 2877)
- Lugar Practical Energy and Climate Plan (S. 3464)
- The Renewable Energy Promotion Act of 2010 (S. 3813)

110TH CONGRESS (2007–2009)

- American Fuels Act of 2007 (S. 133)
- National Energy and Environmental Security Act (S. 6)
- Coal-to-Liquid Fuel Energy Act of 2007 (S. 155)
- Climate Stewardship and Innovation Act of 2007 (S. 193)
- Global Warming Pollution Reduction Act (S. 309)
- Ethanol Infrastructure Expansion Act of 2007 (S. 859)
- Zero-Emissions Building Act of 2007 (S. 1059)
- Clean Fuels and Vehicles Act of 2007
- Biogas Production Incentive Act of 2007 (S. 1154)
- Clean Air/Climate Change Act of 2007 (S. 1168)
- Clean Power Act of 2007 (S. 1201)
- Clean Coal Act of 2007 (S. 1227)
- Save Our Climate Act of 2007 (H.R. 2069)
- Plug-In Hybrid Electric Vehicle Act of 2007 (H.R. 2079)
- Global Warming Education Act (H.R. 1728)

109TH CONGRESS (2005–2006)

- Clear Skies Act (S. 131)
- Clean Power Act (S. 160)
- Climate Stewardship Act (S. 342)
- Clean Coal Power Initiative Act (S. 957)
- Climate Stewardship and Innovation (S. 1551)
- Oil Security Act (S. Amdt. 958)
- The Energy Policy Act (H.R. 1640)
- Clean Air Planning Act (H.R. 1873)
- Keep America Competitive Global Warming Policy Act (H.R. 5049)
- Bioenergy Innovation, Optional Fuel Utilization, and Energy Legaco (BIOFUEL) Act (H.R. 5372)
- Safe Climate Act (H.R. 5642)

108TH CONGRESS (2003–2004)

- Global Climate Security Act of 2003 (S. 17)
- Climate Stewardship Act of 2003 (S. 139)

---

* This is just a partial list.

- Climate Stewardship Act of 2004 (H.R. 4067)
- Clear Skies Act of 2003 (S. 485 and S. 1844)
- National Greenhouse Gas Emissions Inventory Act of 2003 (H.R. 1245)
- Energy Tax Incentives Act of 2003 (S. 597)

107TH CONGRESS (2001–2002)

- Clean Smokestacks Act (H.R. 1256)
- Clean Power Plant Act (H.R. 1335)
- Emissions Reduction Incentive Act of 2001 (S. 1781)
- National Energy Security Act (S. 389)
- Climate Change Risk Management Act (S. 1294)
- International Carbon Conservation Act (S. 769)
- Forest Resources for the Environment and Economy Act (S. 820)
- The Agriculture, Conservation, and Rural Enhancement Act of 2001 (S. 1731)
- Clean and Renewable Fuels Act (S. 892)

The International Political Arena

The majority of the world's GHG emissions are from a relatively small number of countries. In fact, the twenty largest emitters, with 70 percent of the world's population and 95 percent of the global gross domestic product, account for about 85 percent of global GHG emissions. The top six emitting nations/regions—China, the United States, the European Union (EU), India, Russia, and Japan—account for more than 60 percent of global emissions. This value is calculated without taking deforestation input into account. If emissions from land-use change and forestry are also taken into account, Indonesia and Brazil, who have extremely high rates of deforestation, rank among the top five emitters (Figure 8.2).

In absolute terms, China surpassed the United States in 2007 as the largest annual emitter and is currently responsible for 21 percent of global GHG emissions (the United States, with 5 percent of the world's population, is responsible for 17 percent of GHG emissions). The Pew Center projects that emissions will decline from current levels by about 4 percent in the EU and 7 percent in Japan by 2020. Emissions are rising most quickly in developing countries. China's and India's emissions are projected to grow compared with current levels by about 45 percent and 47 percent, respectively, by 2020. Annual emissions from all developing countries surpassed those of developed countries in 2004.

Policies in Key Countries

Significant action is being taken by many countries around the world. Many different countries have policies and programs that help reduce GHG emissions. Some of them deal specifically with climate change, whereas others are managed as a package driven by economic, energy, or development objectives. For example, in the European Union, the following four policies are currently in place:

Important transitions in emitting countries over the coming century

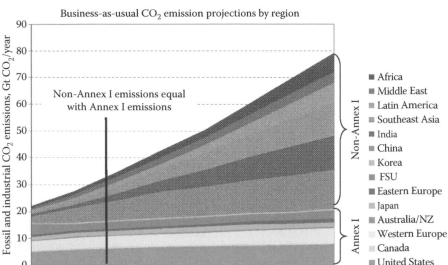

FIGURE 8.2

(See color insert.) This figure shows the "business-as-usual" projections of major GHG-emitting countries throughout the remainder of this century, illustrating why it is imperative that the Annex II countries—those designated in the Kyoto Protocol as not needing to be mandated by emissions control—be held accountable for their emissions. China, for example, is rapidly developing industrially and is now emitting enormous amounts of GHGs into the atmosphere. Without future control of the Annex II countries, the implications of future CO_2 levels (and temperature rise over the 2°C level) are unmanageable.

1. Emissions Trading System: This involves mandatory CO_2 emission limits for 12,000 installations in six major industrial sectors, with emissions trading.

2. Renewable Energy Target: This requires that 20 percent of the energy sources in the EU must be from renewable sources by 2020, including a minimum of 10 percent biofuels in overall fuel consumption.

3. Energy Efficiency Goal: This is a nonbinding goal of energy efficiency improvement of 20 percent from projected 2020 levels.

4. Auto Fuel Economy: This includes mandatory standards to reduce the average CO_2 emissions of new passenger cars from 160 to 120 g/km by 2015.

China also has policies in place to mitigate climate change. They have the National Climate Change Program, which is a comprehensive program that was adopted in 2007, outlining existing and planned policies and programs addressing climate change mitigation and adaptation. The government is currently in the process of developing new energy and climate goals to begin in 2011, with the initial steps of establishing a carbon market. They also have established the following three plans:

1. Energy Intensity Goals: These are national goals designed to reduce energy intensity by 20 percent. This will be achieved through a combination of energy-saving initiatives such as building more efficient coal-fired power plants and shutting down inefficient facilities; appliance standards and consumer subsidies; taxes on petroleum; and mandating provincial and local government action on energy-efficient buildings and public transportation. They estimate that this plan will prevent 1.5 billion tons of emissions.
2. Fuel Economy Standards: These standards are proposed for 2015 and will require all urban cars and light trucks to achieve an average of 36.9 mpg, which is a 27 percent improvement over 2002 levels.
3. Renewable Energy Initiatives: These initiatives include national targets for renewables to provide 15 percent of the primary energy by 2020, including specific targets for wind, solar, biomass, and hydropower capacity.

India is also in the race to lower its emissions and work toward lowering the effects of climate change. They currently have three plans in place to achieve their goals:

1. Renewable Energy: This program deals with solar energy and establishes a target of having a solar capacity of 20 GW by 2022. The country is putting a renewable energy certificate mechanism in place, and they have plans to build a solar capacity of 20 GW by 2022.
2. Energy Efficiency: This is a national program that includes energy-efficiency labels for appliances, mandatory energy audits of large energy-consuming industries, and benchmarks for industrial energy use. One of their goals is to lower energy-consumption rates by increasing energy efficiency.
3. Coal Levy: This program introduced a levy in July 2010 on domestic and imported coal of approximately 50 rupees (about one dollar) per ton. The funds raised will go toward a National Clean Energy Fund.

Canada has pledged to take a constructive approach to achieve real environmental and economic benefits for Canada. Since 2006, the Canadian government has invested more than $10 billion in green infrastructure, energy efficiency, clean-energy technologies, and the production of cleaner energy and fuels. They have developed an Economic Action Plan that includes about $1.8 billion of green investments designed to protect the environment, stimulate the economy, and transform technologies. The Clean Energy Fund, which is part of the Economic Action Plan, is investing $795 million over 5 years in research, development, and demonstration projects to advance Canadian leadership in clean-energy technologies. This includes a large-scale carbon-capture and -storage project as well as smaller-scale projects of renewable and alternative energy technologies. Three carbon-capture and -storage projects have already been announced, costing a total of $466 million from the fund. Up to $146 million will be invested over the next 5 years in projects under the Renewable and Clean Energy portion of the Clean Energy Fund. These projects will support renewable, clean energy, and smart-grid infrastructure.

On October 1, 2010, Environment Canada released the final *Passenger Automobile and Light Truck Greenhouse Gas Emission Regulations* and released a Notice of Intent outlining their commitment to continue working with the United States towards the development of tighter standards for light-duty vehicles for the 2017 and later model years. On May 21, 2010, the Canadian government announced that it will regulate GHG emissions from heavy-duty vehicles. The government is working with the heavy-duty trucking industry, including manufacturers and users, to develop these regulations. On September 1, 2010, the Canadian government released its Renewable Fuel Regulations, which require an average renewable fuel content of 5 percent in gasoline, effective in December 2010.

Table 8.3 illustrates what various countries pledged in the Copenhagen Accord. To date, more than 80 countries have submitted mitigation pledges under the Accord. In creating an effective global response to climate change,

TABLE 8.3

Copenhagen Pledges by Both Developed and Developing Countries

Country	2020 Economy-Wide Emissions Target
Developed countries	
Australia	5–25% below 2000 levels
Canada	17% below 2005 levels
European Union	20–30% below 1990 levels
Japan	25% below 1990 levels
Russia	15–25% below 1990 levels
United States	~17% below 2005 levels
Developing countries	
Brazil	36.1–38.9% below business as usual (BAU) by 2020
China	40–45% emission intensity reduction below 2005 levels by 2020
India	20–25% emission intensity reduction below 2005 levels by 2020 (excludes agricultural emissions)
Indonesia	26% below reference levels by 2020
Korea	30% below BAU by 2020
Mexico	Up to 30% reduction below BAU by 2020 (including 51 million tons CO_2-e by 2012)
South Africa	34% below BAU by 2020; 42% below BAU by 2025

it will require stronger efforts than what has been done to date. For the past 15 years, the main focus of negotiations within the United Nations Framework Convention on Climate Change (UNFCCC) efforts has been the establishment of a legally binding plan to reduce GHG emissions.

This is an important step because binding commitments illustrate a country's commitment to addressing the climate change issue on a global basis. These commitments also provide some assurance to participating countries that other participating countries are also committed and will meet their goals as well. This strategy seemed to work well for the participants of the Kyoto Protocol. The Copenhagen summit presented a more difficult picture, however. It made clear the difficulty of achieving a new set of binding commitments because most of the countries that had binding targets under the Kyoto Protocol are now unwilling to commit to new targets without equal and fair commitments from the United States and the major emerging economies, such as China and India. These countries, however, are not yet prepared to take on the binding commitments required in the eyes of the other countries.

According to the Pew Center, perhaps the best course at this point is an evolutionary one. In this case, it may take a multilateral approach that evolves gradually over time, beginning with initial steps that first build all parties' confidence levels. Once these early commitments are satisfactorily met, then the agreements can evolve into stronger obligations until all parties are on par.

The Pew Center suggests that the countries take incremental steps to strengthen the multilateral architecture in ways that promote stronger action in the near term, while providing a stronger foundation for future binding commitments. By drawing political guidance from both the Bali Action Plan and the Copenhagen Accord, countries could strengthen existing UNFCCC protocols and, if necessary over time, establish new ones. In tandem, countries could also engage in opportunities with programs outside the UNFCCC to mitigate climate change or perhaps work with the World Trade Organization to phase out fossil fuel subsidies. It is building a track record of getting positive results in lowering GHG emissions, lowering fossil-fuel consumption, phasing out substances contributing to climate change, and other measures that can build up the confidence needed within and between countries that will promote a stronger working relationship in order to make further progress under agreements such as the Copenhagen Accord possible. One major concern is that it may take longer to see results with the multilateral approach, and with climate change, this may not be a viable option.

Conclusions

According to the Pew Center for Global Climate Change, addressing the challenge of climate change will ultimately require a comprehensive set of approaches, including support for the development of noncarbon energy sources, the use of market mechanisms to put a price on greenhouse gas emissions, tax incentives, efficiency standards to promote the use of efficient products and technologies, and limits on GHG emissions.

The success of the international effort largely hinges on domestic action by the United States. Action by the United States is critical both because it will promote stronger action by other countries and because it will better position the United States to take on the types of binding commitments needed to ensure a sustained and effective global effort. Although the United States has had a slow start and a defiant past, it is now time to step up to the plate and take part in perhaps one of the most serious global efforts of all time, an effort that reaches toward future generations. What is disturbing is that it appears the Obama administration is trying to achieve progress toward implementing climate change policy and is being hampered by the current actions of Congress.

A specific case in point is that part of the budget debate and resultant decisions during Fiscal Year 2011 resulted in the Department of the Interior getting two-thirds of its climate change funding cut, which resulted in the cancellation of several key projects. Then, 2 weeks later when Congress had cut the funds from various agencies, the climate change line item had been restored to the budget, which leads one to wonder what Congress' stand is on climate change in the first place and what will actually happen to those climate dollar funds.

It is also time for those rapidly developing nations, such as China and India, to realize they are walking into a world much different than the one that industrialization created two centuries ago. This is an era of much greater understanding of the environment and the detrimental effects humans can have on it, and because of it, there has been a collective enlightening—an understanding of what nonrenewability means. With that newfound knowledge of human impact on the environment and the detrimental consequences comes a greater responsibility to protect it, and hence, a new requirement. There can no longer be a "business as usual" approach. Each nation must carry the weight of their due responsibility or not become or remain developed. It is a changing world, and the key is to change it for the better, for future generations. There is no going back when the future is at stake.

Discussion

1. Why do you think climate change became a talking point of the 2008 U.S. presidential elections when it had been repressed in prior ones?
2. Do you agree with President Obama's description of the "shock and trance" cycle of rising gas prices? To what do you attribute this?

3. In President Obama's address to the summit meeting on climate change organized by Governor Schwarzenegger, he commented that "too often Washington has failed to show the same kind of leadership" when referring to other countries and businesses doing their part by investing in clean energy technologies. Do you agree? What examples can you think of that support his claim?

4. Do you believe that in the United States the federal government should set the example for mitigating climate change? If so, what should they do and how should it be implemented to be the least burdensome on the taxpayer?

5. What do you feel the underlying causes are of Congress not being able to come to an easy, workable agreement on their differences of climate change legislation?

6. How much of Congress's problem do you think boils down to the fact that it is just the age-old issue of Democrat versus Republican?

7. What would be your reaction if Congress were able to take away the EPA's right to manage GHGs? How would that affect the future of climate change policy and mitigation progress in the United States?

8. If you were a politician, how would you convince Congress that acting now on climate change is imperative?

9. Why do you think the attitude toward climate change is so different between the 111th and 112th U.S. Congresses?

10. What changes do you think could be made in the U.S. Congress to make it more conducive to getting climate change legislation passed?

11. What is your opinion about having a multilateral approach for countries under the Copenhagen Accord? Do you think giving each country the latitude to choose its own level and rate of reduction will encourage success?

References

Broder, J. M. 2009. Democrats unveil climate bill. *The New York Times*, April 1. http:// www.nytimes.com/2009/04/01/us/politics/01energycnd.html?hp (accessed January 23, 2011). Presents the viewpoint of global warming and politics.

Clips & Comments. 2008. Transcript: President-Elect Barack Obama and Michelle Obama on *60 Minutes*. November 16. http://www.clipsandcomment. com/2008/11/16/transcript-president-elect-barack-obama-and-michelle-obama-on-60-minutes-november-16/ (accessed August 11, 2011).

Energy and Commerce Committee. 2009. Chairman Waxman, Markey release discussion draft of new clean energy legislation. Press Release, March 31. http://waxman.house. gov/News/DocumentSingle.aspx?DocumentID=116749 (accessed August 11, 2011).

Environmental Protection Agency. 2009a. EPA proposes new regulations for the national renewable fuel standard program for 2010 and beyond. EPA-420-F-09-023, May. Discusses the reduction on foreign oil reliance and its implications.

Environmental Protection Agency. 2009b. Endangerment and cause or contribute findings for greenhouse gases under Section 202(a) of the Clean Air Act, final rule. *Federal Register*. 74: 239.

Gore, A. 2006. *An Inconvenient Truth*. Emmaus, PA: Rodale. Presents an excellent overview of the climate change problem, how it has come about, what it means for the future, and why humans need to act now to slow it down.

Greenhouse, L. 2007. Justices say EPA has power to act on harmful gases. *The New York Times*, April 3. http://www.edf.org/documents/ca/NY%20Times%20%20Justices%20Say%20E.P.A.%20Has%20Power%20to%20Act%20on%20Harmful%20Gases%20-%2004.03.07.pdf (accessed November 5, 2010). Discusses the EPA's role in climate change.

Greenway, R. 2009. Greenhouse gas reporting requirements finalized. *Environmental News Network*, September 25. http://www.enn.com/ecosystems/article/40522 (accessed August 11, 2011).

Krauss, C. 2007. As ethanol takes its first steps, Congress proposes a giant leap. *The New York Times*, December 12. http://www.nytimes.com/2007/12/18/washington/18ethanol.html?pagewanted=print (accessed January 23, 2008). Discusses the plans Congress has for renewable energy in order to reduce the nation's heavy reliance on foreign oil.

New York Times. 2008. The one environmental issue. January 1. http://www.nytimes.com/2008/01/01/opinion/01tue1.html (accessed August 11, 2011).

NPR. 2009. Transcript: Barack Obama's inaugural address, January 20. http://www.npr.org/2010/12/02/99590481/transcript-barack-obama-s-inaugural-address (accessed August 11, 2011).

Revkin, R. 2008. Obama: Climate plan firm amid economic woes. *New York Times*, November 18. http://dotearth.blogs.nytimes.com/2008/11/18/obama-climate-message-amid-economic-woes/ (accessed August 11, 2011).

Sanders, B. 2007. Summary of the Global Warming Pollution Reduction Act (S. 309) as introduced January 2007. http://sanders.senate.gov/newsroom/news/?id=f99cc6a4-1b10-4a0e-8d91-75bd0b3ab1bd (accessed August 11, 2011).

Suggested Reading

American Wind Energy Association. 1997. Wind energy and climate change: A proposal for a strategic initiative, October. http://www.ecoiq.com/online resources/anthologies/energy/wind.html (accessed March 20, 2009). Discusses cost-effective methods for supplying electricity to rural villages via renewable wind energy.

Choi, C. Q. 2008. The energy debates: Clean coal. *LiveScience*, December 5. http://www.livescience.com/environment/081205-energy-debates-clean-coal.html (accessed February 22, 2009). Discusses whether or not the clean coal technology performs up to its expectations.

Flook, S. 2007. China set to build 562 new coal plants: Kyoto in perspective. *The Politic*, January 17. http://www.thepolitic.com/archives/2007/01/17/china-set-to-build-562-new-coalplants (accessed January 16, 2009). Discusses air pollution concerns in China and the disastrous effect that will have on global warming if they do not use renewable energy sources but rely on fossil fuels instead as they industrialize.

Gelbspan, R. 1997. *The Heat Is On: The High Stakes Battle over Earth's Threatened Climate*. Reading, MA: Addison Wesley. Offers a look at the controversy environmentalists often face when they deal with fossil-fuel companies.

Gelling, P., and A. C. Revkin. 2007. Climate talks take on added urgency after report. *The New York Times*, December 3. http://www.nytimes.com/2007/12/03/world/asia/03bali.html?pagewanted=pring (accessed January 23, 2008). Discusses the need to cut greenhouse emissions in preparation for the Bali conference, which will discuss what the global plan of action will be after the Kyoto Protocol expires in 2012.

Houghton, J. 2004. *Global Warming: The Complete Briefing*. New York: Cambridge University Press. Outlines the scientific basis of global warming, describes the impacts that climate change will have on society, and looks at solutions to the problem.

McKibben, B. 2007. *Fight Global Warming Now: The Handbook for Taking Action in Your Community*. New York: Holt Paperbacks. Provides the facts of what must change to save the climate and shows how everyone can be proactive in their community to make a difference.

Reuters (London). 2008. Multinationals fight climate change. *The New York Times*, January 21. http://www.nytimes.com/2008/01/21/business/21green.html (accessed April 26, 2009). Looks at the joint efforts of eleven companies using renewable energy.

9

Sociopolitical Impacts of Climate Change

Overview

This chapter addresses some of the sociopolitical impacts associated with climate change. It looks first at the issue of national security and terrorism and how it relates to the freshening of the ocean waters leading to a slowdown of the thermohaline ocean current. In connection with this potential problem, the chapter examinees the very real possibility of abrupt climate-change issues and the possibilities and consequences of a mass migration being triggered as a result. It then takes a look at the positions of several other countries and their views on national security and just how they are preparing their military forces for any future conflicts as a result of a changing climate. Next, this chapter focuses on how humans perceive climate change and how those perceptions could possibly ignite a serious conflict and crisis. It also takes an analytical look at which of the world's countries are the most fragile and which could be facing serious risks in a changing climate. In conclusion, the chapter presents the concept of climate justice and equity, and presents the information from a different point of view, discussing some topics that could be perceived as long-term threats.

Introduction

Peace is a fragile commodity and not to be taken for granted. Throughout history, wars have been fought over all manner of things, and on several occasions, wars have been triggered by climate change. While the gears of climate change are turning and the natural and man-made environments are being altered, it is imperative to stay informed. One of the best ways to stay informed and proactive during the climate change process is through keeping abreast of the ongoing studies in progress.

Climate change is a critical issue to stay current on because every aspect of a changing climate has ramifications on human life. One of the most important ramifications of climate change is the possibility of violent conflict. In order to fully understand the issues, it is important to have a good concept of the consequences of inaction, the ease with which conflict can arise, what a "fragile state" is and which countries have been identified as such, and where other hostilities can creep up that might not be obvious at first glance. All of these issues are important to consider when discussing the sociopolitical impacts of climate change.

Development

Climate Change, National Security, and Terrorism

In October 2003, Andrew Marshall, a highly respected U.S. Department of Defense planner, commissioned a Pentagon study on climate change and U.S. security. The study's principal authors were Doug Randall of the Global Business Network (a California think tank) and Peter Schwartz, former head of planning for Shell Oil. Their conclusion was that climate change could ultimately prove to be a greater risk to the nation than terrorism (Schwartz and Randall, 2003).

Randall and Schwartz, who interviewed leading climate change scientists, conducted additional research, and reviewed numerous climate models with experts in climatology, concluded that climate change could lead to a slowing of ocean currents. Major currents in the ocean carry huge amounts of heat from the equator to the poles, circulating heat energy on the surface and at great depths. One extremely important current is the thermohaline circulation (THC). This global current is significant to major parts of the world; it moves the warm salty Atlantic water that originates near the equator northward toward Greenland and Labrador, where it then cools and sinks (see Chapter 1 for a more in-depth description). The current sinks more than 1.6 kilometers in specific locations, where it then turns over and heads south, making its way back through the Atlantic toward the equator again. From there, the water continues to move south, travels around the southern tip of Africa, and rises to the surface in the Indian and Pacific Oceans, as well as areas near Antarctica. It then heads north toward the equator again, where it picks up heat, and repeats the cycle.

The problem with adding large amounts of freshwater to the ocean through the melting of ice caps and glaciers is that it decreases the salinity of the ocean water and slows the overturning process at the high latitudes. By slowing the process, the freshwater slows down the entire conveyor belt, which means that warmth from the equator will not be brought up into the Northern Hemisphere (Figure 9.1).

The Gulf Stream, which is the current that transports a significant amount of heat northward from the earth's equatorial region toward western Europe, helping to warm its climate, is part of that circulation system. In fact, if it

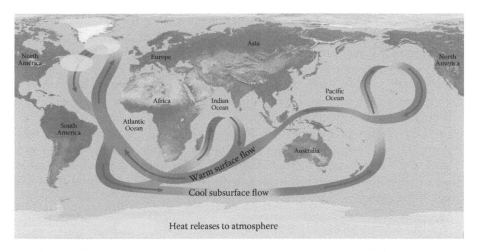

FIGURE 9.1
(See color insert.) Working as a massive conveyor belt of heat, the ocean thermohaline circulation has a significant effect on weather worldwide. As global warming continues to heat up the planet, many scientists are worried that the addition of freshwater to the ocean from the melting of the Greenland ice sheet could stop the North Atlantic conveyor. If it did shut down, or even slow down, it would send colder temperatures to Europe and cause other sudden climate changes around the world. (Courtesy of NOAA, Ocean facts, http://oceanservice.noaa.gov/facts/coldocean.html, 2011).

were not for the Gulf Stream, the North Atlantic and Europe would be on average 5°C cooler. If this extensive current were to shut down, it would have a negative impact on the entire ocean/atmospheric system and cause adverse effects worldwide not only in ocean circulation, but also in the jet stream in the atmosphere that drives storm systems. Based on evidence retrieved from ice cores in Greenland, scientists have determined that the THC has been shut down in the past and that every time it has been shut down, an abrupt climate change has occurred. The chief mechanism for shutting down the THC is the addition of freshwater (Schwartz and Randall, 2003). The report goes on to analyze how an abrupt climate change scenario could

> potentially de-stabilize the geopolitical environment, leading to skirmishes, battles, and even war due to resource constraints such as:
>
> 1. Food shortages due to decreases in net global agricultural production
> 2. Decreased availability and quality of freshwater in key regions due to shifted precipitation patterns, causing more frequent floods and drought
> 3. Disrupted access to energy supplies due to extensive sea ice and storminess (Schwartz and Randall, 2003)

As these conditions persist and global and local carrying capacities are reduced, tensions could mount around the world, leading to two principal strategies: defensive and offensive. Nations that have the resources and are in a position to do so may build fortresses around their countries, protecting and keeping the resources for themselves. Less fortunate nations—especially those who share borders with warring nations—may engage in battle for access to food, clean water, or energy. Unlikely alliances could be formed as defense priorities shift and the goal becomes resources for survival instead of religion, ideology, or national honor.

If these chains of events were to occur, it would pose new challenges for the United States and other developed countries and regions such as Canada, the European Union, and Australia. Randall and Schwartz suggest that in order to be prepared to deal with such changes, it is important that the United States:

- Improve predictive climate models to allow investigation of a wider range of possible scenarios in order to be able to anticipate how and where changes could happen.
- Determine potential impacts of abrupt climate change through modeling and how it could influence food, water, and energy.
- Determine which countries are most vulnerable to climate change and could contribute materially to an increasingly disorderly and potentially violent world.
- Identify "no-regrets" strategies such as enhancing capabilities for water management.
- Rehearse adaptive responses.
- Explore local implications.
- Explore geoengineering options that control the climate.

The authors advised the Department of Defense to look at potential responses now because there is already evidence in place that climate change has reached a threshold where the THC could start to be significantly affected, such as documented measurements of the North Atlantic being freshened by melting glaciers, increased precipitation, and increased freshwater run-off, making it substantially less salty over the past 40 years. Because of this, Randall and Schwartz recommend the report be elevated from a scientific debate to a U.S. national security concern. In their research, they concluded that weather-related events can have an enormous impact on society. They influence food supply, conditions in cities, availability of and access to clean water, and availability of energy.

According to the Climate Action Network of Australia, climate change will probably reduce rainfall in rangeland areas, which would cause a 15 percent drop in grass productivity. This could cause a reduction of the average weight of cattle by about 12 percent, which would significantly reduce the world beef supply. In addition, dairy cows would probably produce 30 percent less milk, and insects may invade new fruit-growing areas. Drinking-water supplies would also be affected, possibly causing a 10 percent reduction in water supply. With this given scenario, several major food-producing regions around the world over the next 15–30 years may not be able to meet demand (Muharam, 2011).

When population numbers are added to the equation, the situation becomes dire. Currently, more than four hundred million people live in the dry, subtropical, overpopulated, and economically poor regions where the negative effect of climate change poses a severe risk to their political, economic, and social stability. In other countries that completely lack resources, the situation will be even worse. In these countries, it is expected that there will be mass emigration as desperate people seek better lives in regions such as the United States that have the resources available to allow them to adapt. This scenario has immediate implications for issues concerning food supply, health and disease, commerce and trade, and their consequences for national security. What the study concluded was that large population movements are inevitable. Learning how to manage populations and border tensions will be critical, and new forms of security agreements dealing specifically with energy, food, and water will be needed. Disruption and conflict will become an everyday fact of life (Townsend and Harris, 2004).

The United States does not stand alone in its view of security issues associated with climate change. Several other countries also support a similar stance. In the past, the potential security implications of climate change received much less attention in Canada than in other countries, but a workshop held in Ottawa in January 2010 changed that stance. The Climate Change and Security Project Team at the University of British Columbia was designed to address this issue by providing a forum where officials from a dozen federal government departments and agencies could come together and discuss possible security-related impacts relevant over the coming two decades; examine the protocols of other countries already addressing these issues; and consult with representatives from Canadian and American universities, think tanks, and nongovernmental organizations, as well as foreign diplomatic missions. The result of this endeavor was the Climate Change and Security Project Report, which identified several themes and policy recommendations for the Canadian government to consider.

A consensus from the meeting was that climate science clearly reveals there are security issues about which Canada needs to be proactive between now and 2030. Scientists confirmed that the changing climate is already generating serious security concerns for Canada, such as changing precipitation patterns, sea-level rise, and extreme weather. Key security-related issues for the government include the knowledge that security-related impacts will occur between now and at least 2030, even if greenhouse gas (GHG) emissions are stopped completely. Of even greater concern is the unknown about specific effects that will occur. Because scientific analysis of regional and local impacts is weak, more research also needs to be done. The take-home message to government officials: Scientific uncertainties do not justify inaction—they point to the need for further research.

Because of this, Canada must begin to address climate-induced security threats, vulnerabilities, and risks immediately. The Canadian Arctic deserves specific security attention, not only because the impacts of climate change are most visible there, but also because of the uncertain intentions and grow-ing presence of other nations in the region. In addition, it is imperative that all regions of Canada be inventoried for risks, because all have been iden-tified as vulnerable. It has been determined that among the most-stressed Canadian public-sector organizations are those responsible for public safety, public health, emergency management, critical-infrastructure protection, and disaster response.

One major issue they also addressed was public awareness. Because cli-mate change is a slow-moving threat, it was suggested that many Canadians are unable to imagine the dramatic set of security-related changes that lie ahead. The participants agreed that actions must be taken to raise pub-lic awareness of the security implications of climate change. In order to accomplish this, it was suggested that government security organizations be open to outside expertise, including that from scientists and others in the academic community, as well as the establishment of robust informa-tion-sharing networks between scientists and security policy makers. A genuine collaboration between the government and the private sector was also identified as being crucial for a successful plan to be developed and implemented.

Concerning military forces, it was recommended that the Canadian Forces start now to prepare for future climate change-related endeavors. A basic premise agreed upon at the Ottawa conference was that climate change would help shape how states form military alliances, why states deploy their military forces, and the environments to which military per-sonnel and equipment will be deployed. In addition, decisions about the role of the military—as a defense force, a resilience builder, or a humani-tarian crisis responder—would affect the size, shape, and structure of the future Canadian Forces. It was acknowledged that climate change would impact all major activities of the Canadian Forces, including deployments, procurement, training, equipment, and energy consumption. Therefore, it is imperative now that climate change be integrated into all policy and planning frameworks within the Department of National Defence and the Canadian Forces. The stance the Canadian government has been advised to take is to accept that the security implications of climate change require a long-term, strategic approach but that planning and preparation needs to begin immediately.

William Hague, the United Kingdom's foreign secretary, recently described climate change as "perhaps the 21st century's biggest foreign policy challenge," in reference to several reports that highlight the threat that climate change poses to "collective security and global order" (Jarvis et al., 2011). In addition, the UK Ministry of Defence stated that "Climate change will amplify existing social, political, and resource stresses and will shift the tipping point at which conflict ignites" (*ScienceDaily*, 2011).

According to Rear Admiral Neil Morisetti, the United Kingdom's climate and energy security envoy and a member of the Royal Navy since 1976, "We see climate change as a threat multiplier, as a catalyst for conflict. We're trying to understand this threat, like any other threat that we look at. It's about trying to reduce risk of the threat of conflict" (*Homeland Security NewsWire*, 2010).

Although Morisetti clarifies that the military is not trying to turn climate change into a military issue, it is important to search for ways to arrest problems before they can surface. For instance, last year, India installed a high-tech fence along its border with Bangladesh in order to deter an influx of illegal immigrants that were displaced by rising sea levels that were flooding their low-lying homeland (*Homeland Security NewsWire*, 2009). In addition there have also been incidences reported by *Scientific American* in March 2009 that "as warming temperatures deplete water supplies and alter land use, military analysts warn, already-vulnerable communities in Asia and Africa could descend into conflicts and even wars as more people clamor for increasingly scarce resources" (Friedman, 2009). Well-known climate change economist Lord Nicholas Stern has cautioned that failing to reduce GHG emissions could bring "an extended world war" (Johnson, 2011).

The United Kingdom acknowledges that the military's role in helping people cope with climate change may not necessarily be only conflict resolution, but also preventing conflict by helping countries grow and protect their resources. In order to do this, the military must reconcile its own contribution to climate change, Morisetti said, acknowledging that "in the military, we burn a lot of gas" (*Homeland Security NewsWire*, 2010). He further clarified that this might be improved by a move to biofuels or an increase in computer-based (rather than field-based) training, although these options would have to be explored further to determine their feasibility.

Australia's military has also warned that climate change could create failed states across the Pacific as sea levels rise and subsequently heighten the risk of conflict over resources. An Australian Defence Force (ADF) analysis conducted in 2009 found that the military could be called on to undertake more security, disaster relief, and reconstruction missions as a result of climate change. The ADF analysis also concluded that the accelerated melting of the Arctic could raise conflict levels if disputes over undersea energy deposits were not resolved peacefully. In addition, the analysis concluded that Australia could face increased illegal migration and fishing as Pacific islands succumb to rising sea levels and climate change impacts key fishing grounds (Agence France-Presse, 2009).

The ADF has been deployed to the Solomon Islands and East Timor in recent years to enforce law and order, also assisting in a 2009 relief operation in Samoa after a tsunami killed 143 people. Air Chief Marshal Angus Houston commented that these types of operations would probably be more frequent in the future: "With the effects of climate change compounding existing pressures, future operations will be more frequent and more intense than those currently underway in East Timor and the Solomon Islands" (*TerraDaily*, 2010).

Houston also believes that rising sea levels would significantly worsen the social problems on the islands. Because the islands have a lot of potential for sustained economic growth, these island nations would struggle to adapt to climate change as changing rainfall patterns, extreme weather, and continually rising sea levels threaten the agriculture and fishing industries on which they depend for their livelihoods. Houston commented on the situation: "From there, it is a small step to political instability and social disorder." Houston also says that although it could take two decades before climate change began to inflict major damage on the South Pacific, Australia would need to be prepared long before that (*TerraDaily*, 2010).

Climate Change, Inaction, and War

Lord Nicholas Stern, one of the world's most prominent climate economists, believes that the failure to address climate change could eventually lead to World War III. Interestingly, he first published his *Stern Review* in 2006 on behalf of the British government and succinctly outlined the potentially catastrophic economic consequences of failing to address climate pollution. Since that time, with the increase in knowledge from the scientific community, Stern has since warned that his report "underestimated the risks" (Johnson, 2011).

Stern explains that there are extreme consequences of inaction—that by not enacting a global policy to cut carbon pollution there will be "potentially immense" consequences of radical transformation of the earth. He further paints the picture that as temperatures increase, sea levels rise, and droughts degrade other areas, there will be serious transformations of the landscape. For example, most of southern Europe will probably be transformed into a desert. Other areas will be flooded, such as Florida and Bangladesh. Stern stresses that no matter where one may live, climate change will change the lives and livelihoods of everyone globally. He points out that "we live where we live because of patterns of climate: where the rivers are, where the seashores are—that's what determines where we are" (Johnson, 2011). The seriousness of what he stresses boils down to the fact that our cost of inaction right now will equate eventually to where we can ultimately live. When you project that in a hundred years or so to hundreds of millions of people—perhaps billions of people—moving, that is a risk for global war, a risk we should be thinking seriously about preventing right now.

Stern also replied to the U.S. military's comment on climate change being a "threat multiplier" to national security. His take on that is that "If we are to learn anything from history, these are the kinds of threats that lead to war, and *geometrically* growing climate change brings threats on a global scale" (Johnson, 2011).

In research conducted by Zhang et al. (2007), they determined that long-term fluctuations of war frequency and population changes followed the cycles of temperature change. Their analyses of historical data illustrated that cooling impeded agricultural production, which then brought about a series of serious social problems, including price inflation and then successively war outbreak, famine, and population decline. Their findings suggest that worldwide and synchronistic war-peace and population cycles in recent centuries have been driven mainly by long-term climate change. Interestingly, their findings imply that the social mechanisms that might mitigate the impact of climate change were not significantly effective during the time period they studied. They looked at the influence of both cyclic and demographic theories. What did appear to be significant in some cases was the presence of a strongly unified society. Because of the overriding uncertainty, however, they concluded that climate change may, therefore, have played a more important role and imposed a wider ranging effect on human civilization than was thought possible.

Zhang et al. did relate the vulnerability today on the agricultural system and the effect it could have on society. In addition, other direct impacts, such as sea-level rise, the spread of tropical diseases, and the increase of extreme weather events, would also add costs to the current economy that is supported by inexpensive energy sources. In severe cases, the economic burden could cause conflict for resources and intensify social contradictions and unrest similar to that seen historically. Where they ultimately see the largest threat, however, is with the impact on ecosystems and whether or not humans could successfully adapt. If adaptation or the costs of adaptation were out of reach, war could be the likely result (Zhang et al., 2007).

Climate Change, Conflict, and State Fragility

According to the Governance and Social Development Resource Centre (GSDRC), there is a wealth of literature on climate change that predicts increased violent conflict as a result of climate-induced changes such as migration, environmental degradation, and resource scarcity (GSDRC, 2007). However, they claim that these references are rarely supported with empirical evidence. Little academic research has been done to date on the connection between climate change and conflict. The most commonly discussed issues are climate change contributing to resource scarcity, which prompts violent conflict, and resource scarcity resulting in migration, which leads to conflict in the receiving area. Sometimes, where there is a clear link between climate and conflict, climate change is just one factor contributing to conflict, but not necessarily the primary factor. There is a consensus that climate change does pose a very serious threat to human security and that increased global competition for resources will lead to increased international tension, which could spark violent, interstate conflict, as previously mentioned.

Analysts Dan Smith and Janani Vivekananda conducted an assessment of the social and human consequences of climate change for International Alert in the United Kingdom. They claim that many of the world's poorest places face two very real problems: climate change and violent conflict. Their conclusion was that in fragile states the consequences of climate change could interact with existing sociopolitical and economic tensions, compounding underlying violent conflict. They believe that conflict-sensitive climate change policies could promote peacebuilding if they were put in place (Smith and Vivekananda, 2007).

Smith and Vivekananda also conclude that those who would feel the effects of climate change the most severely are those living in poverty, in unstable states under poor governance. In order to understand how the effects of climate change will interact with socioeconomic and political problems, it is necessary to understand the consequences of climate change. Smith and Vivekananda have identified four key elements of primary risk: political instability, economic weakness, food insecurity, and large-scale migration.

They contend that states and communities need to adapt in order to handle the challenges of climate change. The most vulnerable communities with the weakest adaptive capacity are those in fragile states. The conclusions they came to as a result of present world conditions include:

- There are currently forty-six countries where governments will struggle to handle climate change. This creates a high risk of political instability, with the potential for violent conflict in the long term. The forty-six countries at risk are listed in Table 9.1.
- There are an additional fifty-six countries where governments will struggle to handle climate change. This creates a high risk of political instability, with the potential for violent conflict in the long term. The countries involved here are listed in Table 9.2.
- Most of the countries facing climate change and violent conflict cannot be expected to adapt alone. Some of them lack the desire, but more lack the capacity. Some lack both.
- Enhancing the ability of communities to adapt to the consequences of climate change reduces the risk of conflict. Peacebuilding activities, addressing socioeconomic instability and weak governance, and enhancing the ability of communities to adapt to climate change are necessary.

TABLE 9.1

Countries Facing a High Risk of Armed Conflict as a Consequence of Climate Change

Afghanistan	Iran
Algeria	Iraq
Angola	Israel and occupied territories
Bangladesh	Jordan
Bolivia	Lebanon
Bosnia and Herzegovina	Liberia
Burma	Nepal
Burundi	Nigeria
Central African Republic	Pakistan
Chad	Peru
Colombia	Philippines
Congo	Rwanda
Côte d'Ivoire	Senegal
Democratic Republic of the Congo	Sierra Leone
Djibouti	Solomon Islands
Eritrea	Somalia
Ethiopia	Somaliland
Ghana	Sri Lanka
Guinea	Sudan
Guinea-Bissau	Syria
Haiti	Uganda
India	Uzbekistan
Indonesia	Zimbabwe

Source: Smith, D., and Vivekananda, J. A climate of conflict: The links between climate change, peace and war. London: International Alert, 2007.

TABLE 9.2

Countries Facing a High Risk of Political Instability as a Consequence of Climate Change

Albania	Fiji	Maldives	Taiwan
Armenia	Gambia	Mali	Tajikistan
Azerbaijan	Georgia	Mauritania	Thailand
Belarus	Guatemala	Mexico	Timor-Leste
Brazil	Guyana	Moldova	Togo
Cambodia	Honduras	Montenegro	Tonga
Cameroon	Jamaica	Morocco	Trinidad and Tobago
Comoros	Kazakhstan	Niger	Turkey
Cuba	Kenya	North Korea	Turkmenistan
Dominican Republic	Kiribati	Papua New Guinea	Ukraine
Ecuador	Kyrgyzstan	Russia	Vanuatu
Egypt	Laos	Saudi Arabia	Venezuela
El Salvador	Libya	Serbia (Kosovo)	Western Sahara
Equatorial Guinea	Macedonia	South Africa	Yemen

Source: Smith, D., and Vivekananda, J. A climate of conflict: The links between climate change, peace and war. London: International Alert, 2007.

- International cooperation is required to support local action, both as a way of strengthening international security and to achieve the goals of sustainable development.
- Policies and strategies for development, peacebuilding, and climate change are often disconnected. Peacebuilding and adaptation, as a result, need to be accomplished in tandem.

Based on their conclusions, Smith and Vivekananda made several recommendations on how to address climate change in fragile states. For example, they suggest:

- Conflict and climate change must be moved to a more important level on the international agenda. International guidelines should be developed regarding adaptation.
- More research is necessary to fully understand the ramifications of indirect local consequences. This needs to be accomplished before effective policies and programs can be developed.
- Adaptation needs to be prioritized over mitigation in fragile states to address the consequences of climate change in order to prevent conflict.
- National adaptation plans of action must be conflict-sensitive, taking into account the sociopolitical and economic context. Migration must be managed through research to identify likely migration flows.
- Governments must engage the private sector and develop guidelines to help companies identify how their commercial operations can support adaptation.

The work of Smith and Vivekananda is seen as an important contribution to the sociopolitical management of climate change. It breaches a topic on which not much research has been done to date (Smith and Vivekananda, 2007).

Another issue of state fragility is when climate change and politics threaten water supplies. In Lebanon's Bekaa Valley, for example, a historic feud over irrigation was refueled in February 2009 because of climate change and an increased scarcity of water in the region. The lack of water has been triggering increasing social conflict in Lebanon because farmers have been unable to grow crops in the region. The residents blame rising temperatures, spiraling population growth, and inefficient irrigation as the chief reasons for the economic and social breakdown the region is currently experiencing.

According to Randa Massad, an irrigation expert at the Lebanese Agricultural Research Institute, experts estimate that demand for water in Lebanon will have increased by more than 80 percent by 2025 as Lebanon's population is expected to nearly double in size—from 4 to 7.6 million (*IrinNews*, 2009). During this same period, climate scientists predict that, because of climate change, average summer temperatures will increase by 1.2°C. This equates to greater evapotranspiration, which could increase the demand for irrigation in the Bekaa's farmlands by as much as 18 percent. "Water shortage is not a new phenomenon in the Bekaa region. What is new is that it is occurring in an increasingly changed environment and this makes it more serious and long-lasting," says Massad.

The government is indeed taking it seriously. They are currently implementing and developing a 10-year water strategy designed to promote integrated water-resources management, including controlling unlicensed wells, updating antiquated distribution mechanisms, such as exposed canals that lose water to evaporation, and addressing the lack of wastewater-treatment facilities for water reuse. Hussein Amery, an expert on water conflict in the Middle East, said, "Lebanon will be very seriously affected by climate change. When you have bad blood and a history of conflict, water becomes a trigger" (*IrinNews*, 2009).

The Concept of Climate Justice and Equity

Climate justice has to do with the issue of who should carry the more significant portion of the weight in controlling greenhouse gas emissions in order to mitigate the effects of climate change. Historically, in climate change negotiations, there has been concern that negotiations and corresponding responsibilities expected of, and assigned to, countries are not allocated and recognized on a fair basis. There are those who hold the belief that it is the developed countries that are responsible for emitting the bulk of the GHGs—far more than developing nations. Therefore, these countries should face the biggest responsibility for taking action to address climate change. In addition, it is the wealthier countries that should support developing nations and help them adopt cleaner technology through financing and technology transfer.

Supporters of climate justice often accuse developed nations of ignoring these responsibilities and choosing merely to point fingers at newly developing countries, such as China and India. Developed nations sometimes argue that if they must comply, so must developing nations. Climate justice contends, on the other hand, that the burden of reductions must lie with industrialized countries. Many supporters of climate justice claim that developed nations choose to sit back, do very little, and then point fingers at newly industrializing nations, expecting them to make a difference by lowering their own emissions targets.

Proponents of climate justice refer to the principle of *common but differentiated responsibilities* when talking about developing nations and climate change. This principle, which was acknowledged at the ratification of the United Nations Framework Convention on Climate Change, recognizes that the largest share of GHGs originated in developed countries, per capita emissions in developing countries are still relatively low, and the share of global emissions originating in developing countries will grow to meet their social and development needs. What this equates to is that developed countries are the most responsible for climate change and that it is unfair to expect third-world countries to make emissions reductions in the same way that the developed countries are expected to make them.

Another important distinction between developed and developing countries is that the GHG emissions from developed countries are generally connected to luxuries, and emissions from third-world countries are for basic needs. An example of GHG emissions from a luxury in a developed country would be those generated as a result of choosing to drive a personal car rather than ride public transportation. An example of GHG emissions from a basic need in a developing country would be those generated as a result of cooking. As explained in a World Resources Institute report, "This is exemplified by the large contrasts in per capita carbon emissions between industrialized and developing countries. Per capita emissions of carbon in the U.S. are over 20 times higher than India, 12 times higher than Brazil, and seven times higher than China" (Shah, 2009).

Supporters of climate justice have also brought up another point that rarely hits the news. Many emissions in countries such as India and China are from wealthy country corporations outsourcing production to these countries. Products are then exported or sold to the wealthy. Currently, however, the "blame" for such emissions are put on the producer, not the consumer. Companies who try to avoid tighter regulations and higher wages in wealthier countries attempt to outsource such production. The issue has been raised that the country responsible for the production of the item should be responsible for the emissions produced as a result of its production.

When it comes to addressing climate justice and equity at climate negotiations, supporters of the cause also claim that, although social justice and equity are part of the climate change issues, they are largely ignored in the discussions because the discussions are generally dominated by the wealthy nations and oil-producing countries, who talk more about economic effectiveness. The explanation for the traditional lack of focus on climate justice and equity is that the topic is only of real concern to the third-world countries. Without as strong a voice as the wealthy countries when it comes to discussion and negotiation, developing nations have their concerns glossed over or passed by completely. As far as to how to fix the inequity, a solid solution remains elusive.

Conclusions

One aspect of climate change that needs to be addressed by all governments is how to handle security issues as climate change continues. This problem requires long-term plans that must be put into effect now in order to account for various scenarios and contingencies. Oftentimes, it is the problems that planners and analysts did not give as much thought to that turn out to be the most difficult to manage and have the furthest-reaching ramifications. Whether the issues are rising sea levels, drought, failed agriculture, or just a homeland that is safer than the one a refugee leaves, serious thought needs to be given, just as much as any safety or emergency plan that is developed for another application or contingency. This is not the time to be remiss or procrastinate. Climate change is already in progress.

There is also validity in the climate-justice and equity theory. Responsibility should be taken in a like manner as to the damage initially done to create the problem. Without a just arrangement, tensions and feelings of hostility arise that could also potentially lead to conflicts that could otherwise be prevented. There are so many issues causing political strife globally, that adding to the mix with climate change could very well be in itself a tipping point the world community would do better without.

Discussion

1. What do you see as the highest security threats for developed nations? How does climate change relate to them?

2. If you were in charge of developing a nation's security plan, how would you go about designing it and whom would you include as your advisors? Why? How would you specifically deal with climate-change issues?

3. For potential security risks, how do you rank public safety, public health, emergency management, critical infrastructure protection, disaster response, communication networks, economic facilities, government/public infrastructure, and natural resources? Why?

4. At the Canadian conference, it was stated that because climate change is a slow-moving threat, many Canadians are unable to imagine the dramatic set of security-related changes that lie ahead. Participants agreed that actions must be taken to raise public awareness of the security implications of climate change. Do you agree with this philosophy? Why or why not? If so, how would you devise an effective public awareness campaign?

5. Lord Nicholas Stern, one of the world's most prominent climate economists, believes that the failure to address global warming could eventually lead to World War III. Do you agree with his assessment? Why or why not?

6. Water is one point of contention for causing conflict as a result of climate change. What other potential triggers can you think of?

7. Do you agree with the concept of climate justice and equity? If so, what role do you think developed countries should take to assist developing nations adopt clean technologies and better prepare for climate change?

8. Do you believe that the countries that outsource their production to other countries should be held responsible for the GHGs produced as a result of the production process? Why or why not? If so, what kind of implementation process could be put in place to fairly manage it?

References

Agence France-Presse. 2009. Australian military warns of climate conflict. *COSMOS Magazine,* January 7. http://www.cosmosmagazine.com/news/2460/australian-military-warns-climate-conflict (accessed April 6, 2011). Discusses Australia's view on military implications with climate change.

Friedman, L. 2009. How will climate refugees impact national security? *Scientific American,* March 23. http://www.scientificamerican.com/article. cfm?id=climage-refugees-national-security (accessed August 11, 2011).

Governance and Social Development Resource Centre. 2007. Climate change, conflict, migration and fragility. http://www.gsdrc.org/go/topic-guides/climate-change-adaptation/climate-change-conflict-migration-and-fragility (accessed August 12, 2011).

Homeland Security NewsWire. 2009. Tensions simmer between India and Bangladesh over dwindling water sources. October 26. Discusses military issues concerning climate change.

Homeland Security NewsWire. 2010. U.S., U.K. military leaders address climate change's role as a global threat multiplier. July 7. http://homelandsecuritynewswire.com/us-uk-military-leaders-address-climate-changes-role-global-threat-multiplier (accessed April 5, 2011). Discusses military issues concerning climate change.

IrinNews. 2009. Lebanon: Climate change and politics threaten water wars in Bekaa. February 1. http://www.irinnews.org/Report.aspx?ReportId=82682 (accessed September 15, 2010). Provides information about the effects climate change is having in Lebanon's Bekaa Valley and a historic feud over irrigation.

Jarvis, L., H. Montgomery, N. Morisetti, and I. Gilmore. 2011. Climate change, ill health, and conflict. *ScienceDaliy,* April 5. Reprinted from *BMJ,* April 5: 342. http://www.sciencedaily.com/releases/2011/04/110405194110.htm (accessed April 6, 2011). Discusses military security issues and climate change in the United Kingdom.

Johnson, B. 2011. Climate inaction risks a new world war. *Grist,* March 10. http://www.grist.org/climate-policy/2011-03-10-nicholas-stern-climate-inaction-risks-new-world-war (accessed April 2, 2011). Discusses Nicholas Stern's views on climate policy.

Muharam, H. 2011. The top 100 effects of global warming. *Global Climate Change News Brief,* January 11. http://hendrawanm.wordpress.com/2011/01/11/global-warming-effects-and-causes-a-top-10-list-climate/ (accessed April 11, 2011). Discusses the effects global warming will have on everyday life.

Schwartz, P., and D. Randall. 2003. An abrupt climate change scenario and its implications for United States national security. http://www.climate.org/topics/PDF/clim_change_scenario.pdf (accessed August 11, 2011).

ScienceDaily. 2011. Climate change threatens global security, warn medical and military leaders. April 5. http://www.sciencedaily.com/releases/2011/04/110405194110.htm (accessed August 13, 2011).

Shah, A. 2009. Climate justice and equity. *Global Issues*. http://www.globalissues.org/article/231/climate-justice-and-equity (accessed November 12, 2010). Discusses the gap in climate change negotiations, responsibilities, and benefits between developed and undeveloped nations.

Smith, D., and J. Vivekananda. 2007. A climate of conflict: The links between climate change, peace and war. London: International Alert. PDF available at http://reliefweb.int/node/22990 (accessed October 5, 2011). Discusses the consequences of climate change on fragile states.

TerraDaily. 2010. Australia military head warns of pacific climate instability. November 3. http://www.terradaily.com/reports/Australia_military_head_warns_of_Pacific_climate_instability_999.html (accessed April 6, 2011). Discusses the preparedness of Australia's military regarding climate change.

Townsend, M., and P. Harris. 2004. Now the Pentagon tells Bush: Climate change will destroy us. *The Observer*, February 22. http://www.guardian.co.uk/environment/2004/feb/22/usnews.theobserver (accessed October 5, 2010). Discusses abrupt climate change and natural security.

Zhang, D. D., P. Brecke, H. F. Lee, Y.-Q. He, and J. Zhang. 2007. Global climate change, war, and population decline in recent human history. *Proceedings of the National Academy of Sciences*. 104(49). http://www.pnas .org/cgi/doi/10.1073/pnas.0703073104 (accessed April 5, 2011). Discusses the impacts of long-term climate change on social unrest and population collapse.

Suggested Reading

Barnett, J., and N. W. Adger. 2007. Climate change, human security and violent conflict. *Political Geography*, 26: 639–655. Discusses the vulnerability of local places and social groups to climate change.

Kahl, C. 2006. *States, Scarcity and Civil Strife in the Developing World*. Princeton, NJ: Princeton University Press. Discusses environmental and demographic stresses.

Smith, P. J. 2007. Climate change, weak states, and the war on terrorism in South and Southeast Asia. *Contemporary Southeast Asia: A Journal of International and Strategic Affairs*, 29(2): 264–285. http://muse.jhu.edu/login?uri=/journals/contemporary_southeast_asia_a_journal_of_international_and_strategic_affairs/v029/29.2smith01.pdf (accessed July 10, 2011). Argues that climate change threatens to undermine the U.S. government's objective of working with well-governed states that have the capacity to cooperate on counterterrorism efforts.

10

Military Issues and Climate Change

Overview

This chapter takes a glimpse into the operations of the U.S. military and shows where it is picking up the slack in leadership that Congress has created and leading in creative and innovative ways of its own. Through consulting with scientists and research organizations, the Department of Defense (DoD) has apparently listened closely and heeded the messages it has been told because it has recognized the seriousness and urgency of climate change and the very real implications it has, and will have, on the security of not only the United States but also the entire world community. In response, it has implemented programs and protocols to respond in a proactive way in its energy choices and direction of the future. This chapter provides insight as to where the military is currently headed on the topic of climate change action. First, it discusses how the military views climate change, the effect it is having on military effectiveness, and what that means in terms of security for the future. Next, it outlines how the Pentagon is taking the lead on cutting back on its enormous use of fossil fuels in order to lower its dependence on hostile nations, keep troops out of harm's way, and build toward a sustainable future. The chapter then illustrates how the DoD is retrofitting its operations to embrace the use of green energy and provides concrete examples of the many successes it has already had and the noteworthy goals it has set for itself for the short and long term.

Introduction

On September 30, 2009, Carol Browner, director of the White House Office of Energy and Climate Change Policy, and Kathleen Hicks, Deputy Under Secretary of Defense for Strategy, Plans and Forces, joined leading military and defense experts to discuss the critical links between climate change, energy, and national security. A key topic discussed was how climate change and the energy situation threaten the country's national security and the ways the U.S. military is preparing to meet these challenges.

According to Senator John Warner (R-VA), "Leading military and security experts agree that global warming could increase instability and lead to conflict in already fragile regions of the world. We ignore these facts at the peril of our national security and at great risk to those in uniform who serve this nation" (MacGillis, 2009). Vice Admiral Dennis V. McGinn, U.S. Navy (Ret.), Member, CNA Military Advisory Board, added, "The Armed Services realize that America's growing dependence on oil isn't just expensive, it can be dangerous. That is why the Department of Defense has taken on the leadership challenge to reduce energy use across the board in order to increase mission effectiveness and save millions of dollars" (MacGillis, 2009).

As a result of these observations, the DoD is currently helping to develop alternative fuel and power sources, which began in earnest in 2008, when they either procured or produced the equivalent of almost 10 percent of their electricity from renewable energy sources. Nellis Air Force Base, for example, has commissioned a 14.2-megawatt solar power array, which was the largest in the Americas (Figure 10.1). In addition, the DoD has reduced its energy use by more than 10 percent since 2003 and aims to improve energy efficiency by 30 percent by 2015.

The DoD says that climate change is one of the key components being examined for its Quadrennial Defense Review. Kathleen Hicks also adds,

FIGURE 10.1
The solar array at Nellis Air Force Base, just outside Las Vegas, Nevada, is one of the largest in North America. It currently provides one-fourth of the base's energy needs. (Courtesy of Airman First Class Nadine Y. Barclay, U.S. Air Force, http://www.af.mil/photos/media_search.asp?q=solar&page=5.)

"Climate change can have critical implications for U.S. national security and for the role and use of U.S. military forces, including by serving as a threat multiplier, exacerbating tensions, and potentially leading to migration or conflict within or between states over scarce resources" (MacGillis, 2009). The military also believes that, in addition, climate change provides opportunities for the United States and U.S. armed forces to enhance their cooperation with regional and international organizations.

In addition, the National Intelligence Council recently reported that climate change could directly impact the United States by threatening energy supplies, damaging military bases, increasing food and water shortages, and stressing the economy. According to Dennis Blair, director of National Intelligence, "The Intelligence Community judges global climate change will have important and extensive implications for U.S. national security interests over the next 20 years" (Blair, 2009).

Because of these numerous implications, the DoD is currently leading the way in becoming more efficient and effective in their use of energy and working to reduce their carbon "boot print," so to speak. This chapter traces the path of the DoD's recognition and growing concern over climate change and the subsequent action the Pentagon has taken to cut back its enormous consumption of fossil fuels and turn to green energy and renewable technology instead.

Development

Climate Change and Military Effectiveness

The Center for a New American Security, a Washington-based think tank that analyzes the interrelationship of natural resources and national security, issued a report in April 2010 citing the impacts from extreme drought, heat waves, desertification, flooding, and extreme weather events as a result of climate change as reasons why the U.S. military needs to be prepared now for a climate change-impacted world of the future (Carmen, Parthemore, and Rogers, 2010). Their report, *Broadening Horizons: Climate Change and the U.S. Armed Forces*, cites that the effects of these environmental events will be amplified by existing sociopolitical factors. It advises, "Countries and regions of strategic importance—from Afghanistan to the Arctic, China to Yemen—are likely to confront major environmental pressures on both their societies and ecosystems." The report also pointed out that phenomena such as desertification leading to humanitarian situations like mass migrations are happening now, and they are currently impacting national security issues.

The armed forces acknowledges that adapting to climate change plays a significant role in day-to-day operations as well, but even with advice from outside organizations, the armed forces have been aware of the need to deal with climate change issues. As part of the effort already started, the DoD released its Quadrennial Defense Review in February 2010, which for the first time ever identified climate change as having an impact on its operations around the world. The report acknowledges that although climate change alone does not cause conflict, it may act as an accelerant of instability or conflict, placing a burden to respond on civilian institutions and militaries around the world. In addition, extreme weather events may lead to increased demands for defense support to civil authorities for humanitarian assistance or disaster response both within the United States and overseas.

The report also details how the military is addressing climate-related issues, both in its own operations—in terms of reducing the military's reliance on fossil fuels—and in helping develop energy efficient and renewable technologies. The Pentagon defines energy security as "assured access to reliable supplies of energy and the ability to protect and deliver sufficient energy to meet operational need." They view this as a strategic priority and one that greener energy sources can help better secure (Berger, 2010).

The report also acknowledges that one key difficulty the military faces in bringing itself up to date with the realities of a changing climate is that they feel they currently lack the data necessary to generate requirements, plans, strategies, training, and material to prepare for future challenges that may be related to climate change. They believe that although the scientific information and data has improved over recent years, it needs to be presented in a more usable form to the decision makers who need it.

At a semi-public forum at Minnesota State University in July 2010, Senator John Warner (R-VA), local activist Leigh Pomeroy, civil rights attorney Ashwin Madia, and St. Paul schoolteacher Alec Timmerman made the case that finding sustainable forms of energy can keep soldiers safer, improve the nation's economy, and keep the country strong for future generations. One consensus that emerged from the meeting was that science was already at the point to support a major overhaul in how the United States thinks about energy. In addition, any citizen who honestly looks at the nation's energy situation would have to conclude that something needs to be done proactively right now. The specialists represented at the forum voiced that an intelligent move was to target national security as a means to bridge the political gap that currently exists in this country that is crippling the progress being made toward revamping the nation's energy policy and views on climate change. According to Pomeroy: "To oppose things that need to be done, frankly, is unpatriotic. Anyone who does is doing so for selfish reasons" (Murray, 2010).

The consensus of the panel was that if climate change action was framed in the context of national security issues in the United States, that would be the most efficient and productive method of getting more people active and on board toward supporting sustainable energy and finding meaningful long-term solutions to the problem. According to Madia, "We know what to do. Many wonder why it's not being done. Politics is at the center of it. It's undeniable that there's a group of our population that do not find the science as compelling as we do. We've got to address the political problem" (Murray, 2010).

One military official believes that climate change hinders the military's effectiveness. Dorothy Robyn, Deputy Under Secretary of Defense for Installations and Environment of the DoD, said during an event at the United Nations climate negotiations in Cancun, Mexico, "Energy dependence and climate change are both threats to our effectiveness as war fighters" (Restuccia, 2010).

The DoD is currently the largest consumer of energy in the United States. In 2009 alone, the DoD spent $13 billion on fuel. Robyn confirmed the DoD is currently working to reduce its greenhouse gas (GHG) emissions and reliance on oil. As an example, she cited two efforts under way; Navy Secretary Ray Mabus is working on options for a fleet of nuclear-powered vessels and biofuel-powered planes.

In an article that appeared in *Scientific American* in March 2011, the U.S. military's heavy reliance on oil—much of it imported—presents a real challenge. According to Rep. Steve Israel (D-NY), "We are reliant on our adversaries for our national security" (Biello, 2011).

For example, consider these facts: an F-16 fighter jet burns 105 liters of jet fuel a minute with its afterburners engaged; the C-17 cargo consumes 11,350 liters an hour (Figure 10.2). The DoD fuel bill came to a whopping $14 billion

FIGURE 10.2
The F-16 burns 28 gallons of jet fuel a minute when its afterburners are engaged—most of it obtained from imported oil. (Courtesy of Master Sgt. Andy Dunaway, U.S. Air Force.)

in 2010. Secretary of the Navy Ray Mabus said, "For every dollar the price of a barrel of oil goes up, the Navy spends $31 million more for fuel. Our dependence on fossil fuels creates strategic, operational and tactical vulnerabilities for our forces" (Biello, 2011).

Because of these enormously high fuel-consumption rates, the DoD and Department of Energy have now partnered on initiatives to further develop and test energy-storage technologies, and the DoD is looking at alternative fuel sources. The navy has taken a lead and set a goal of deriving half of its energy from non-fossil-fuel sources by 2020, as well as making half of its bases self-sufficient in terms of energy. Currently, the navy is engaged in the following:

- They have ordered approximately 150,000 liters of jet fuel derived from *Camelina*—an oil-seed plant like canola.
- They have obtained more than 75,000 liters of diesel-like fuel for ships from algae (an order the U.S. Air Force has matched by requisitioning 150,000 liters of bio-jet fuel).
- They have obtained their first hybrid electric-drive ship, which uses electric motors for speeds under 12 knots. It already saved $2 million in fuel costs on its maiden voyage from Pascagoula, Mississippi, to San Diego.

According to Mabus, concerning the navy's history of fuel switches—from wind to coal in the nineteenth century and coal to oil supplemented by nuclear over the course of the twentieth century, "I am confident—as we lead again in changing the way we power our ships and aircraft—that the naysayers who say, 'it's too expensive, the technology is not there,' are going to be proven wrong again" (Biello, 2011). Mabus is confident that the switch to alternative fuels will make the troops more efficient and better fighters.

The Pentagon Takes the Lead on Cutting Back on Fossil Fuels

The U.S. Navy celebrated Earth Day 2010 by test flying its main attack aircraft, the F/A-18 Super Hornet, on a biofuel blend, which was part of an ambitious push by the Pentagon to increase U.S. security by using less fossil fuels. Although Congress may be waffling about moving forward and enacting climate change policy, the U.S. military appears to be moving full speed ahead in making transitions to renewable energy and cutting back on their tremendous use of fossil fuels. In addition to the electric-drive ships that are saving enormous fuel costs, they are also utilizing solar-based water purification in Afghanistan, which is reducing the need for dangerous convoys and keeping the troops safer, as well as employing solar and geothermal energy sources on home soil.

There are many innovations that we all take for granted that originated with the military—usually offshoots from wartime. One advantage civilians may benefit from is that the size of the military's investment will create economies of scale that will help bring down the costs of renewable energy, and military innovations in energy technologies could spread to civilian uses, in a similar way that the Internet and GPS technology did. An added bonus is that by using alternative fuels, the military will help the United States cut back on the overall emissions of GHGs. The U.S. military could just be the route to lead the United States down the road to energy independence.

One of the big concerns the military faces, according to Amanda J. Dory, Deputy Assistant Secretary of Defense for Strategy in the Office of the Secretary of Defense, is the ability of adversaries to attack energy supplies. She also expresses concern that climate change trends that will cause resource scarcity, environmental destruction, and other problems will occur even under conservation initiatives. The DoD's concern is that they do not have the luxury of waiting for a "100 percent certainty" before making their decisions. In their favor, they are veterans at dealing with both complexity and uncertainty and are ready and able to take measures to deal with climate change issues and prepare the troops to respond now (Schoof, 2010).

DoD Embraces Green Energy: Turning Goals into Examples

The U.S. military defines itself as having a broad mission, which includes managing ambiguous or incomplete pieces of information, anticipating threats, and keeping Americans safe. They are also responsible, in tandem with the intelligence community, for monitoring and analyzing information and factors that could destabilize foreign states or that may require humanitarian assistance. It is in this context that defense specialists and the military categorize climate change and the U.S. "energy posture"—how the DoD approaches its energy use, consumption, costs, and sources and how these patterns, in turn, affect the readiness of the armed forces. This, in turn, also affects how the troops are called on for support domestically and internationally. At home, the armed forces could be needed to support civil authorities, such as during Hurricane Katrina. Abroad, the military's services could be required for a variety of humanitarian and security endeavors, from natural-disaster response to assisting nations that are suffering from famine or desertification.

For these reasons, the Pew Charitable Trusts teamed up with the DoD and produced a report called "Reenergizing America's Defense: How the Armed Forces Are Stepping Forward to Combat Climate Change and Improve the U.S. Energy Posture," which provides an overview of the important initiatives the DoD has undertaken in order to lead in energy strategies and technologies. It includes endeavors in energy conservation, renewable energy investments, and digital-grid research. The military is currently working to better understand the challenges involved in transitioning to a greener future in order to obtain a more secure energy future (Pew Trusts, 2010). Some of the highlights in the report include the following:

- The army plans to have four thousand electric vehicles in the next 3 years, which is one of the biggest electric fleets in the world (Figure 10.3).

- The air force plans to provide 25 percent of the energy at its bases with renewable energy by 2025 and use biofuel blends for half of its aviation fuel by 2016.

- The navy plans to launch a strike group, called the Great Green Fleet, by 2016 that runs entirely on non-fossil-fuel energy, including nuclear ships, combat ships that run on hybrid electric power systems using biofuels, and aircraft that fly only on biofuels.

- In Afghanistan, the navy is moving toward more solar and wind energy so that it can reduce reliance on the fuel convoys, keeping troops out of danger. The solar-powered water-purification units are reducing the need for fossil fuel to clean water and for purified water brought in by truck.

- The navy is planning to have half of its bases generate all their own energy by 2020.

- A 30-megawatt geothermal project is being developed at Hawthorne Army Depot in Nevada.

FIGURE 10.3

The army's transition to four thousand electric vehicles within the next 3 years is part of its energy-efficiency initiative. (Courtesy of the U.S. Army.)

- The army is investing American Recovery and Reinvestment Act funds in lighter, more deployable power systems and microgrids for use in forward operating bases.

The air force was spending $10 million *per day* on energy and consumes 2.5 billion gallons of aviation fuel a year, making it the DoD's largest energy user. To help reduce its energy use, it created a comprehensive energy program and policy, the "Air Force Energy Plan 2010." The object of the plan is to "make energy a consideration in everything they do" (United States Air Force, 2010). The plan focuses on three main goals:

- Reduce demand across aviation, ground operations, and installations. For example, flight simulators will be used for training purposes to save fuel, and flight routes will be reconfigured when possible.
- Increase the supply of renewable and alternative energy for aviation, ground operations, and installations. They plan to meet 25 percent of base energy needs with renewable energy sources by 2025 and obtain 50 percent of their aviation fuels from biofuel blends by 2016.
- Change the culture by changing the mindset of air force members. This includes energy-awareness training and education.

Navy Secretary Ray Mabus commented that the biggest obstacles in general concerning the implementation of cleaner energy are the lack of infrastructure in place and the high price of alternative fuels. He also iterated that both hurdles would fall as the navy helps build a demand for them. He believes that once the military puts the infrastructure in place, the use of alternative fuels will take off with the general public (Schoof, 2010).

Each service branch has established institutional capabilities for developing the plans and policies needed to reduce energy demand, increase supplies of alternative-energy sources, and ensure that U.S. troops have the best technologies to complete their missions. The DoD is determined to meet these challenges and keep the United States strong and lead by example.

Conclusions

Because the DoD is the world's single largest consumer of fossil fuels, its move toward green technology could not come at a better time. As Congress quibbles and defeats measure after measure, the DoD's progress is welcome news indeed. It is even better that they are not just making promises and projections for the future—they are accomplishing those goals right now, and where they do make future projections, they have included near-, mid-, and long-term goals, not the long-term time frames Congress always seems to plan.

There is no hiding the fact that the military currently uses an enormous amount of fossil fuels in conjunction with their efforts in the Middle East and elsewhere. In fact, according to the CIA Factbook, if the U.S. military were a nation state, it would be ranked number thirty-seven in terms of oil consumption—ahead of the Philippines, Portugal, and Nigeria (CIA, 2011). There are many critics of the military for using the amount of energy they do, which as the statistics show is extremely excessive, but that is where the debate begins to branch into multiple issues. Oftentimes, those on the negative side of the argument are critics of the military's activities to begin with. However, with the issue at hand here—climate change—the steps the military is taking are positive ones. At the rate of fuel consumption they presently use, when they have transitioned to 25–50 percent alternative fuels, that adds up and significantly cuts back GHG emissions.

Currently, the DoD uses 93 percent of all U.S. government fuel consumption. The air force is the largest consumer of fuel energy in the federal government and also uses 10 percent of the nation's aviation fuel. By 2016, it plans to fuel half of its domestic transportation by U.S.-produced synthetic blends. The air force is currently the leading purchaser of renewable energy within the federal government and has been a longtime pioneer of renewable energy development. The U.S. Army has recently prioritized renewable energy programs in Iraq. Examples of the programs they are involved in include the Tactical Garbage to Energy Refinery Program, which currently converts 0.91 metric tons of waste to 42 liters of JP-8 fuel; an insulating-foam technology; hybrid-electric manned ground vehicles; and highly efficient portable cells (Vogel, 2009).

Promising for consumers are the innovations that come from the military, which is where many of our health, safety, communications, and even entertainment items have originated, such as pacemakers, water filters, scratch-resistant lenses, memory foam, shoe insoles, cordless tools, safety grooving in pavement, adjustable smoke detectors, satellite radios, GPS, and international calling. Although the military operates under a healthy budget, which serves as a point of contention for many people, there are more benefits realized by the general public than what the military often gets credit for, and perhaps the efforts of the military will give renewable energy infrastructure the boost it needs to benefit the world and help halt climate change at the same time.

Discussion

1. Ashwin Madia made the following comment at the public forum in Minnesota about climate change action, national security, and mitigation inaction: "We know what to do. Many wonder why it's not being done. Politics is at the center of it. It's undeniable that there's

a group of our population that do not find the science as compelling as we do. We've got to address the political problem." Do you believe that the reason the United States is behind other countries in controlling their GHG emissions is largely political in nature? Why or why not? If so, what should be done to change this?

2. Madia also said maybe the United States should spend more money on climate change research instead of giving "it to countries who don't like us." Do you agree? Why or why not?

3. Why is the military a likely candidate to lead the way with the implementation of alternative fuels and green energy?

4. Besides the Internet and GPS technology, can you think of any other military-based technology from which the general public has benefited?

5. Do you think the military is a likely candidate to be a major leader in alternative energy?

6. What concerns do you think other countries have about the work and progress that the U.S. military is making along these lines?

References

Berger, M. 2010. Climate concerns spur changes in U.S. military. *IPS Inter Press Service*, April 29. http://www.ipsnews.net/print.asp?idnews=51259 (accessed April 6, 2011). Outlines the thoughts of the U.S. military on the effects of climate change and the need for action.

Biello, D. 2011. U.S. military links alternative energy research to lives—and dollars—saved. *Scientific American*, March 3. https://www.scientificamerican.com/article.cfm?id=alternative-energy-research-saves-lives (accessed April 6, 2011). Explores how the military looking at the future of energy may be key to improving U.S. national security, economic prosperity, and health.

Blair, D. C. 2009. Annual threat assessment of the intelligence community for the Senate Select Committee on Intelligence. Senate Select Committee on Intelligence, February 12. http://www.dni.gov/testimonies/20090212_testimony.pdf (accessed April 11, 2011). Discusses the threat climate change poses to military bases, food supplies, and the economy in the United States.

Carmen, H., C. Parthemore, and W. Rogers. 2010. *Broadening Horizons: Climate Change and the U.S. Armed Forces*. Washington, DC: Center for a New American Security. http://www.cnas.org/node/4374 (accessed August 13, 2011).

Central Intelligence Agency. 2011. *The World Factbook*. https://www.cia.gov/library/publications/the-world-factbook/ (accessed August 11, 2011).

MacGillis, B. 2009. Climate change poses critical risks for national security. The Pew Charitable Trusts, September 30. http://www.pewtrusts.org/news_room_detail.aspx?id=55256 (accessed April 7, 2011). Discusses the risks associated with national security caused by climate change.

Murray, R. 2010. Energy, national security intertwined? *The Free Press*, July 15. http://mankatofreepress.com/local/x540035009/Energy-national-security-intertwined (accessed April 6, 2011). Outlines a forum held that discussed the benefits of sustainable energy for the country's future.

Pew Charitable Trusts. 2010. Reenergizing America's defense: How the armed forces are stepping forward to combat climate change and improve the U.S. energy posture. The Pew Project on National Security, Energy and Climate. http://www.pewtrusts.org/our_work_report_detail.aspx?id=58542&category=919 (accessed November 2, 2010). Discusses the actions the DoD is taking to rekey their strategies and technologies and gear toward sustainable energy sources.

Restuccia, A. 2010. DoD official: Climate change hinders military's effectiveness. *The Hill*, December 9. http://thehill.com/blogs/e2-wire/677-e2-wire/132961-dod-official-climate-change-hinders-militarys-effectiveness (accessed April 6, 2011). Presents information concerning the U.S. military's stance on climate-change and energy-dependence ramifications.

Schoof, R. 2010. Pentagon taking the lead on cutting back on fossil fuels. *McClatchy Newspapers*, April 21. http://www.mcclatchydc.com/2010/04/20/v-print/92532/pentagon-taking-the-lead.html (accessed April 6, 2011). Provides an overview of the Pentagon's plans to increase national security by using less fossil fuels.

United States Air Force. 2010. Air Force energy plan. Air Force Energy Program Management Office. http://www.safie.hq.af.mil/shared/media/document/AFD-091208-027.pdf (accessed August 14, 2011).

Vogel, S. 2009. Pentagon prioritizes pursuit of alternative fuel sources. *The Washington Post*, April 13.

Suggested Reading

Ackerman, S. 2011. Afghanistan's green Marines cut fuel use by 90 percent. *Wired*, January 13. http://www.wired.com/dangerroom/2011/01/afghanistans-green-marines-cut-fuel-use-by-90-percent (accessed April 6, 2011). Provides news coverage about the U.S. Marines in the Helmand Province in Afghanistan building a solar-energy generator.

Beidel, E. 2011. Navy takes biofuels campaign into uncharted waters. *National Defense Industrial Association*, January. Discusses the Navy's endeavors into biofuel use, research, and implementation.

Drummond, K. 2010. Green Marines phase out fuel, take solar generators to war. *Wired*, September 22. http://www.wired.com/dangerroom/2010/09/green-marines-phase-out-fuel-take-solar-generators-to-war (accessed April 6, 2011). Discusses how marines are phasing out fuel in favor of solar generators.

Gardner, T. A. 2010. U.S. military leads climate change combat. The Pew Charitable Trusts, April 20. http://www.pewtrusts.org/news_room_detail.aspx?id=58610 (accessed April 7, 2011). Outlines how the military, as the government's largest fuel consumer, is transitioning over to a renewable fleet.

Gerson, N. L. 2007. National security and the threat of climate change. CAN Corporation. http://securityandclimate.cna.org/report (accessed April 11, 2011). Discusses the national security challenges that global climate change presents.

National Intelligence Council. 2008. Global trends 2025: The National Intelligence Council's 2025 project. http://www.dni.gov/nic/NIC_2025_project.html (accessed April 6, 2011). Provides information about how key global trends might develop over the next 15 years to influence world events.

Perlin, J. 2010. Greener battlefields would be safer for troops. Miller-McCune, October 28. http://www.miller-mccune.com/science-environment/greener-battlefields-would-be-safer-for-troops-24716 (accessed April 6, 2011). Discusses the hazards of driving refueling tankers in the field and the necessary transition to alternative fuels.

Pew Charitable Trusts. 2010. Reenergizing America's defense: How the armed forces are stepping forward to combat climate change and improve the U.S. energy posture. The Pew Project on National Security, Energy and Climate. http://www.pewtrusts.org/our_work_report_detail.aspx?id=58542&category=919 (accessed November 2, 2010). Discusses the actions the DoD is taking to rekey their strategies and technologies and gear toward sustainable energy sources.

Rosenthal, E. 2010. U.S. military orders less dependence on fossil fuels. *New York Times*, October 4. http://www.nytimes.com/2010/10/05/science/earth/05fossil.html?pagewanted=pring (accessed April 6, 2011). Discusses the military's new stance on fossil fuel dependence and the state of vulnerability in which it puts the U.S.

Sabin-Wilson, L. 2009. Turning seawater into jet fuel. Impact Lab, August 20. http://www.impactlab.net/2009/08/20/turning-seawater-into-jet-fuel (accessed April 10, 2011). Outlines a process the navy is working on to convert seawater into a kerosene-based jet fuel.

Snider, A. 2010. Navy is sailing along with climate change planning. *Politics Daily*, May 28. http://www.politicsdaily.com/2010/05/28/navy-is-sailing-along-with-climate-change-planning (accessed April 6, 2011). Discusses the navy's progress on climate change issues.

11

Economics of Climate Change and Socioeconomic Implications

Overview

Two of the most significant components of climate change that must be kept near the forefront of the issue are economics and the socioeconomic implications associated with any relative economic decisions made. This chapter discusses those components. It begins by looking at how climate change relates to economic considerations overall and then delves into specific sectors of the economy to see how they will be affected as temperatures begin to change, rainfall and wind patterns shift, and seasonal timing is altered. The chapter will explore the impacts of drought and desertification on the economy, what can be expected in the future for the fishing and forestry industries, how changes in the world's monsoon patterns will affect the lives of millions of people, what it means for agriculture and food production, how significant the impacts to the recreation and tourism industry are expected to be, how it will now affect the healthcare system, and what it means for the transportation and energy industries. Finally, this chapter examines the economic effects that all these impacts will have—and are currently having—on the insurance industry and who is carrying that economy burden, the Stern Review and what it foretells, and finally how we, as a world community, can meet the challenges of climate change impacts.

Introduction

As the scientific community continues to make advances in the scientific aspect of climate change, which in turn assists policy makers in their decisions for national and global responses, it also fine-tunes the economic perceptions and options available to adapt, mitigate, and plan for the future by being able to evaluate a range of beneficial and educated outcomes. It is too easy for inaction to be motivated by the perceived high cost of mitigation and adaptation, which in the past has been the default policy option for many.

The direct costs of not taking on the challenges posed by climate change are often neglected or ignored—and unfortunately, in the past, have not been calculated into a long-term solution. The indirect effects of climate change are considered even less frequently in many circumstances, yet they are generally substantial in the long run. In addition, the true economic impact of climate change is full of "hidden" costs. In reality, it is not just the replacement value of infrastructure, as it may appear at first glance—there are also costs of components, such as workdays and productivity lost, provisions of temporary shelter and supplies, potential relocation and retraining costs, transportation costs, and other issues specific to the situation.

Added to these costs, the range of various climate change possibilities— from rising sea levels to stronger and more frequent extreme weather—will also have impacts on the natural environment as well as human-made infrastructure and their ability to contribute to economic activity and quality of life. Impacts will vary globally and across different sectors of the economy and are often linked with each other, making the chain reactions and effects difficult to predict. Because of all these interrelationships, it is imperative to gain a working knowledge of the economic interactions of the various components involved in global climate change. Through an understanding of these complicated interactions, it is possible to plan for the future in the most efficient way possible.

Development

Climate Change and Economic Considerations

The economic effects of climate change touch virtually every sector of the economy and have far-reaching effects. Climate change affects agriculture and food production, which affect domestic and global trade. It affects the fishing and forestry industries. The issues of sea-level rise and intense weather events affect tourism and the recreation, construction, and insurance industries. Climate change also impacts transportation networks, has energy considerations, and relates to healthcare and labor issues. Because of its far-reaching impacts, it cannot be ignored. Instead, it must be dealt with now in order to plan for, mitigate, and adapt as efficiently and wisely as possible in order to reduce negative impacts, conserve resources, and provide for future generations.

As an economic issue, standard cost-benefit analysis can be applied to the climate change issue. In order to do this, it is necessary to establish a method of determining the costs and benefits and a method of determining whether a cost-benefit analysis is appropriate through the willingness of a society to place a value on goods and services. With climate change, however, it can be difficult to assign values to some of the costs and benefits because of the nature of some of the impacts. For example, how do you assign a value to an ecosystem? Or to a particular species? Or to human health? Based on research by economist S. J. DeCanio, it is also impossible to know the value future generations may place on a particular asset, which also affects the valuation of costs and benefits (DeCanio, 2007).

Another issue to consider is compensation between regions and populations. If some countries benefit from future climate change but others experience hardship and losses, there is no guarantee that the winners would compensate the losers (Intergovernmental Panel on Climate Change, 2001).

Therefore, the inherent uncertainties of the future potential impacts of climate change present a major problem in its management and the resultant costs and benefits of actions taken in response. Two possible ways to handle the situation include iterative risk management and sequential decision making. The iterative risk-management approach may involve looking at the situation in a "low-probability/worst-case scenario" light. An approach-based method looks at the situation with the mindset that over time, decisions related to climate change can be revised as new, improved information is developed. This approach works well with climate change because climate change is such a long-term problem, and scientists are still learning about it; this approach offers a greater degree of flexibility and also allows for improved information over time, enabling the approach to adjust for that.

The Climate Change Science Program, in a study conducted in 2009, suggested two related decision-making management strategies that may work well when faced with an issue that has high uncertainty, such as climate change. The first were resilient strategies, which strive to identify a range of possible future circumstances and then choose approaches that work reasonably well across the entire spectrum. The second were adaptive strategies, which strive to choose methods that can be improved as more is learned (Climate Change Science Program, 2009).

According to Smit et al. (2001) climate change impacts can be measured as economic costs, especially those that are connected to market transactions and that directly affect the gross domestic product (GDP). Monetary measures of nonmarket impacts on human health and ecosystems are more difficult to calculate. Smit et al. have also identified the following issues as difficulties:

- Vulnerability: There is a limited understanding of the potential market sector impacts of climate change in developing countries.
- Adaptation: The future level of adaptive capacity in human and natural systems to climate change will affect how society will be impacted by climate change. As a result, assessments may under- or overestimate adaptive capacity.
- Socioeconomic trends: Future predictions of development affect estimates of future climate change impacts.

- Knowledge gaps: Calculating distributional impacts requires detailed geographical knowledge, but this is a major source of uncertainty in climate models.

The Intergovernmental Panel on Climate Change (IPCC) defined adaptation to climate change as "initiatives and measures to reduce the vulnerability of natural and human systems against actual or expected climate change effects." They defined vulnerability to climate change as "the degree to which a system is susceptible to, and unable to cope with, adverse effects of climate change, including climate variability and extremes" (IPCC, 2007).

Smit et al. (2001) describe adaptive capacity as the ability of a system to adjust to climate change. They describe the determinants of adaptive capacity:

- Economic resources: Wealthier nations are better able to bear the costs of adaptation to climate change than the poorer ones.
- Technology: Lack of technology can impede adaptation.
- Information and skills: Information and trained personnel are required to assess and implement successful adaptation options.
- Social infrastructure: The availability, type, and condition of infrastructure—such as housing, drainage systems, roads, and water-storage systems—affects a community's ability to adapt.
- Institutions: Nations with well-developed social institutions have greater adaptive capacity than those with less effective ones.
- Equity: Nations with government institutions in place that allow an equitable access to resources have a greater adaptive capacity than nations that do not.

Based on these criteria, Smit et al. (2001) concluded that developing countries with limited resources, low levels of technology, inadequate infrastructure, unstable institutions, and inequitable access to resources cannot adapt well and are highly vulnerable to climate change. Developed nations, on the other hand, have a much greater adaptive capacity (Smit et al., 2001).

The work done by Richard S. J. Tol (2009) shows that climate change affects nearly every component of the economy, directly or indirectly. For instance, he says weather affects agriculture, energy use, health, and many aspects of nature—which in turn affects everything and everyone. As a result, the causes and consequences of climate change are very diverse, and those in low-income countries who contribute least to climate change are most vulnerable to its effects. Climate change is also a long-term problem. Some greenhouse gases have an atmosphere lifetime measured in tens of thousands of years, and each year humans release billions of metric tons of greenhouse gases into the atmosphere. As a graphic representation of how much is released, he said "if all the emissions were priced, as of January 2009, the amount of

CO_2 released into the atmosphere equaled 1.5 percent of the world income" (Tol, 2009). Tol also explains that many factors influence the economic aspect of climate change, and although scientists do not have it all figured out yet, they at least can identify their areas of ignorance on the subject.

According to Tol, any study of the economic effects of climate change begins with some assumptions of future emissions, the extent and pattern of warming, and other possible aspects of climate change such as sea-level rise and changes in rainfall and storm patterns. Any studies done must then translate from climate change to economic consequences. There are many methods available in which to accomplish this: statistical analysis, natural science papers and research, climate models, and lab experiments. The physical impacts must then each be given a price and totaled. For example, the effect of sea-level rise is comprised of additional coastal protection and land lost, which includes the cost of dike construction, the value of land, decisions about which properties to protect, health and welfare needs, and other human needs. Climate models may also be used to estimate the future effect of climate change, which can then be extrapolated to other countries and then be added up. This method assumes that spatial patterns will remain constant, however.

What is interesting about Tol's research on the estimates of the welfare impact of climate change is that he looked at several studies done of climate change using various research methods: literature search, interviews, models, and statistical analysis, and the results were surprisingly similar, as shown in Table 11.1. In Table 11.1, the second column shows the long-term increase in temperature. The third column shows the effect on welfare at that future time, expressed as percentages of income. For example, the entry for Nordhaus and Boyer (2000) estimates that the effect of 2.5°C temperature increase is as bad as losing 1.5 percent of income. The next column shows the percentage change in annual GDP of the regions hardest hit by climate change, with the adjacent column showing the region associated with it. The last two columns show the percentage change in GDP for regions that are least hurt by climate change (and in most cases would even benefit from a warmer climate) and where those regions are.

Tol's conclusions from his study are that the welfare effect on the current economy is relatively small—a few percentage points of GDP—and that this loss of output can look large or small, depending on context. From one perspective it is about equivalent to a year's growth in the global economy, which might suggest that over a century or so, the economic loss from climate change is not all that significant. He points out, however, that the damage is not negligible. An environmental issue that causes a permanent reduction of welfare, lasting into the indefinite future, is significant. In addition, models suggest that in places where there may be some initial benefits of warmer temperatures, they are short-lived, and those positives quickly turn into negatives, causing prolonged environmental damage. Based on research, the turning point in terms of economic benefits occurs at about 1.1°C warming. In addition, low-income countries are typically less able to adapt to climate

TABLE 11.1

Estimates of the Impact of Climate Change on Society

Study	Warming (°C)	Impact (% of GDP)	Region Worst Off (% of GDP)	Region Worst Off (Region)	Region Best Off (% of GDP)	Region Best Off (Region)
Mendelsohn et al. (2000)	2.5	0.0 0.1	−3.6 −0.5	Africa	4.0 1.7	Eastern Europe and the former Soviet Union
Nordhaus and Boyer (2000)	2.5	−1.5	−3.9	Africa	0.7	Russia
Tol (2002)	1.0	2.3 (1.0)	−4.1 (2.2)	Africa	3.7 (2.2)	Western Europe
Maddison (2003)	2.5	−0.1	−14.6	South America	2.5	Western Europe
Rehdanz et al. (2005)	1.0	−0.4	−23.5	Sub-Saharan Africa	12.9	South Asia
Hope (2006)	2.5	0.9 (−0.2 to 2.7)	−2.6 (−0.4 to 10.0)	Asia (without China)	0.3 (−2.5 to 0.5)	Eastern Europe and the former Soviet Union

Source: Adapted from Tol, R., *Journal of Economic Perspectives*, 23, 2, 2009.
Mendelsohn et al. (2000) only include market impacts. Rehdanz et al. (2005) only consider market impacts on households. The numbers used by Hope (2006) are averages of previous estimates by Fankhauser and Tol (1996).

change because of a lack of resources, and they have less capable facilities to support a changing climate.

Economic Impacts of Drought and Desertification

Two topics brought into the public light because of climate change are drought and desertification. No longer are these abstract problems heard occasionally on the news that only plague Africa and Asia. These are now global problems experienced by developed nations, as well, such as the massive, lengthy droughts that have stuck the southwestern United States in recent years. In fact, it is estimated that 6 million hectares are lost to desertification, along with 24 billion metric tons of topsoil lost to erosion each year.

The arid Southwest in the United States, for example, is projected to become even drier in this century. To date, there is emerging evidence that the process has already begun (Barnett et al., 2008). The existing deserts in the United States are projected to expand to the north and east, as well as upward in elevation as the climate changes. Increased drying contributes to several changes that lead to desertification. Increased drought conditions cause perennial plants to die because of water stress and low tolerance to plant diseases. Simultaneously, non-native grasses have invaded the region,

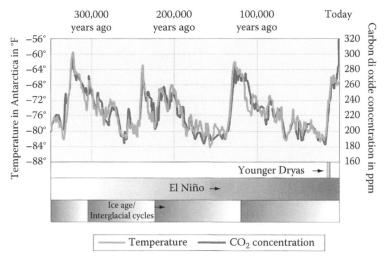

FIGURE 1.1

Fluctuations in temperature (orange line) and in the amount of carbon dioxide concentrations in the atmosphere (blue line) over the past 350,000 years. The temperature and carbon dioxide concentrations at the South Pole run roughly parallel to each other, showing the strong correlation between the two. (From Casper, J. K., *Global Warming Trends: Ecological Footprints,* Facts on File, New York, 2009.)

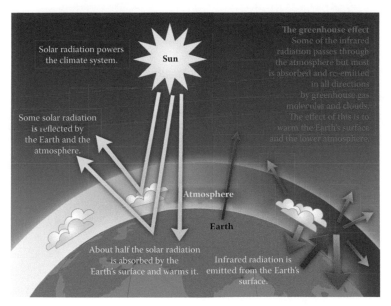

FIGURE 1.2

This diagram illustrates the earth's natural greenhouse effect. The natural greenhouse effect is what makes earth a habitable planet—if the greenhouse effect did not exist, the earth would be too cold, and life could not exist. As more greenhouse gases are added to the atmosphere through the burning of fossil fuels, deforestation, and other measures, however, less heat is able to escape to space, warming the atmosphere unnaturally—causing a situation called the enhanced greenhouse effect. This is what is contributing to climate change today.

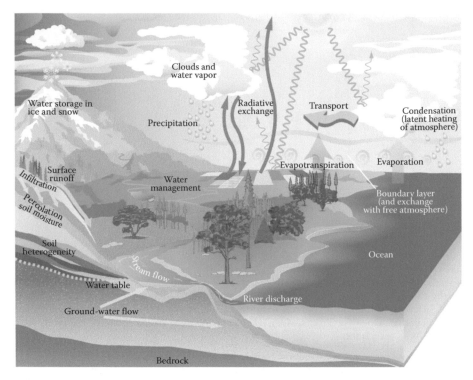

FIGURE 1.3
The earth's hydrologic cycle. (Courtesy of NASA, NASA energy and water cycle study, http://
news.cisc.gmu.edu/NEWS%2005%20Discovery%20and%20Product%20projects.htm, 2011.)

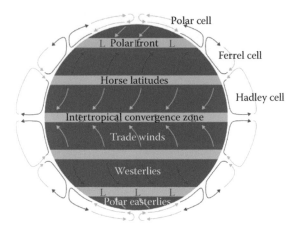

FIGURE 1.4
This diagram represents the earth's major atmospheric circulation patterns. Major wind systems, such as the trade winds and westerlies lie between permanent bands of high or low
pressure located at specific latitudes. (Courtesy of NASA, The water planet: Meteorological,
oceanographic and hydrographic applications of remote sensing, Section 14 of *Remote Sensing
Tutorial*, http://rst.gsfc.nasa.gov/Front/tofc.html, 2011.)

FIGURE 1.5
On August 5, 2010, an enormous chunk of ice, approximately 251 square kilometers in size, broke off the Petermann Glacier along the northwestern coast of Greenland. According to climate experts at the University of Delaware, the Petermann Glacier lost about one-fourth of its 70-kilometer-long floating ice shelf. The recently calved iceberg is the largest to form in the Arctic in 50 years. (Courtesy of NASA, Ice island caves off Petermann Glacier, http://earthobser vatory.nasa.gov/NaturalHazards/view.php?id=45207, 2010.)

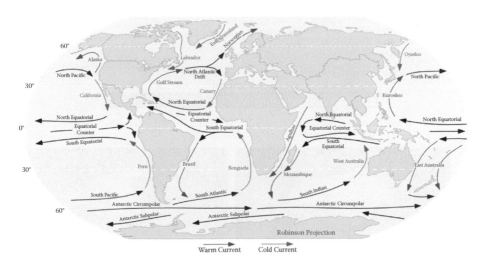

FIGURE 1.6
The earth's major ocean currents are responsible for the global transport of heat. Without them, many areas would be much cooler than they currently are.

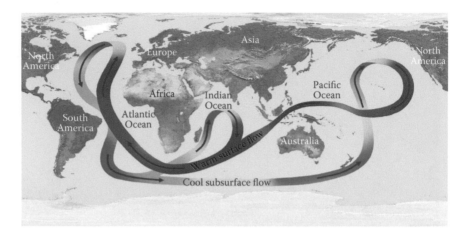

FIGURE 1.7
The Great Ocean Conveyor Belt is the major transport mechanism of heat in the ocean. If its flow were disrupted, it could trigger an abrupt climate change, such as an ice age in western Europe. (Courtesy of NOAA, Ocean facts, http://oceanservice.noaa.gov/facts/coldocean.html, 2011.)

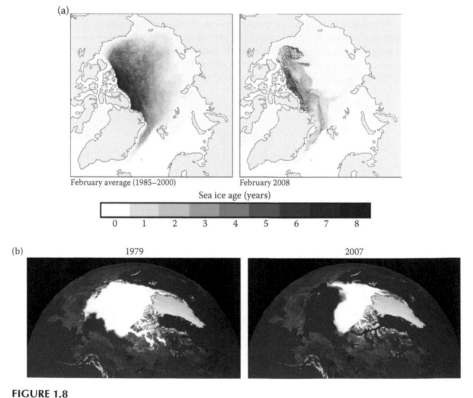

FIGURE 1.8
In the Arctic, sea-ice extent fluctuates with the seasons. It reaches its peak extent in March, near the end of Northern Hemisphere winter, and its minimum extent in September, at the

end of the summer thaw. In September 2007, Arctic sea-ice extent was the smallest area on record since satellites began collecting measurements about 30 years ago. (a) Although a cold winter allowed sea ice to cover much of the Arctic in the following months, this pair of images shows the drastic change in conditions. On the right (February 2008), the ice pack contained much more young ice than the long-term average (left). In the past, more ice survived the summer melt season and had an opportunity to thicken over the following winter. The area and thickness of sea ice that survives the summer has been declining over the past decade. (Courtesy of NASA, NASA and the International Polar Year, http://www.nasa.gov/mission_pages/IPY/multi media/ipyimg_20080326.html, 2008.) (b) This shows the comparison between the September annual minimum of sea ice in 1979 and the September image from 2007, illustrating the drastic decline in ice. (Courtesy of NASA, "Remarkable" drop in Arctic Sea ice raises questions, http://www.nasa.gov/vision/earth/environment/arctic_minimum.html, 2008.) (c) Since 1880 and the rise of industrialization, the land and ocean temperatures have steadily climbed upward, emphasizing the anthropogenic effect of climate change. (Courtesy of NASA/GISS, Global annual mean surface air temperature change, http://data.giss.nasa.gov/gistemp/graphs/, 2006.) (d) The average monthly Arctic sea-ice extent has steadily decreased by 2.5 million square kilometers since 1979. (Courtesy of National Snow and Ice Data Center, Arctic Sea ice extent remains low; 2009 Sees third-lowest mark, http://nsidc.org/news/press/20091005_minimumpr.html, 2009.)

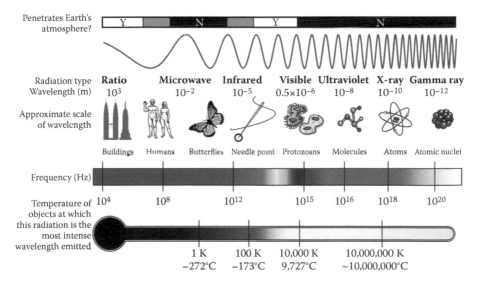

FIGURE 2.1
The sun's electromagnetic spectrum ranges from short wavelengths, such as x-rays, to long wavelengths, such as radio waves. The majority of the sun's energy is concentrated in the visible and nearly visible portion of the spectrum—the wavelengths located between 400 and 700 nm.

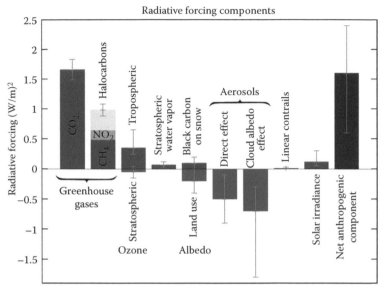

FIGURE 2.3
This graphic illustrates the concept of radiative forcing. In order to understand climate change, it is necessary to understand the components in the atmosphere that control the overall warming and cooling on short- and long-term bases. (Adapted from NASA/GISS, Forcings in GISS climate model, http://data.giss.nasa.gov/modelforce/.)

FIGURE 3.1
This is the Muir Glacier located in Alaska. The left image was taken in 1941, and the right image was taken in 2004. The massive melting that has taken place is attributed largely to anthropogenic warming of the atmosphere. (Courtesy of National Snow and Ice Data Center; left photo by William O. Field; right photo by Bruce F. Molnia.)

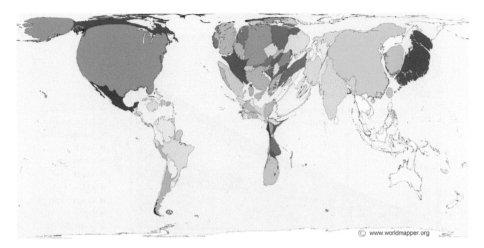

FIGURE 5.7
A cartogram depicting world global warming. This illustrates the world in terms of carbon emissions. The United States, Europe, and China are disproportionately large, whereas Africa is barely visible. (Courtesy of SASI Group, University of Sheffield, and Mark Newman, University of Michigan, http://www.worldmapper.org/display.php?selected= 295, 2006.)

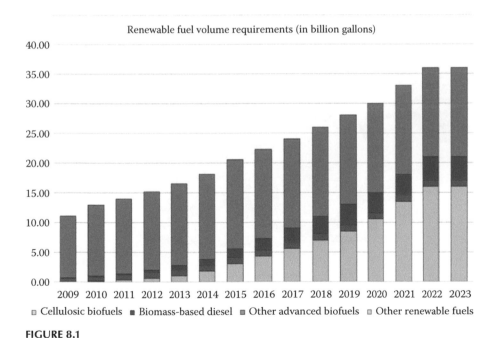

FIGURE 8.1
This schematic illustrates the working plan for the Renewable Fuel Standard 2 (RFS2), which contains a four-part mandate for life-cycle GHG-emissions levels relative to the 2005 baseline of petroleum for renewable fuel, advanced biofuel, biomass-based diesel, and cellulosic biofuel.

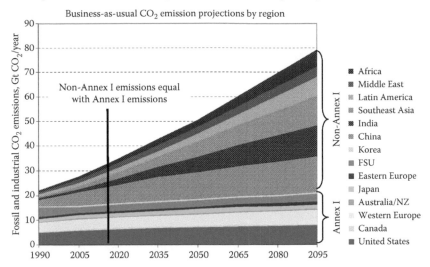

FIGURE 8.2

This figure shows the "business-as-usual" projections of major GHG-emitting countries throughout the remainder of this century, illustrating why it is imperative that the Annex II countries—those designated in the Kyoto Protocol as not needing to be mandated by emissions control—be held accountable for their emissions. China, for example, is rapidly developing industrially and is now emitting enormous amounts of GHGs into the atmosphere. Without future control of the Annex II countries, the implications of future CO_2 levels (and temperature rise over the 2°C level) are unmanageable.

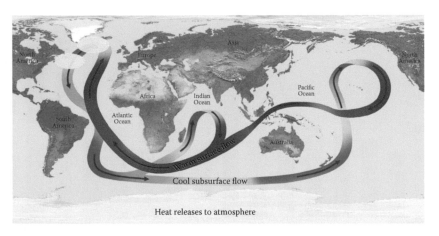

FIGURE 9.1

Working as a massive conveyor belt of heat, the ocean thermohaline circulation has a significant effect on weather worldwide. As global warming continues to heat up the planet, many scientists are worried that the addition of freshwater to the ocean from the melting of the Greenland ice sheet could stop the North Atlantic conveyor. If it did shut down, or even slow down, it would send colder temperatures to Europe and cause other sudden climate changes around the world. (Courtesy of NOAA, Ocean facts. http://oceanservice.noaa.gov/facts/cold ocean.html, 2011).

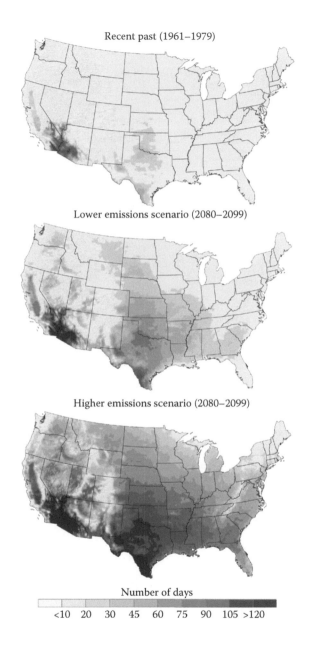

FIGURE 11.1

The number of days in which the temperature exceeds 100°F by late this century, compared to the 1960s and 1970s, is projected to increase strongly across the United States. For example, parts of Texas that recently experienced about 10–20 days per year over 100°F are expected to experience more than 100 days per year in which the temperature exceeds 100°F by the end of the century under the higher-emissions scenario. The lower-emissions scenario assumes a CO_2 concentration of 550 ppm, and the higher-emissions scenario assumes a concentration of 850 ppm. (Courtesy of U.S. Global Change Research Program; image found in Brennan, P., *Orange County Register*, May, 12, 2011).

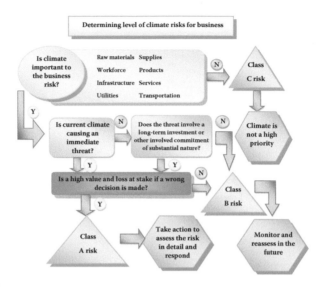

FIGURE 11.6
It is an imperative part of any successful business implementation plan today to include the influence of any climate change impacts that may be significant to the operation of a business.

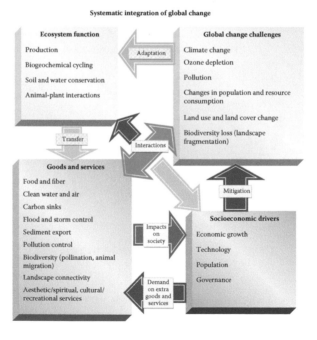

FIGURE 11.7
Economics are influenced interactively by ecosystems and their resultant goods and services, as well as socioeconomic drivers, and global change. These components can act independently, dependently, or interdependently to form the complete economic picture, and a working knowledge of all components is necessary in order to make sound, long-term business decisions.

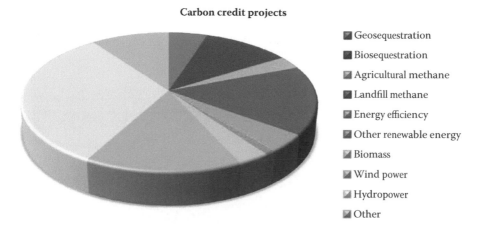

Carbon credit projects

- Geosequestration
- Biosequestration
- Agricultural methane
- Landfill methane
- Energy efficiency
- Other renewable energy
- Biomass
- Wind power
- Hydropower
- Other

FIGURE 12.1
Trading carbon credits is one way to share the burden of reducing CO_2 emissions globally. Current trading ratios are as follows: geosequestration (5%), biosequestration (11%), agricultural methane (3%), landfill methane (16%), energy efficiency (4%), biomass (3%), wind power (15%), hydropower (32%), other renewable energy (1%), and other (10%).

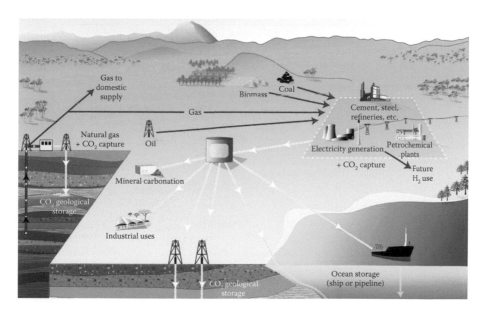

FIGURE 12.2
The IPCC's schematic diagram of possible CCS systems, showing the sources for which CCS might be relevant, transport of CO_2, and storage options. (Courtesy of Rubin E. et al., IPCC special report: Carbon dioxide capture and storage technical summary, http://www.ipcc.ch/pdf/specialreports/srccs/srccs_technicalsummary.pdf, 2011.)

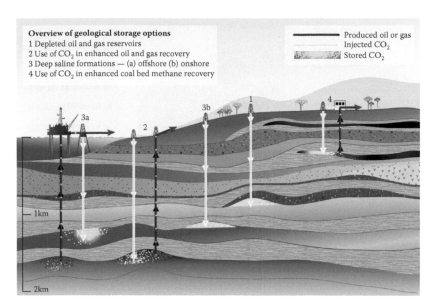

FIGURE 12.3
CO₂ can be sequestered in deep underground geological formations. (Courtesy of Rubin E. et al., IPCC special report: Carbon dioxide capture and storage technical summary, http://www.ipcc.ch/pdf/special-reports/srccs/srccs_technicalsummary.pdf, 2011.)

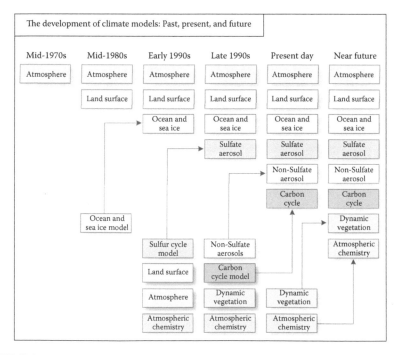

FIGURE 13.1
The evolution of climate models beginning in the mid-1970s and extending into the near future. (From Casper, J. K., *Global Warming. Climate Management: Solving the Problem*, Facts on File, New York, 2010.)

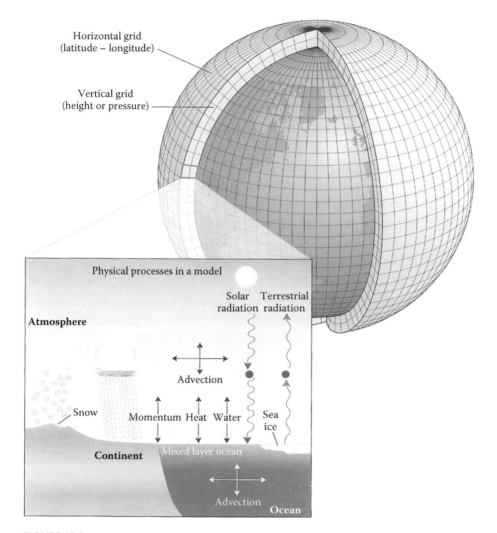

FIGURE 13.2

A climate model is comprised of a set of x/y/z points placed around the globe at specified intervals in a netlike structure, called its resolution. A small grid with lots of points close together has a high resolution and is more detailed; a large grid with points spread farther apart has a low resolution and less detail. In the model, each point x/y/z intersection has a value associated with it—one value for each variable represented in the model. In this example, each grid point would have a distinct value for solar radiation, terrestrial radiation, heat, water, advection, atmosphere, and so on. (From Casper, J. K., *Global Warming. Climate Management: Solving the Problem*, Facts on File, New York, 2010.)

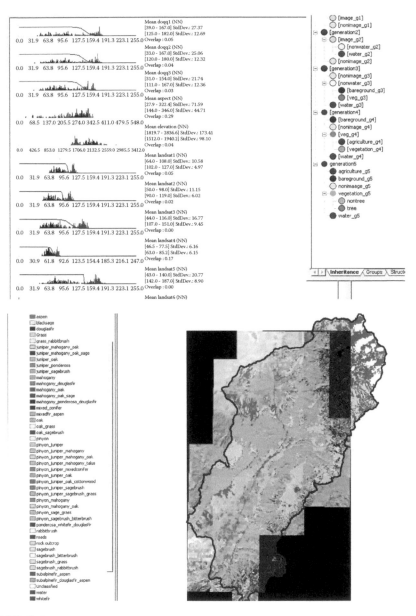

FIGURE 13.6

The top portion illustrates how the different classes are determined during a classification relying on spectral signatures. In this more complicated approach, classes are determined by finding natural breaks in spectral signatures and fitting curves to the resultant spectral data. The bottom approach is another option used that is less complicated, called supervised classification, in which field data points of known information are identified on the digital map and attributed. Once those have been coded, the image-processing software will compare the spectral signatures of all the other pixels to the known ones and group them according to those that are most similar. Both approaches were used in the project.

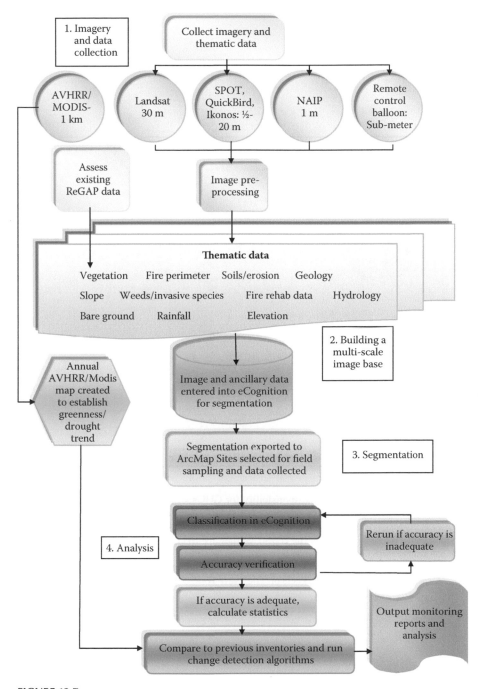

FIGURE 13.7
Summary of the range-monitoring process. It is imperative to have a working model in place before beginning a project of this nature to avoid costly or repetitive mistakes.

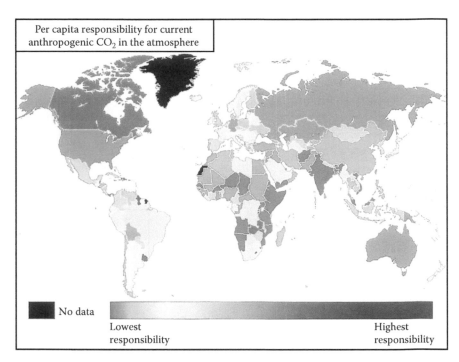

FIGURE 14.2

This map shows the per capita responsibility for GHGs worldwide. In many cases, the largest offenders are often the most wealthy, industrialized nations that are likely to encounter the least in losses overall because they have many of the resources necessary to mitigate the negative effects. Unfortunately, the countries that are likely to encounter the greatest losses are the undeveloped countries and those located close to sea level that have not contributed significantly to the climate change problem. (From Casper, J. K., Global Warming. *Climate Management: Solving the Problem,* Facts on File, New York, 2010.)

such as cheatgrass. One of the downsides with these grasses, besides being invasive species, is that they provide additional fuel for wildfires, which leads to a more rapid loss of natural vegetation. In addition, any precipitation that does occur in the area comes in the form of downpours, and when there is less vegetation to protect the soil, water erosion increases. Once this starts, higher air temperatures and decreased soil moisture reduce soil stability, further contributing to erosion and degrading the landscape.

The conditions of the lands are affected by other factors and their interaction with climate on a local scale. Large-scale, unregulated livestock grazing in the Southwest during the later 1800s and early 1900s contributed greatly to desertification. By 1920, grazing was at an all-time high on the public lands in the West. Then, by the 1970s, grazing had been reduced by about 70 percent, but the arid lands are very fragile ecosystems, and they have been very slow to show signs of recovery. As climate change causes these lands to become even drier and more fragile, their recovery will slow down even more.

Current management practices also need to be reassessed. Today, managers tend to focus more on water resources than protection of the land's productivity. As temperatures rise and the landscape further dries out, the condition of the land will need to become a priority to keep it from becoming further degraded.

In addition to mismanagement of the land, climate change caused by the burning of fossil fuels has led to temperature increases, shifting population patterns, and drying up of aboveground water supplies, which have all added to the problem. The economic impact is a global as well as a regional and national issue. When a region has lower harvests, less consumable goods are produced for internal use or export, which in turn stunts economic growth. With the lack of food in developing nations, especially when they already have excessive foreign debts, to have to pay for imported food further inhibits their economic growth capabilities. In addition to that, the reliance on imports may lead to increased influence from outside political forces. The lending nations could also be impacted economically because their own resources are being drained. With fewer goods to sell, farmers have less money with which to purchase consumer goods, causing those industries to decline, creating a ripple effect in the economy. Once foodstuffs are in short supply, it can upset international futures markets. According to Edouard Saoma, Director of the United Nations Food and Agricultural Organization, if the situation is not changed, it "could completely deplete supplies and cause a food crisis" (Anselmo, 2011).

As a result of this possibility, several solutions have been proposed, and a few nations already have plans in effect to solve the problems. Such plans to combat drought and desertification include:

1. Encouraging the development of new technology
2. Constructing dams and storage reservoirs in monsoon areas and others prone to regular flooding

3. Constructing canals

4. Designing better irrigation

5. Managing water and forestry more efficiently

6. Implementing tree and grass planting programs

7. Reducing fossil fuel burning

All of these plans, however, represent a significant investment of capital.

The Fishing and Forestry Industries

Climate change is already impacting the world's oceans and will have serious consequences for the hundreds of millions of people who depend on fishing for their livelihoods, according to the United Nations Food and Agricultural Organization (2008). Impacts on fisheries that have already been observed include an increase in the frequency and intensity of extreme weather events, such as the El Niño phenomenon in the South Pacific; the warming of the world's oceans, with the Atlantic in particular showing signs of warming deep below the surface; and warmer-water species increasing toward the South and North Poles.

There has also been an increase in salinity in near-surface waters in hotter regions, while the opposite is happening in colder areas because of more precipitation, melting ice, and other processes. The oceans are also becoming more acidic, which is harmful to coral reefs. The fishing communities in the high latitudes, as well as those that rely on the coral reef systems, will be most exposed to the impact of climate change. Fisheries located in deltas, coral atolls, and ice-dominated coasts will be vulnerable to flooding and coastal erosion caused by subsequent rises in sea level. Based on information from the Food and Agricultural Organization, about forty-two million people work directly in the fishing sector, the great majority in developing countries. Counting those who work in processing, supply, marketing, and distribution, the fishing industry supports several hundred million jobs. In addition, fish is also the world's most widely traded foodstuff and a key source of export earnings for many poorer countries. The sector has particular significance for small island states.

The distribution of marine fish and plankton are predominantly determined by climate, and true to expectations, marine species have been documented moving northward in U.S. waters. The timing of plankton blooms is also shifting. Extensive shifts in the ranges and distributions of both warm- and cold-water species of fish have also been documented (Janetos et al., 2008). The polar regions are key areas where climate change has already made a significant impact, including on the fishing industry. The waters around Alaska are already undergoing significant alterations in marine ecosystems with important implications for fisheries and the residents there who depend on them. Alaska is one of the leading areas for salmon, crab, halibut, and herring catch. In addition, many native communities depend on local harvests of fish, walrus, seals, whales, seabirds, and other marine species for their food supply. Climate change causes significant alterations in marine ecosystems with important impacts on the fisheries economy. Ocean acidification associated with increasing amounts of CO_2 concentrations is a serious threat to cold-water marine ecosystems (Feely et al., 2004; Orr et al., 2005).

One case in point is the northern Bering Sea. Situated in an extremely productive fishing area off of Alaska's west coast, the Bering Sea pollock fishery is the world's largest single fishery. It has experienced major declines over the past decade as air and water temperatures have risen and sea ice has rapidly declined. Resident populations of fish, seabirds, seals, walrus, and other species there depend on plankton blooms that are regulated by the extent and location of the ice edge in the spring. As sea ice retreats, however, the location, timing, and species composition of the plankton blooms change, reducing the amount of food reaching the living creatures on the ocean floor, which radically changes the species composition and populations of fish and other marine life forms. This, in turn, is causing devastating results for the fishing industry (Grebmeier et al., 2006).

As climate change continues, the fishing industry will be hit even harder economically. Throughout this century, changes already observed on the shallow shelf of the northern Bering Sea will most likely affect a much broader portion of the Pacific-influenced sector of the Arctic Ocean. As changes occur, the most productive commercial fisheries are expected to become more distant from existing fishing ports and processing infrastructure, requiring either relocation or greater investment in transportation time and fuel costs. These changes will also affect the ability of native peoples to successfully hunt and fish for the food they need to survive. Coastal communities are already noticing a displacement of walrus and seal populations. Bottom-feeding walrus populations are threatened when their sea-ice platform retreats from the shallow coastal feeding ground in which they depend.

Forests provide many services that are critical to the well-being of humans. They provide air and water quality maintenance, watershed protection, and water flow regulation; wildlife habitat and biodiversity conservation; recreational opportunities and aesthetic fulfillment; raw materials for wood and paper products; and climate regulation and carbon storage. As climate change continues, it will change the balance of the forest ecosystems, and the changes are not projected to be for the better. According to Brent Sohngen of the Department of Agricultural, Environmental, and Development Economics at Ohio State University, terrestrial ecosystems store approximately one trillion tons of CO_2 in the biomass of living trees and plants (Sohngen, 2009). He says that current estimates suggest that it would be possible to increase carbon efficiently in order to reduce future damage caused by impending climate change. He has proposed several methods, which include afforestation (planting old agricultural land with trees), reduced deforestation, and proper forest management. Sohngen says that current estimates in the literature accounting for opportunity costs and implementation and management costs imply that an additional 6.8 billion tons of CO_2 per year may be sequestered in forests by 2030 for $30 per ton of CO_2. About 42 percent of this would come from avoided deforestation, with the rest roughly equally split between afforestation and forest management plans.

Sohngen also states that analysis indicates that if society follows an "optimal" carbon-abatement plan, forestry could accomplish about 30 percent of the total abatement needed throughout the remainder of the century. If society instead chooses to place strict limits on emissions in order to meet the 2°C temperature increase limitation, then with what forestry alone contributes, it could lower overall abatement costs by as much as 50 percent. The benefits Sohngen projected represent a potentially large, additional benefit of a forestry carbon-sequestration program. In the United States, forest growth currently offsets about one-fifth of the fossil fuel carbon emissions, serving as a carbon sink—a highly beneficial service that forests provide to store CO_2 and offset climate change.

Wildfires are a significant impact as a result of climate change, some of them anthropogenically caused. In a study that appeared in *Science*, as reported in *USA Today* on July 7, 2006, not only the number, but the size of large forest fires in the West has grown "suddenly and dramatically" during the past 20 years partly because of climate change (O'Driscoll, 2006). Since 1987, the wildfire season has been extended 2.5 months longer than traditionally because springtime was seeing higher temperatures, and winter snow pack was melting more quickly. With earlier snowmelt, the soil and vegetation dries out sooner, which leads to more fires that burn bigger and longer. In a study conducted by the University of California-San Diego and the University of Arizona, 1,166 large forest fires were analyzed from 1970 to 2004 on national forest and parkland in the western United States. From 1987 to 2004, there were four times as many forest fires and 6.5 times as much land burned as there was from 1970 to 1987.

According to Tom Swetham of Arizona's Laboratory of Tree-Ring Research, "Climate is the principal reason. The rising temperatures and dryness due to warming are the main factors in the northern Rockies" (O'Driscoll, 2006). Each year large amounts of federal funding are allocated toward fighting fires on public lands. As climate change continues, this effort is expected to intensify, representing a significant financial investment in forested areas (Sedjo and Sohngen, 2009).

Changes in Monsoon Patterns

According to a recent study presented in *USA Today*, the annual monsoon cycle—a part of the southern Asian summer and also critical to agricultural production in Bangladesh, Nepal, Pakistan, and India—could be weakened and delayed as climate change causes temperatures to rise in the future, according to the results of a model constructed by a Purdue University research group. The group discovered through computer modeling that they could influence monsoon dynamics and cause there to be less summer precipitation, a delay in the start of the monsoon season, and longer breaks between rainy periods.

What is critical about this is that half of the world's population lives in areas that are currently affected by monsoons, where even slight deviations from the norm can have significant effects on the sensitive monsoonal patterns and climate. The summer monsoon affects water resources, agriculture, economics, ecosystems, and human health throughout southern Asia, representing a very important climate modulator.

Effects may be drastic if climate change delays the arrival of expected monsoons. At the Purdue Climate Change Research Center, researcher Noah Diffenbaugh noted that "agricultural production, water availability and hydroelectric power generation could be substantially affected by delayed monsoon onset and reduced surface runoff. Alternatively, the model projects increases in precipitation over some areas, including Bangladesh, which could exacerbate seasonal flood risks." The summer monsoons are a critical source of water supply. They are responsible for about 75 percent of the total annual rainfall in major parts of the region and produce nearly 90 percent of India's water supply (Associated Press, 2009).

General circulation models have been used for projections of what may happen to monsoon patterns for this region, but it has been difficult to model southern Asia and try to determine what will happen in specific areas because of its extremely diverse topography. Diffenbaugh's team used the highest-resolution climate-model data available and simulated a monsoon circulation under the effects of projected climate change. The model projected a delay in the beginning of monsoon season of 5–15 days by the end of the twenty-first century and an overall weakening of the summer monsoon precipitation over south Asia. The team noted that increasing temperatures in the future strengthen some aspects of large-scale monsoon circulation but weaken the fine-scale interactions of the land with the moisture in the atmosphere, which, they believe, could lead to reduced precipitation over the Indian subcontinent.

According to Moetasim Ashfaq, lead author of the study and a graduate student in earth and atmospheric sciences at Purdue University, "It is the more subtle, local-scale processes that are key in this case. Our model shows a decrease in convective precipitation, which is critical for summer precipitation in this region. Our findings show it is not just a question of whether monsoon circulation is stronger or weaker. Even with a strong monsoon system, if circulation changes enough to change where and when rain is delivered, then that could have an impact that has not been captured in the large-scale evaluations" (Associated Press, 2009).

From observation, the atmospheric conditions that lead to reduced precipitation can also lead to an intensification of extremely hot conditions. The model in Ashfaq's study also illustrated an eastward shift in monsoon circulation, which would indicate more rainfall over the Indian Ocean, Bangladesh, and Myanmar but significantly less over India, Nepal, and Pakistan. Less moisture over the land in combination with dry summer air would lead to less moisture in the atmosphere and, therefore, less rainfall.

To give credence to the study, monsoon moisture flow circulates from the ocean to land. In the summer months, the land warms much faster than the ocean. This, in turn, creates a pressure gradient that draws air masses from the ocean to the continent, bringing moist air that encourages the formation of a large-scale monsoon system. The monsoon system usually starts in early June and ends in late September and generally starts at the southeast tip of India and moves northwest to the rest of India and Pakistan. The climate model used by the research team accurately recreated the monsoon season of past years, and its future projections are consistent with what has been seen in recent drought years over the region, which lends credibility to their results (Associated Press, 2009).

The significance of the changing monsoon circulation patterns caused by climate change is the effect it will have on millions of people. A delay in the timing of the monsoon would cause India's economic growth to be negatively impacted. According to Shravya Reddy (2009) at the Natural Resources Defense Council, it could affect food production and cause India's GDP growth rate to slip by 1 to 1.5 percentage points, and the GDP growth could slip as low as 5 percent this year from the current 7 percent. It will also negatively affect the farming industry. Farming currently constitutes 18 percent of India's GDP, and up to 70 percent of Indians depend on farm incomes for their living. In addition, 60 percent of India's farms depend on rains for irrigation. Reduced crop yields—especially food crops—will raise prices higher and higher. In addition, consumer prices rose 10.21 percent in May from a year earlier, after gaining 9.09 percent in the previous month.

The IPCC's fourth assessment report lays out how a rise in global average temperature will cause more variability in the monsoon, leading initially to both an increase in precipitation in some areas and a decrease in others, which will cause increases in both floods and droughts. They also add to that prediction that eventually the monsoon will bring less rain, over fewer parts of the country, and the rainfall will be confined to short, intense bursts. Because there will be more runoff, there will also be a decrease in groundwater aquifers, which will lead to widespread water shortages for irrigation and human consumption.

Agriculture and Food Production

The agricultural and food production industries will experience some of the greatest disruptions and changes of all from climate change. Although climate change affects agriculture, climate is also affected by agriculture, which currently contributes 13.5 percent of all human-induced greenhouse gas emissions globally. In the United States, agriculture represents 8.6 percent of the country's total greenhouse gas emissions, including 80 percent of its nitrous-oxide emissions and 31 percent of its methane emissions (EPA, 2011).

Increased agricultural production will be necessary in the future in order to feed growing global populations, and agricultural productivity is directly tied to the climate and land resources. Climate change can have both beneficial and detrimental impacts on plants. As greenhouse gas (GHG) levels rise in the atmosphere, projected climate changes are likely to increasingly challenge farmers' capacity to efficiently produce food, feed, fuel, and livestock products. Crops in the future, as climate change progresses, are expected to respond to three factors: rising temperatures, changing water resources, and increasing CO_2 concentrations. Generally, warming causes plants that are below their optimum temperature to grow and produce faster. Although this can be a good thing, it is not always. For instance, when cereal crops grow faster, there is less time for the grain itself to grow and mature, which reduces the yields. For some annual crops, late-season stress can be avoided by adjusting their planting date.

The time period during which wheat and other small grain grow and mature (called the grain-filling period) shortens dramatically with rising temperatures. Moderate increases in temperature will decrease yields of corn, wheat sorghum, beans, rice, cotton, and peanut crops. In addition, some crops are sensitive to high nighttime temperatures, which have been rising even faster than daytime temperatures. Nighttime temperatures are expected to continue to rise in the future. These changes in temperature are critical to the reproductive phase of growth because warm nights increase the respiratory rate and reduce the amount of carbon that is captured during the day by photosynthesis to be released in the fruit or grain. As temperatures continue to rise and drought periods increase, crops will be more frequently exposed to temperature thresholds at which pollination begins to fail and the quality of the vegetable crops decreases. Grain, soybean, beans, rice, cotton, peanut crops, and canola crops have relatively low optimal temperatures and will have reduced yields and will begin to fail as warming increases (Hatfield et al., 2008).

Temperature also affects the length of the growing season and water availability. For crops that do better in higher temperatures, such as sweet potatoes, this means the growing season will be longer, but for those that are temperature sensitive, such as lettuce and broccoli, their growing season will be truncated. As temperatures rise, plants will also be required to use more water to keep cool, but that only works to a certain point, and when that level is reached and temperatures exceed the threshold for that species, it will not produce seed anymore, which means it will no longer reproduce (Hatfield et al., 2008). Temperature change will also cause a migrational shift in species. If temperatures in a region warm, species will shift poleward; if they cool, they will migrate toward the equator. Therefore, temperature will be one of the major factors determining where crops will be able to be efficiently grown.

Higher CO_2 levels typically cause plants to grow larger. This can also pose a problem, because for some crops, this makes them less nutritious. On the other hand, CO_2 also makes some plants more water-use efficient, so that they produce more plant material with less water. This could be a benefit in dry or drought areas, depending on the type of plant and the specific location.

In some cases, there may be adaptation measures to accommodate for these changed conditions. In the most simple of cases, changing the planting dates and planting a crop later in the year could overcome the problems of excessive heat with the least economic impact. The economic feasibility of this, for the farmer, would depend on several factors, such as the optimum planting date for maximum profits and timing delivery of produce with market demand and competition.

In some cases, the typical fruits grown in a region may have to be discontinued if they require longer winter chilling periods than what the area receives at that point. Another noted issue on the increase with climate change is the increase in heavy downpours. Another event associated with climate change is less frequent, but more intense, precipitation episodes. This pattern is predicted to increase in areas across the United States. One of the early consequences of excessive rainfall is delayed spring planting, which in turn lowers profits for farmers who typically produce high-value early season crops such as melon, sweet corn, and tomatoes. When cultivated fields flood during the growing season, it causes crop losses because of low oxygen levels in the soil, increased susceptibility to root diseases, and increased soil compaction because of the heavy farm equipment used on wet soils.

For an example of the magnitude of losses felt by farmers, in the spring of 2008, when heavy rains caused the Mississippi River to rise two meters above flood stage, it inundated hundreds of thousands of hectares of cropland, which coincided with the time farmers were harvesting wheat and planting corn, soybeans, and cotton. The farmers suffered economic losses of about $8 billion, with many farmers going bankrupt and others taking many years to recover (National Oceanic and Atmospheric Administration, 2008). Heavy downpours can also reduce the quality of the crops, causing significant economic losses. When vegetable and fruit crops are sensitive to short-term minor stresses, extreme weather events can cause significant economic damage.

According to the U.S. Global Change Research Program, drought frequency and severity are projected to increase in the future over large areas of the United States, especially if GHG levels in the atmosphere continue to rise. Unfortunately, increased drought will occur at a time when crops will require additional water in order to survive as temperatures climb. To put this in perspective, for many high-value crops, only hours or days of moderate heat stress at critical growth stages can reduce grower profits by negatively affecting visual or flavor quality, even when total yield is not reduced (Peet and Wolf, 2000).

Another major concern with rising temperatures is the northward expansion of invasive weeds. Today, southern farmers lose more of their crops to weeds than do northern farmers. Based on the 2009 U.S. Global Change Research Program Report, southern farmers in the United States lose 64 percent of their soybean crop to invasive weeds, whereas northern farmers lose 22 percent. One especially problematic, very aggressive invasive weed that plagues the southern regions of the United States—kudzu—has traditionally been confined to areas where winter temperatures do not drop below certain thresholds, but as temperatures continue to rise with the changing climate, it will expand its range (along with other similar invasive weeds) northward into key agricultural areas, causing economic losses in expanded areas. Currently, kudzu has invaded approximately one million hectares of the southeast United States and is a carrier of the fungal disease soybean rust, which seriously threatens the soybean production business, playing havoc with that part of the economy (Karl et al., 2009).

Controlling weeds represents a huge economic expense. In the United States alone, it currently costs more than $11 billion annually, with the majority of that expenditure on herbicides, which suggests that both herbicide use and costs will increase as temperatures and CO_2 levels rise with climate change. In addition, the most commonly used herbicide in the United States, glyphosate (commonly known as RoundUp®), actually loses its efficacy on weeds grown at the CO_2 levels that are projected to occur in the coming decades. This means that higher concentrations of the chemical and more frequent usage will be necessary, which in turn increases both the economic and environmental costs associated with the use of chemicals (Kiely et al., 2004).

Insects and diseases also thrive in warming temperatures, necessitating an increased use of pesticides. As temperatures climb, both insects and pathogens will expand their ranges northward. In addition, as temperatures warm, it will allow more insects to survive over the winter, whereas cold winter temperatures in the past served to control their populations. Insects present two negative aspects to crop production: They not only damage crops but also carry disease, which kills crops and vegetation. Crop diseases are expected to increase as earlier springs and warmer winters promote proliferation and higher survival rates of disease pathogens and parasites. Consequently, the longer growing season will also allow some insects to produce more generations during a single season, which will greatly increase the insect population. In addition, plants that grow in higher CO_2 concentrations may grow larger, but they tend to be less nutritious, so insects need to consume more in order to meet their protein requirements, which causes greater crop destruction (Hatfield et al., 2008).

Rangeland grazing for cattle and other livestock will also take a toll under rising temperatures. In the United States, eastern pasturelands are planted and managed, and western rangelands are native pastures, not seeded and much more arid. These rangelands are already experiencing the effects of climate change caused by increasing levels of CO_2. Specific grasses are beginning to dominate that do better under higher CO_2 levels, and the quality of the forage is beginning to drop, meaning that more acreage is needed to provide the animals with the same nutritional value, resulting in an overall decline in livestock productivity. Woody shrubs and invasive cheatgrass are encroaching into once-fertile grasslands, which further reduces the forage value of the rangeland. The combination of these factors leads to a decline in livestock productivity and economic decline for the rancher.

It is also expected that livestock production will fall as temperatures rise. Temperature and humidity interact to cause stress in animals, just as in humans, and the higher the heat and humidity, the greater the stress and discomfort, and the larger the reduction in the animals' ability to produce milk, gain weight, and reproduce. Milk production will decline in dairy operations, the number of days it takes for cows to reach their target weight will grow longer in meat operations, the conception rate in cattle is projected to fall, and swine growth rates are predicted to decline because of increased heat. As a result, swine, beef, and milk production are all projected to decline as temperatures rise (Hatfield et al., 2008).

Impacts to the Recreation and Tourism Industry

Recreation and tourism play important roles in the economy and are very important components in the lives of many people worldwide. Tourism is one of the largest economic sectors in the world, and it is also one of the fastest growing (Hamilton and Tol, 2004). The employment opportunities created by recreational tourism provide economic benefits not only to individuals but also to communities. In the United States alone, more than 90 percent of the population participates in some form of outdoor recreation, which equates to nearly 270 million people, and several billion days spent annually in a wide variety of outdoor recreation activities (Cordell et al., 1999). In some regions tourism and recreation are major employment centers, creating jobs for many in primary and secondary industries. For example, in the United States, in areas where popular national parks are located, such as Yellowstone National Park in Wyoming or Zion National Park in Utah, there are not only employment opportunities at the parks, but opportunities also in the regions around the parks in a variety of sectors, including lodging facilities, restaurants, other tourist attractions, retail stores, and so forth. These businesses bring billions of dollars to the regional economies in which they are located. The same applies globally, making tourism and recreation one of the world's key industries. Recreational activities, such as fishing, hunting, skiing, snowmobiling, diving, beach-going, and other outdoor activities, make important economic contributions and are a part of family traditions that have value that goes beyond financial returns.

According to the Outdoor Industry Foundation, the outdoor recreation industry has an average annual $730 billion impact on the U.S. national economy. The report indicated that this amount factors in the amount Americans spend on outdoor trips and gear, the companies that provide that gear and related services, and the companies that support them. The outdoor industry also supports 6.5 million jobs. This equates to one in 20 U.S. jobs, and generates about $88 billion in federal and state tax revenue while stimulating 8 percent of all consumers spending (Outdoor Industry Foundation, 2006).

Because most recreation and tourism activities occur outside, increased temperature and precipitation have a direct effect on the enjoyment of these activities and on the desired number of visitor days and the associated level of visitor spending as well as tourism employment. Weather conditions are a critical factor influencing tourism. The availability and quality of natural resources is another important factor that determines the quality of outdoor recreation and tourism, such as the condition of beaches, wetlands, forests, wildlife, and snow—all of which are affected by climate change. Climate change, however, is currently changing the outcome of that for many areas and will continue to do so in the future.

Climate change will equate to reduced opportunities for some activities and locations and expanded opportunities for others. According to the IPCC, the effects of climate change on tourism in a particular area will depend in part on whether the tourist activity is summer- or winter-oriented and the elevation of the area and the impact of climate on alternative activities. Therefore, the range of effects on each area will differ depending on the climate change effects at that particular location. In the short term, the length of the season for, and desirability of, several of the most popular outdoor activities, such as hiking, beach-going, backpacking, horseback riding, and camping, will most likely be enhanced, because of warmer temperatures and a longer summer season. Other activities, however, are likely to be negatively impacted by even slight increases in warming, such as snow- and ice-dependent activities: skiing, snowboarding, ice climbing, snowmobiling, and ice fishing. Therefore, areas today that serve as ski resorts may not have adequate snow in the future, so they may have to become mountain resorts used for other activities, such as hiking, mountain biking, or horseback riding (IPCC, 2007). This is already occurring in some of the alpine areas in Europe, where some ski resorts have now become "mountain beach" resorts. According to an article in the UK *Telegraph* (de Quetteville, 2008), many of Europe's picturesque alpine towns are transforming their venues into golfing, cycling, and cross-country trekking to offset economic losses from lack of adequate snow cover to support their dwindling skiing industry.

The results of a Swiss study reveal that snow cover suitable for winter tourism has fallen by 60 percent in the past 60 years (de Quetteville, 2008). As a result, instead of their traditional snow-themed activities, these destinations are moving into golfing, cycling, and cross-country trekking in order to offset losses caused by declining snowfall.

According to Stephan Lerendu, the director of the Tourist Office at Avoriaz, France, "People think snow is a resource that will never run out but it's difficult right now. Tennis courts, saunas, and gyms are must-haves. Golf courses are indispensable for ski stations. We are adapting ski schools for the summer, with downhill biking and Nordic walking instead. We are creating green, blue, red, and black pistes for cycling. It's the future" (de Quetteville, 2008).

Currently in Crans-Montana, Switzerland, golf courses already host European PGA tournaments. It also offers canyoning and paragliding. The statistics for snow days in the area are shocking: at low altitudes—up to 800 meters—the number of snow days (days with enough snow to make a snowman) has fallen from 28 days to 13. At mid-altitudes—up to 1,300 meters—days suitable for cross-country skiing have gone from 55 to 38. At high altitudes, for downhill skiing, snow days have declined from 93 to 74 (de Quetteville, 2008). According to Dr. Christoph Marty, who led the research, "There is nothing similar in 1,000 years of records. Since this impact takes time to filter through the environment, what we are seeing now is only the fruit of what we did 20 years ago."

If climate change negatively impacts the recreational uses of mountain areas, these changes will have a ripple effect of economic consequences throughout the local, national, and international economy, particularly in business sectors such as the travel industry, the transportation industry, sports equipment industry, food industry, and others related to leisure activities. For example, according to the U.S. Global Change Research Program, some western ski resorts, such as those in Utah, Colorado, and California, could face a 90 percent decrease in snow pack, making the country's most iconic ski locations facing obliteration.

Some of the possible identified effects that will occur in the recreation and tourism industries include:

- Declines in cold-water and cool-water fish habitat may affect recreational fishing.
- Shifts in migratory bird populations may affect recreational opportunities for bird-watchers and wildlife enthusiasts.
- Coastal regions may lose pristine beaches because of sea-level rise.
- Winter recreation (skiing, snowmobiling, ice fishing) is likely to be affected by reduced snow pack and fewer cold days, and costs to maintain these industries will rise.
- Tourism and recreation in locations such as Alaska will most likely undergo climate-driven transformation through loss of wetlands and a reduction in habitat for migratory birds.
- Arctic breeding and nesting areas for migratory birds will be lost, affecting bird-watching activities.
- Melting permafrost in the Arctic will pose economic damage to structures and other infrastructures to residents there.

- Hunting and fishing will change as animals' habitats shift and as relationships among species in natural communities are disrupted by their different responses to rapid climate change.
- Water-dependent recreation in areas projected to get drier and beach recreation in areas that are expected to see rising sea levels will suffer.
- Some regions will see an expansion of the season for warm weather recreation such as hiking and bicycle riding.

In summary, the net economic effect of short-term climate change on recreational activities is probably going to be positive. In the longer term, however, as the effect on ecosystems and seasonality become more pronounced, the net economic effect on tourism and recreation is not known with clear certainty and will vary from location to location.

Healthcare and Labor-Related Impacts

Climate change presents a unique challenge to human health. Unlike specific health threats caused by one particular disease vector, there are several different ways that climate change can lead to potentially serious health effects. For instance, there are the directly measurable health effects that occur as a result from incidents such as heat waves and severe storms. There are also health issues that result from, or are made worse because of, pollution and airborne allergens. The third class of health issues includes climate-sensitive infectious diseases.

When trying to assess the economic impacts of the health-related issues of climate change, it is necessary to also factor in the need to manage new and changing climate conditions. Potential health risks must be recognized so that they can be avoided, minimized, or mitigated effectively. Diseases being introduced from elsewhere globally must also be considered, because society is global in nature. It is currently expected that developing nations will suffer even greater health consequences from climate change. With global travel so common, however, there are no real barriers or dividing lines anymore, requiring all areas to plan accordingly.

In addition, climate extremes—such as severe storms and drought—can undermine public-health infrastructure, further stress environmental resources, destabilize economies, and potentially create security risks anywhere in the world. An increase in the risk of illness and death related to extreme heat and heat waves are very likely in the future. Some reduction in the risk of death related to extreme cold is also expected. Also of note, the population—for example in the United States—is aging, and older people are more vulnerable to hot weather and heat waves. Today, the U.S. population over age 65 is 12 percent. By 2050, however, it is projected there will be 21 percent over 65, or more than 86 million people. Diabetics are also at greater risk of heat-related death.

According to a report issued by the thirteen-agency U.S. Global Change Research Program released in 2009, which discusses extreme weather and how climate change will affect life in the United States, both the current and projected effects through 2099 under various GHG–emissions scenarios, that portion of the earth will see some drastic changes (Figure 11.1). For example, by the end of the century, summers will be much more severe across the country. In the New England area, summers will resemble those of present-day North Carolina. The Great Lakes region can expect to feel more like Oklahoma. Texas, typically hot now with 10 to 20 days a year over 38°C, will jump to approximately 100 days of that temperature range—over 3 months instead of 3 weeks. Chicago, which has experienced severe heat waves and resulting high death tolls in the past (they experienced a heat wave in 1995 that killed more than 700 people), can expect that kind of relentless heat up to three times a year. The southwest portion of the country, which has experienced multiple droughts and resultant wildfires in the recent past, can expect to face even more frequent droughts, as spring rains decline by as much as half. Cities such as Los Angeles, Salt Lake City, and Phoenix, which have large populations, will face difficult times as snow packs shrink and melt earlier and water evaporates more rapidly, resulting in reservoirs remaining at only partial capacity.

Another major health concern as temperatures rise is the urban heat-island effect. Large amounts of concrete and asphalt in cities absorb and hold heat. Tall buildings prevent heat from dissipating and reduce airflow. Simultaneously, there is generally little vegetation to provide shade and evaporative cooling. As a result, parts of cities can be up to 6°C warmer than the surrounding rural areas, compounding the temperature increases that people experience as a result of human-induced warming, such as heat given off from automobile traffic, machinery, and other means. As human-induced warming is projected to raise average temperatures by approximately 3.5–6.5°C by the end of this century under the higher-emissions scenario, heat waves are expected to continue to increase in frequency, severity, and duration. As an illustration, by the end of the century, the number of heat-wave days in Los Angeles is projected to double, and the number in Chicago is projected to quadruple if emissions are not reduced. This means that cities like Chicago can expect that the average number of deaths caused by heat waves could more than double by 2050 under a lower-emissions scenario and quadruple under a higher-emissions scenario (Figure 11.2).

The U.S. Global Change Research Program recognizes that the full effect of climate change on heat-related illness and death involves a number of factors including actual changes in temperature and human population characteristics, such as age, wealth, and fitness. In addition, adaptation at the scale of a city includes options such as heat-wave early warning systems, urban design to reduce heat loads, and enhanced services during heat waves, which all have an economic impact on the supporting society.

Climate change and warming temperatures will also make it more challenging to meet air-quality standards necessary to protect public health.

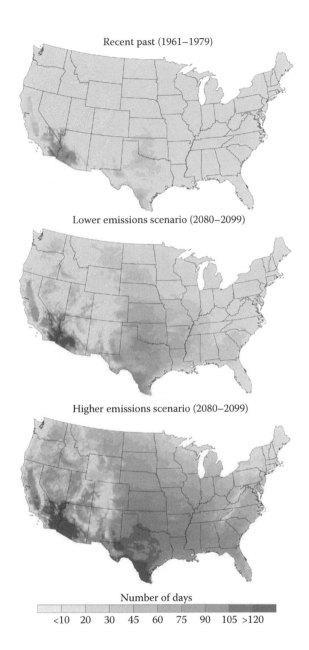

FIGURE 11.1
(See color insert.) The number of days in which the temperature exceeds 100°F by late this century, compared to the 1960s and 1970s, is projected to increase strongly across the United States. For example, parts of Texas that recently experienced about 10–20 days per year over 100°F are expected to experience more than 100 days per year in which the temperature exceeds 100°F by the end of the century under the higher-emissions scenario. The lower-emissions scenario assumes a CO_2 concentration of 550 ppm, and the higher-emissions scenario assumes a concentration of 850 ppm. (Courtesy of U.S. Global Change Research Program; image found in Brennan, P., *Orange County Register*, May, 12, 2011).

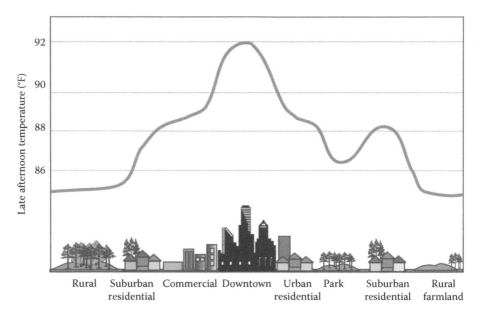

FIGURE 11.2
The urban heat-island effect is projected to be responsible for many heat-related deaths as temperatures climb in urban areas throughout the remainder of the century. (Courtesy of U.S. Global Change Research Program, Human health sector, http://www.globalchange.gov/publi cations/reports/scientific-assessments/us-impacts/full-report/climate-change-impacts-by-sector/human-health, 2011.)

Higher temperatures and associated stagnant air masses are expected to make it more challenging to meet air-quality standards, especially for ground-level ozone (a component of smog). According to the U.S. Global Change Research Program (USGCRP), it has been established that breathing ozone results in short-term decreases in lung function and damages the cells lining the lungs (USGCRP, 2009). It also increases the incidence of asthma-related hospital visits and premature deaths. Vulnerability to ozone effects is greater for those who spend time outdoors, especially with physical exertion, because this results in higher amounts reaching the lungs. As a result, those most at risk include children, outdoor workers, and athletes. Ground-level ozone concentrations are affected by several factors, including weather conditions, emissions of gases from vehicles and industry that lead to ozone formation (especially nitrogen oxides and volatile organic compounds [VOCs]), natural emissions of VOCs from plants, and pollution blown in from other locations. Warmer temperatures are projected to increase the natural emissions of VOCs, accelerate ozone formation, and increase the frequency and duration of stagnant air masses that allow pollution to accumulate, which will exacerbate health symptoms. The formation of ground-level ozone occurs under hot and stagnant conditions, which are the same weather conditions that accompany heat waves, which are expected to increase as climate change continues. Oftentimes, heat waves and air pollution occur together because they both occur under the same weather conditions. These interactions will more than likely increase as climate change continues, as well.

Extreme weather events are also responsible for emotional trauma, injury, and illness, and even death is known to result from extreme weather occurrences. The health impacts of extreme storms can also contribute to physical and mental health problems, which are projected to increase.

When looking at severe weather events—such as Hurricane Katrina—it is important to realize that injury, illness, emotional trauma, and death result from extreme weather. The number and intensity of severe events are already increasing and are projected to increase further in the future. Human health impacts are expected to be more severe in poorer countries where the emergency preparedness and public-health infrastructure is less developed. Early warning and evacuation systems and effective sanitation lessen the health impacts of extreme events, and developing countries often lack these. Indirect health impacts of extreme storms go beyond direct injury and death to problems such as carbon-monoxide poisoning from portable electric generators in use following hurricanes and mental-health impacts such as depression and post-traumatic stress disorder.

Several disease-causing agents that are commonly transmitted by water, food, or animals are susceptible to changes in replication, survival, persistence, habitat range, and transmission as a result of climate change factors such as increasing temperature, precipitation, and extreme weather events. Issues that will become problems as temperatures rise will include:

- Cases of waterborne *Cryptosporidium* and *Giardia* increase following heavy downpours. These parasites can be transmitted in drinking water and through recreational water use.

- Heavy rain and flooding can contaminate certain food crops with feces from nearby livestock or wild animals, increasing the likelihood of food-borne disease associated with fresh produce.

- Climate change affects the life cycle and distribution of the mosquitoes, ticks, and rodents that carry West Nile virus, equine encephalitis, Lyme disease, and hantavirus. Moderating factors, such as housing quality, land-use patterns, pest-control programs, and a robust public-health infrastructure, are likely to prevent the large-scale spread of these diseases in the United States.

- As temperatures rise, tick populations that carry Rocky Mountain spotted fever are projected to shift from south to north.

In addition, rising temperatures and carbon dioxide concentrations increase pollen production and prolong the pollen season in several plants with highly allergenic pollen, presenting health risks for those with allergies.

Climate change has caused an earlier onset of the spring pollen season in some areas such as the United States, because these areas warm up earlier in the year now. Based on data from Ebi et al. (2008), laboratory studies suggest that increasing carbon dioxide concentrations and temperatures increase ragweed pollen production and prolong the ragweed pollen season.

Another plant whose growth and toxicity is increased by CO_2 is poison ivy. In this case, the vines grow twice as much per year in air saturated with a doubled preindustrial CO_2 concentration as they do in unsaturated air. Similar effects also occur with the growth of stinging nettle and leafy spurge, which are two noxious weeds that cause rashes when they come into contact with human skin (Ziska, 2003).

The segment of the population who is at the highest risk for health-related problems because of climate change are infants and children, pregnant women, the elderly, people with chronic medical conditions, outdoor workers, and people living in poverty. Children's small ratio of body mass to surface area and other factors make them vulnerable to heat-related illness and death. They are also more sensitive to air pollution because of their increased breathing rate relative to body size, developing respiratory tracts, and the additional amount of time they spend outside. Children also have immature immune systems, which greatly increases their risk of serious problems developing from food-borne or waterborne diseases. In addition, developmental factors make them more likely to have complications from infections such as *E. coli* or *Salmonella* (Ebi, 2008).

The elderly are at greater risk to have debilitating chronic diseases or limited mobility. They are also more sensitive to extremes because they have a reduced ability to regulate their own body temperature or sense when they are too hot. They are also at greater risk of heart failure, which adds complications to the fact that the heart must work harder to cool the body down when it is too hot. The poor will also feel the negative effects of the heat because they oftentimes lack shelter with air conditioning.

Future reduction in the vulnerability of populations to the effects of climate change on human health will require putting together a comprehensive working long-term plan that takes into account several issues, such as municipal- and public-service planning, including components like water, energy, and health services. It must also take into account the growth pattern projections of future populations.

Transportation and Energy Considerations

Although transportation is a significant source of GHG emissions worldwide and has been identified as having a major impact on climate, the effects of climate change will also have major impact on the transportation sector. Climate change is slated to cause numerous disruptions to transportation networks. Some of this has already been demonstrated. For example, when major flooding occurred in the Midwest of the United States in 2008, it restricted regional travel and disrupted freight and rail shipments across the entire country. In most countries, transportation networks are vital to nations' economies, safety, and quality of life.

Extreme events present major challenges for transportation, and these events are becoming more frequent and intense. According to the National Research Council, what is necessary at this point is for transportation planners to now take into account climate change when they do their long-term planning and project development, because extreme events are becoming so much more frequent and transportation networks are long-term projects, climate change must be one of the major considerations when building new systems or upgrading old ones (National Research Council, 2008).

In order to be prepared to withstand the effects of climate change, it is important to have a plan in place that addresses a range of adaptation responses that can be employed to reduce risks through either the redesign or relocation of infrastructure. Adapting to climate change is not an easy process: It is an evolutionary process that takes time through the adoption of a series of plans, risk-management designs, and adaptive responses. If these are followed, vulnerable transportation infrastructure can be made more resilient (Potter et al., 2008).

One thing that has been identified is that sea-level rise and storm surge will increase the risk of major coastal impacts, including both temporary and permanent flooding of airports, roads, rail lines, and tunnels. Transportation infrastructure along coastal areas is becoming increasingly vulnerable to sea-level rise. Because most high-density cities are located in coastal areas, the potential exposure of this transportation infrastructure to flooding is enormous. Another factor is that populations increase dramatically in these areas during the summer months because the beaches in coastal locations are generally important tourist destinations.

For example, the Gulf Coast of the United States area has an estimated 3,860 kilometers of major roadway and 395 kilometers of freight rail lines at risk of permanent flooding within 50–100 years as climate change and land subsidence combine to produce an anticipated sea-level rise of approximately 1.2 meters. The potential for disruption is enormous (Kafalenos, 2008).

Storm surge will also be another major impact. More intense storms, when coupled with sea-level rise, will result in destructive storm surges. According to the National Research Council, in the Gulf Coast area of the United States, an estimated 97,000 kilometers of coastal highway are already exposed to periodic flooding from coastal storms and high waves. Some of these routes are currently used as evacuation routes during hurricanes and could become seriously compromised in the future as climate change escalates. The coastal areas are also major centers of economic activity. In the United States, for example, six of the nation's top ten freight centers are currently threatened by sea-level rise. Seven of the ten largest ports (by tons of traffic) are located on the Gulf Coast. In addition, approximately two-thirds of all U.S. oil imports are transported through this region. Sea-level rise would potentially affect commercial transportation activity valued in the hundreds of billions of dollars annually through inundation of area roads, railroads, airports, seaports, and pipelines (Kafalenos et al., 2008, and National Research Council, 2008).

More intense storm surges are also expected. Currently, 97,000 kilometers of coastal highway are subject to destructive storm surges. Despite all of this, the tourism markets still continue to grow and apply pressure to these locations, encouraging continued growth and development. Any growth will generate demand for more transportation infrastructure and services, which will continue to challenge transportation planners to meet their demand. It is extremely important, however, that they take climate change into account as one of the most important, pressing variables.

Impacts on harbor infrastructure from wave damage and storm surges are projected to increase. Subsequent changes will be required in harbor and port facilities to accommodate higher tides and storm surges. Changes in the navigability of channels are expected—some will become more accessible and extend further inland because of deeper waters, whereas others will be restricted because of changes in sedimentation rates and sandbar locations. There will also be reduced clearance under some waterway bridges for boat traffic. Some channels may have to be dredged more frequently.

Airports are another consideration. Those located in coastal cities are often located adjacent to rivers, estuaries, or open ocean. Runways in coastal areas may become inundated with water unless effective protective measures are taken. Several of the world's busiest airports may face closure or restrictions.

Energy source, supply, and use are major components of the climate-change issue. It is the primary contributing factor to climate change, and in return, climate change will eventually affect our production and use of energy. According to the Energy Information Administration, about 87 percent of U.S. GHG emissions originate from energy production and use (Energy Information Administration, 2008).

Rising temperatures are expected to increase energy requirements for cooling and reduce energy requirements for heating. Changes in precipitation have the potential to affect prospects for hydropower, either positively or negatively. Increases in hurricane intensity are expected to cause further disruptions to oil and gas operations in the Gulf, like those experienced in 2005 with Hurricane Katrina and 2008 with Hurricane Ike (Wilbanks et al., 2007).

Resulting Insurance Costs and Impacts

The insurance industry—one of the world's largest industries—is one of the principal means through which the costs of climate change are distributed across society. Most of the economic consequences of climate change are processed through the public and private insurance markets, which is principally the mechanism that aggregates and distributes all of society's risk with climate change. This is one of the key ways in which the measurable costs of climate change are manifested, reported, and acknowledged by society.

In an average year, approximately 90 percent of insured catastrophic losses worldwide are weather-related. In the United States, for example, about half of all these losses are insured, which amounted to $320 billion between 1980 and 2005 (values are inflation adjusted to 2005 dollars). Surprisingly, the bulk of the cost was not for the headline-grabbing incidents such as hurricane damage, either. It was for other things caused by climate change, such as flood damage, that people do not often hear about on the news. In the United States, this is not helping the federal government, either. Because catastrophic weather events are on the rise, coupled with private insurers' withdrawal from various markets, it is placing the federal government at increased financial risk as the default last-resort insurer. Of note, the National Flood Insurance Program would have gone bankrupt after the storms of 2005 had they not been given the ability to borrow $20 billion from the U.S. Treasury (Wilbanks et al., 2007) (Figures 11.3, 11.4, and 11.5).

A LOOK AT HURRICANE KATRINA

According to the Insurance Information Institute, Hurricane Katrina was the costliest disaster in the history of insurance. For this single storm, the insurance industry paid $41.1 billion ($45.1 billion in 2009 dollars) and more than 1.7 million claims—covering a six-state area. About 15,000 claims adjusters from across the country were involved in handling the record number of claims for damage to homes, businesses, and vehicles. Louisiana alone accounted for 63 percent of insured losses, and Mississippi accounted for approximately 30 percent. In addition to that, the federally operated National Flood Insurance Program also paid out $16.1 billion in

FIGURE 11.3
Storm surge in New Orleans, Louisiana, during Hurricane Katrina. (Courtesy of Jocelyn Augustino, FEMA, http://www.fema.gov/photolibrary/.)

FIGURE 11.4
An aerial photo of destroyed Mississippi Gulf Coast Highway I-90 on October 4, 2005, as a result of winds and tidal surge from Hurricane Katrina. This section of the bridge connected Pass Christian, near Gulfport, to Bay St. Louis. (Courtesy of John Fleck, FEMA, http://www.fema.gov/photolibrary/.)

claims from flooding. Damage from flooding, including storm surge from a hurricane, is one of the events covered by the National Flood Insurance Program, but it is not covered under standard homeowner's insurance policies. The Insurance Information Institute also reported that an additional $2 to $3 billion of insured damages occurred at off-shore energy facilities.

Both public and private insurance are experiencing losses that require a combination of risk-based premiums and improved loss prevention. What has been identified to be a certain trend is that future increases in losses will be attributable to climate change as it increases the frequency and intensity of several different types of extreme weather events, such as severe thunderstorms and heat waves (Trapp et al., 2007).

FIGURE 11.5
After Hurricane Katrina, demolition was the only choice for many buildings such as this one along Highway 90 near Biloxi, Mississippi. After seven months (this picture was taken April 1, 2006), it was still difficult to comprehend the degree of devastation the Mississippi coast area sustained. (Courtesy of George Armstrong, FEMA, http://www.fema.gov/photolibrary/.)

Because of population increases and the fact that world populations are expanding and moving into harm's way, people are at greater risk of having climate-related insurance claims in the future. In the past, it is been easier for insurance companies to plan for future growth in claims. According to the 2009 U.S. Global Climate Change Research Program Report, this will no longer be the case: The past will no longer be able to be used as a reliable basis for the future as climates continue to change. It is now a challenge to design insurance systems that adequately price risks, reward loss prevention, and do not foster risk taking—such as repeatedly rebuilding flooded homes. This is especially an economic challenge when factoring in the insurers' ability to recover rising losses, combined with information gaps on the impacts of climate change and adaptation strategies. In fact, rising losses are already affecting the availability and affordability of insurance. In the United States, for example, several million customers can no longer economically purchase private insurance coverage and are turning to state-mandated insurance pools or going without any insurance coverage. Offsetting rising insurance costs is one benefit of mitigation and adaptation investments to reduce the impacts of climate change (Peterson et al., 2008).

There are no sectors of the insurance industry that will escape the effects of climate change. Segments that will be affected include:

- Property
- Crops
- Forest products
- Transportation infrastructure
- Livestock
- Business and supply-chain interruptions caused by weather extremes
- Water shortages
- Electricity outages
- Legal consequences
- Compromised health or loss of life

Because insurance companies can see the future consequences of climate change, some are actively working as partners in climate science and participating in the formulation of public-policy and adaptation strategies. Some companies have even promoted adaptation measures by providing premium incentives for their customers who are willing to fortify their properties, engaging in the process of determining building codes and land-use plans, and participating in the development and financing of new technologies and practices. For example, some insurance companies have been working with the Federal Emergency Management Agency (FEMA) Community Rating System, which is a point system that rewards communities that undertake floodplain management activities to reduce flood risk beyond the minimum requirement set by the National Flood Insurance Program. Everyone in these communities is rewarded with lower flood-insurance premiums (–5 to –45 percent) (Federal Emergency Management Agency, 2008). Other insurance agencies have acknowledged that mitigation and adaptation work together with a coordinated climate risk-management strategy and are now offering "green" insurance products that are designed to capture those associated benefits (Mills, 2006 and 2009).

Expanding populations, especially high concentrations in urban areas, is occurring globally but is most prevalent in lower-income countries. Even worse, many large cities are located in vulnerable areas such as floodplains and coasts. In the majority of these cities, the poor often live in the most marginal of the environments, in areas that are most susceptible to extreme events, and their ability to adapt is limited by their lack of financial resources.

Another major problem caused by current climate change is that over half of the world's population—including most of the world's major cities—depends on glacier melt or snowmelt to supply water for drinking and municipal uses. Although many locations today may have abundant water supplies and even have frequent floods caused by increases in glacier melt rates because of increased temperatures worldwide, this will not remain the case. The trend is projected to reverse as temperatures continue to rise and resulting glacier mass reduces. More winter precipitation is also projected to fall as rain rather than snow in the future, as well (IPCC, 2007). One consequence of this will be increased global migration. In addition, climate change may also alter trade relationships by changing the comparative trade advantages of regions or nations. Where climate change limits access to scarce resources or increases incidences of damaging weather events, consequences are highly likely to be felt in the global economy and security.

The overall energy economy is enormous and has far-reaching implications. It covers both financial and managerial resources and must take both into account to be adaptive. Mitigation projects, however, will most likely have to be taken on a regional scale because of the regional effects of extreme weather events and reduced water availability and the effects of increased cooling demands on especially vulnerable places and populations.

Most of the research that has been done to date has been on the effects of climate change on energy production and use and has focused on the impacts of production and use in buildings—specifically heating and cooling buildings. Studies so far indicate the demand for cooling energy increases from 5 to 20 percent per 1.8°F or warming, and the demand for heating energy drops by 3 to 15 percent for each 1.8°F of warming (Scott and Huang, 2007). As temperatures rise, the demand for electricity is expected to increase. An increase in peak demand can lead to a disproportionate increase in energy infrastructure investment. This in turn implies a greater demand for electricity because most cooling of buildings is provided by electricity, whereas heating is provided by a combination of natural gas and fuel oil. What is even worse is that the majority of electricity is currently generated from coal; therefore, producing more electricity to keep up with a growing demand in the future has the potential to increase the CO_2 emissions entering the atmosphere if more energy efficient technology is not introduced, noncarbon energy sources are not employed, or carbon capture and storage (discussed in Chapter 12) is not utilized as an option (Scott and Huang, 2007).

Another reason to wean ourselves off of fossil fuels is because generation of electricity in thermal power plants (coal, gas, or oil) is water intensive. For example, power plants rank only slightly behind irrigation in terms of freshwater withdrawals in the United States (Bull et al., 2007). In some regions, reductions in water supply caused by decreases in precipitation and/or water from melting snowpack are expected to be significant, causing a competition for water among various sectors, including energy production. Because of this, it is expected that water shortages will limit power-plant electricity production in many areas of the world. According to Bull et al., locations in the United States that can expect future water constraints on electricity production in thermal power plants by 2025 include Arizona, Utah, Texas, Louisiana, Georgia, Alabama, Florida, California, Oregon, and Washington state.

Energy production and delivery systems are also exposed to sea-level rise and extreme weather events in vulnerable regions. Sea-level rise will damage equipment as it floods and erodes. It will also increase the costs of energy generation if facilities have to be relocated farther inland. Extreme events, such as hurricanes, are expected to become more common and will damage and destroy energy facilities. The electricity grid is also vulnerable to climate change, including temperature changes to severe weather events. Severe weather events, such as ice storms, thunderstorms, and hurricanes, can destroy power lines. Electric power transformers can fail under extremely high temperatures. In addition, climate change is likely to affect some renewable energy sources across the nation, such as hydropower production in regions subject to changing patterns of precipitation or snowmelt. Hydroelectric generation is very sensitive to changes in precipitation and river discharge. For instance, every 1 percent decrease in precipitation results in a 2–3 percent drop in streamflow, and every 1 percent decrease in streamflow can result in a 3 percent drop in power generation. Such magnifying sensitivities can occur when water flows through multiple power plants in a river basin (Bull et al., 2007).

Climate change may also affect renewable energy sources. For example, changing cloud cover affects solar-energy resources, changes in winds affect wind power, and temperature and water availability affect biomass production (particularly related to water requirements for biofuels) (Bull et al., 2007). It has been recommended that more research be done in this area to get a better idea about where specific impacts would occur and how significant they would be (Scott and Huang, 2007).

The Stern Review: The Economics of Climate Change

Sir Nicholas Stern, Head of the UK Government Economic Service, chair of the Grantham Research Institute on Climate Change and the Environment at the London School of Economics (LSE), chair of the Centre for Climate Change Economics and Policy at Leeds University and LSE, and a former Chief Economist of the World Bank, compiled and released the Stern Review on October 30, 2006, for the British government, which provided clear scientific evidence that emissions from economic activity—particularly the burning of fossil fuels for energy—are causing changes to the earth's climate. The report discussed the effects of climate change on the global economy. It is one of the most widely known economic reports on climate change. The executive summary stated that "the [r]eview first examines the evidence on the economic impacts of climate change itself, and explores the economics of stabilizing greenhouse gases in the atmosphere." The second half of the review considered the "complex policy challenges involved in managing the transition to a low-carbon economy and ensuring that societies can adapt to the consequences of climate change that can no longer be avoided" (Stern, 2007).

The central message of the review was that if immediate action is not taken to reduce GHG emissions, the overall costs and risks of climate change can be equated to losing on the order of 5 percent of the GDP each year, forever. Even worse, if a wider range of risks and impacts is considered, the estimates of damage may rise to 20 percent or more of the global GDP. Stern concluded in his report that if action is taken immediately to reduce climate change, the cost could be as little as 1 percent of the annual global GDP. He also advised that actions taken globally by nations must be based on mutually reinforcing international frameworks. The recommendations he proposed in his review to cut GHG emissions are by the following methods:

- Increased energy efficiency
- Adoption of clean power, heat, and transport technologies
- Reduction of deforestation
- Supporting innovation and deployment of low-carbon technologies
- Remove barriers to energy efficiency—inform, educate, and persuade individuals to change their behavior
- Carbon pricing and budgeting through tax, trading, or regulation
- Carbon-capture technologies
- Changes in demand for energy-intensive technologies

Stern also proposes the following global-based approaches:

- International emissions trading
- Actions to reduce deforestation
- Technology cooperation
- Adaptation

His review stated that climate change is the greatest and widest-ranging market failure ever seen, presenting a unique challenge for economics. From an economics point of view, he urged the use of market mechanisms to offer incentives to change the global economy. The review provided recommendations, including environmental taxes, to minimize the economic and social disruptions that climate change will cause. It stated, "our actions over the coming few decades could create risks of major disruption to economic and social activity, later in this century and in the next, on a scale similar to those associated with the great wars and the economic depression of the first half of the 20th century. And it will be difficult or impossible to reverse these changes. Tackling climate change is the pro-growth strategy for the longer term and it can be done in a way that does not cap the aspirations for growth of rich or poor countries." Along with this, Stern also advocated greater personal and social responsibility in every individual's life that is then supported by sustained public and private investment in environmental change. In June 2008, Stern updated his report and increased the estimate for the annual cost of achieving stabilization between 500 and 550 ppm CO_2e to 2 percent of GDP to account for faster than expected climate change (Jowit and Wintour, 2008).

Stern's major conclusions concerning climate change include the following:

- The benefits of strong, early action on climate change outweigh the costs.
- Scientific evidence indicates increasing risks of serious, irreversible impacts from climate change associated with business-as-usual paths for emissions.
- Climate change may initially have small positive effects for a few developed countries, but it is likely to be very damaging for the much higher temperature increases expected by mid- to late century under "business as usual" scenarios.
- Transition to a low-carbon economy will bring challenges for competitiveness but also opportunities for growth. Policies to support the development of a range of low-carbon and high-efficiency technologies are critical now.
- Establishing a carbon price, through taxation, trading, or regulation, is an essential foundation for climate change policy. Creating a broadly similar carbon price signal around the world and using carbon finance to accelerate action in developing countries are urgent priorities for international cooperation.
- Climate change threatens the basic elements of life for people around the world—access to water, food production, health, and use of land and the environment.
- The impacts of climate change are not evenly distributed—the poorest countries and people will suffer earliest and most. If and when the damages appear, it will be too late to reverse the process. Thus we are forced to look a long way ahead.
- Integrated assessment modeling provides a tool for estimating the total impact on the economy; our estimates suggest that this is likely to be higher than previously suggested.
- Emissions have been, and continue to be, driven by economic growth, yet stabilization of greenhouse gas concentration in the atmosphere is feasible and consistent with continued growth.
- Adaptation policy is crucial for dealing with the unavoidable impacts of climate change, but it has been underemphasized in many countries.
- An effective response to climate change will depend on creating the conditions for international collective action.

- There is still a window of time in which to avoid the worst impacts of climate change if strong collective action starts immediately.

The Stern Review has had positive responses from several sources, as interviewed by the BBC. Pia Hansen, a European Commission Spokeswoman, said that doing nothing was not an option, that action must be taken now. Simon Retallack of the UK think tank IPPR reported, "This [review] removes the last refuge of the 'do-nothing' approach on climate change, particularly in the U.S." Richard Lambert, Director General of the Confederation of British Industry, said that a global system of carbon trading is "urgently needed." Charlie Kronick of Greenpeace reported, "Now the government must act and, among other things, invest in efficient decentralized power stations and tackle the growth of aviation" (BBC, 2006).

Although there was much praise for the report, there was also criticism. For example, Ruth Lea, Director of the Centre for Policy Studies, questioned the scientific consensus on climate change on which the Stern Review was based. In an article in the *Telegraph*, she commented that "Authorities on climate science say that the climate system is far too complex for modest reductions in one of the thousands of factors involved in climate change—such as carbon emissions—to have a predictable effect in magnitude, or even direction" (Lea, 2006). She also questioned the long-term economic projections made in the review, saying that economic projections of even a year or two were often wrong.

Professor Bill McGuire of Benfield UCL Hazard Research Centre commented that Stern may have greatly underestimated the effects of climate change (BBC, 2006). In addition, David Brown and Leo Peskett of the Overseas Development Institute, a UK think tank on international development, argued that the key proposals on how to use forests to combat climate change may prove difficult to implement (Brown and Peskett, 2006). In addition, on the BBC radio program *The Investigation*, several economists and scientists argued that Stern's assumptions in the review were more pessimistic than those made by most experts in the field and that the review's conclusions are at odds with the mainstream view (Cox and Vadon, 2007).

From an economic point of view, one of the issues that has been debated among economists is the discount rate used in the review. The concept of "discounting" is used by economists to compare economic impacts occurring at different times. It is used in Stern's calculation of the costs of mitigation, as well as the costs of "business as usual" climate change. Economists commonly state four main reasons for placing a lower value on consumption occurring in the future rather than in the present (Quiggin, 2007):

- Future consumption should be discounted simply because it takes place in the future and people generally prefer the present to the future (inherent discounting).

- Consumption levels will be higher in the future, so the marginal utility of additional consumption will be lower.
- Future consumption levels are uncertain.
- Improved technology of the future will make it easier to address climate change concerns.

Using a high discount rate decreases the assessed benefit of actions designed to reduce greenhouse gas emissions. The Stern Review's discount rate for climate change damages is approximately 1.4 percent, which was lower than that used in most previous studies on climate change damages.

Her Majesty's (HM) Treasury has issued a document where several economists are quoted praising the Stern Review; it includes favorable reviews written by Robert Solow, James Mirrlees, Amartya Sen, Joseph Stiglitz, and Jeffrey Sachs. Economists have varying views over the cost estimates of climate change mitigation given in the review, as well. Paul Ekins of King's College London commented that Stern's central mitigation cost estimate is "reasonable," but economists Robert Mendelsohn and Dieter Helm have commented that the estimate is probably too low (Cox and Vadon, 2007; Helm, 2008).

Stern's response to the various criticisms his review has received sticks with the conviction that early and strong action toward climate change is critical. His response is as follows (Dietz et al., 2007): "The case for strong and urgent action set out in the [r]eview is based, first, on the severe risks that the science now identifies (together with the additional uncertainties . . . that it points to but that are difficult to quantify) and, second, on the ethics of the responsibilities of existing generations in relation to succeeding generations. It is these two things that are crucial: risk and ethics. Different commentators may vary in their emphasis, but it is the two together that are crucial. Jettison either one and you will have a much-reduced programme for action—and if you judge risks to be small and attach little significance to future generations you will not regard global warming as a problem. It is surprising that the earlier economic literature on climate change did not give risk and ethics the attention they so clearly deserve, and it is because we chose to make them central and explicit that we think we were right for the right reasons."

Later in 2008, in an official letter, Joan Ruddock, MP of the UK government, dismissed the criticisms of the review made by several economists, which in her view show "a fundamental misunderstanding of the role of formal, highly aggregated economic modeling in evaluating a policy issue" (Ruddock, 2008).

Meeting the Challenges of Climate Impacts

Now that the public is becoming more educated (and seeing firsthand some of the effects of climate change), businesses are beginning to take a stronger role, attempting to project and take steps to adequately prepare for climate change. Some of the action taken so far has been reactive—businesses responding and adapting to changes that have already taken place. An example of this—reacting, such as getting out of harm's way—would be a business relocating somewhere further inland (in a safer zone) after losing their place of business to Hurricane Katrina. Another example, illustrated by the Pew Group, is farmers that are currently interested in obtaining drought- and flood-resistant seeds in response to increased weather extremes. The Pew Center identifies the fact that successful adaptation over the long term requires recognizing and acting on threats from an early stage—ideally before they occur—and identifying appropriate responses (Sussman, 2008).

One thing Sussman points out is that with the changes currently happening with the climate, it may be necessary to alter traditional business practices, requiring companies to put themselves through a paradigm shift, no longer relying on historical trends and decisions or keeping a "business as usual" mindset. Long-term mission assessments must now be made with climate change a central theme in order to plan ahead. Sussman cites a major reason to complete a risk assessment is to determine the likelihood of changes, with the best approach being proactive adaptation where future climate change is consciously anticipated and appropriate options incorporated into decision making. For example, in locating and building a new facility, it will be important to take into account the location of rivers if a water draw for electric power generation is necessary or cooling water is needed. Sussman believes that screening to identify the potential risks of short- and long-term climate change is the first step in determining whether or not a risk assessment is necessary to identify further actions.

The purpose of the screening is to determine whether the business might be at risk; what aspects are at risk and from what; and whether a more complete risk assessment is needed to determine exactly what, if any, further actions are needed. The goal of the screening is to classify/screen risks into one of three categories, as illustrated in Figure 11.6:

- Class A: What to assess now
- Class B: What can wait and be studied at a later date
- Class C: What requires no action

It is important to look at the business from a long-term perspective and determine the role climate change may have on it. This means identifying the projected climate and physical effects of climate change that could possibly threaten the success of the business operation, such as flooding, sea-level rise, or drought. Looking at the immediacy of an issue is also important, because it can prevent costly business decisions and potential mistakes and

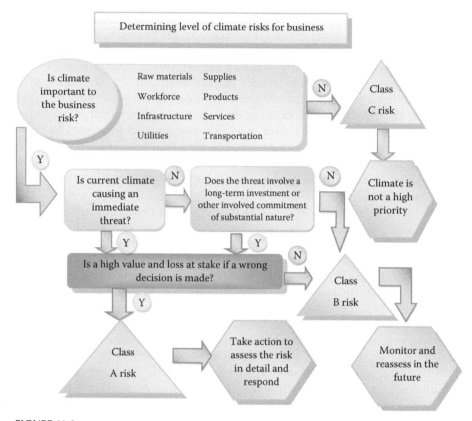

FIGURE 11.6
(See color insert.) It is an imperative part of any successful business implementation plan today to include the influence of any climate change impacts that may be significant to the operation of a business.

poor investment choices. For example, choosing a location of a business that may later need to be relocated could prove costly enough to threaten the survival of the business. Any sizable cost of a poor decision, whether it be a significant change in cash flow, a downturn in an investment, sizable restriction of growth goals, negative impacts on the firm's reputation, or any other significant measure of success, necessitates comprehensive risk assessment and management (Sussman, 2008).

Making long-term decisions based on developing carefully thought-out future plans is a wise economic move. Decisions that are made today as to where to locate a business and what design standards to build to will all have implications decades down the road. What Sussman emphasizes is that decisions made now will help plan for and avoid future vulnerabilities. It is important to gain a firm understanding in a long-term business solution of the systematic integration of global change in general and the relationships between and among ecosystem functions, socioeconomic drivers, goods and services, and global change challenges (Figure 11.7). Everything human populations do in their production of goods and services, which are made available through ecosystem functions, affect global systems. Those effects cannot only be far-reaching, but can cascade through various ecosystem functions. Although we may not always directly see the effects of our actions on the environment—or see them manifested immediately—the effects are still felt, not only by the human component but by the natural world as well.

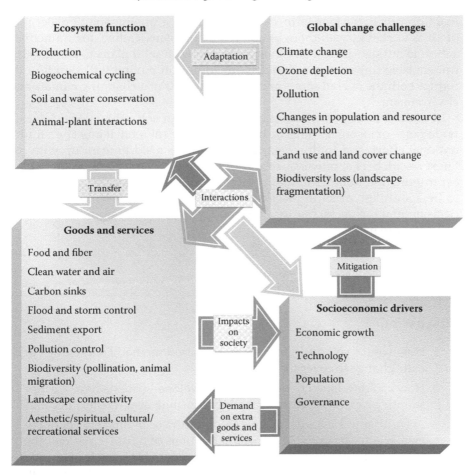

FIGURE 11.7
(See color insert.) Economics are influenced interactively by ecosystems and their resultant goods and services, as well as socioeconomic drivers, and global change. These components can act independently, dependently, or interdependently to form the complete economic picture, and a working knowledge of all components is necessary in order to make sound, long-term business decisions.

Conclusions

Scientific evidence continues to mount that climate change will directly or indirectly affect all sectors and regions of the global economy, although not equally. While there may be temporary benefits from a changing climate, the costs of climate change will rapidly exceed any benefits briefly gained in the few areas of the world that may realize any short-term benefits, and the costs of climate change will rapidly exceed any benefits, placing major stress on public sector budgets, personal income, and job security. Because of the extreme total economic costs of climate change, any delayed action or inaction will, in the long term, be the most costly economic solution. Because of this, workable, well-thought-out economic plans calling for immediate action to mitigate GHG emissions combined with solid plans to adapt to the unavoidable impacts from damage already set in motion will significantly reduce both the overall short- and long-term costs of a changing climate and environment.

In addition, not just one country or region can solve this global issue. A wide range of resources must be brought into the solution. It must be a multinational effort, and in this regard, governments need to team up with the latest researchers at universities and private research institutions for the latest information in order to make the most informed decisions. There needs to be a seamless mesh in place between the scientific community and the decision-making sector, so that the most prudent policies and investment decisions are made.

Discussion

1. How would you design an approach-based method concerning climate change to cover the next decade?
2. How can developed nations help undeveloped nations have a better adaptive capacity to deal with climate change over the next decade?
3. According to Richard Tol, any study of the economic effects of climate change begins with some assumptions of future emissions, the extent and pattern of warming, and other possible aspects of climate change such as sea-level rise and changes in rainfall and storm patterns. Any studies done must then translate from climate change to economic consequences. He states there are many methods available through which to accomplish this: statistical analysis, natural science papers and research, climate models, and lab experiments. If you were an economic analyst, which would you choose and why?

4. How many economic ramifications can you think of that could be a result of drought and desertification? Are they short or long term, and how much of society would they ultimately affect?

5. Using an atlas, locate at least twelve major seaports and discuss the economic and socioeconomic ramifications of climate change, both short and long term.

6. In addition to the impacts discussed, what other impacts can you think of that farmers will face in a warming world?

7. In addition to the impacts discussed, what other impacts can you think of that ranchers will face in a warming world?

8. What mitigation steps can farmers and ranchers begin to take now to minimize the economic damage they will face in the coming years?

9. What types of tourism industries will be negatively affected by climate change? Can you name any that might be affected positively?

10. Should healthcare benefits be changed to reflect issues relating to climate change? Should any increased costs related to illness connected to the climate raise insurance premiums so that the cost is passed on to the patient? What if someone cannot afford insurance? Who should be responsible for payment in that case?

11. If you were a transportation planner and were instructed to take into account climate change in the design of a long-term planning and project development, how would you go about it and what elements would you include? Why?

12. Concerning homeowner's insurance for a home destroyed by a hurricane: Should a homeowner be allowed to rebuild in the same place again, knowing the risks involved? Why or why not?

13. If you were a land manager, what kind of plan would you put together to protect the fragile arid lands of the American Southwest or any other similar fragile arid region?

14. The Stern Report says: "Establishing a carbon price, through taxation, trading, or regulation, is an essential foundation for climate change policy. Creating a broadly similar carbon price signal around the world, and using carbon finance to accelerate action in developing countries are urgent priorities for international cooperation." Do you agree with this philosophy? Why or why not? If not, what would you propose?

References

Anselmo, R. 2011. Social and economic impact of drought and desertification. *Economic and Social Council: 41st Agenda.* http://www.munfw.org/archive/41st/ecosoc1.htm (accessed April 4, 2011). Discusses how drought and desertification has gone from being a topic only addressed by environmentalists to one posing a serious threat to the global ecological, economic, and political future of society.

Associated Press. 2009. Study: Climate change to affect monsoon in South Asia. *USA Today*, February 27. http://www.usatoday.com/weather/climate/globalwarming/2009-02-27-climate-change-asian-monsoon_N.htm (accessed April 15, 2011). Discusses the impacts to the monsoon patterns caused by climate change.

Barnett, T. P., D. W. Pierce, H. G. Hidalgo, C. Bonfils, B. D. Santer, T. Das, G. Bala, A. W. Wood, T. Nazawa, A. A. Mirin, D. R. Cayan, and M. D. Dettinger. 2008. Human-induced changes in the hydrology of the western United States. *Science*. 319(5866): 1080–1083.

BBC. 2006. Expert reaction to stern review. October 30. Provides professional reactions to the Stern Report.

Brennan, P. 2011. Report: Strong U.S. action needed on warming. *Orange County Register*, May 12. http://sciencedude.ocregister.com/tag/national-academy-of-sciences/ (accessed August 19, 2011).

Brown, D., and L. Peskett. 2006. The challenge of putting Stern's prescriptions into practice. ODI Weblog, December 5. http://blogs.odi.org.uk/blogs/main/archive/2006/12/05/David_Brown.aspx. Offers some criticisms and questions about the Stern Review.

Bull, S. R., D. E. Bilello, J. Edmann, M. J. Sale, and D. K. Schmalzer. 2007. Effects of climate change on energy production and distribution in the United States. In: *Effects of Climate Change on Energy Production and Use in the United States.* Edited by T. J. Wilbanks, V. Bhatt, D. E. Bilello, S. R. Bull, J. Edmann, W. C. Horak, Y. J. Huang, M. D. Levine, M. J. Sale, D. K. Schmalzer, and M. J. Scott. Synthesis and Assessment Product 4.5. Washington, DC: U.S. Climate Change Science Program, pp. 45–80.

Climate Change Science Program. 2009. *Best Practice Approaches for Characterizing, Communicating, and Incorporating Scientific Uncertainty in Decision Making: A Report by the U.S. Climate Change Science Program and the Subcommittee on Global Change Research.* Washington, DC: National Oceanic and Atmospheric Administration. http://www.globalchange.gov/publications/reports/scientific-assesments/saps/311 (accessed October 22, 2010). Discusses methods for dealing with economic considerations of climate change.

Cordell, H. K., B. McDonald, R. J. Teasley, J. C. Bergstrom, J. Martin, J. Bason, and V. R. Leeworthy. 1999. Outdoor recreation participation trends. In: *Outdoor Recreation in American Life: A National Assessment of Demand and Supply Trends.* Edited by H. K. Cordell and S. M. McKinney. Campaign, IL: Sagamore Publishing, pp. 219–321.

Cox, S., and R. Vadon. 2007. Running the rule over Stern's numbers. *BBC News*, January 26. http://news.bbc.co.uk/1/hi/sci/tech/6295021.stm. Discusses the mitigation analysis in the Stern Review.

DeCanio, S. J. 2007. Reflections on climate change, economic development, and global equity, October 17. http://www.stephendecanio.com/Stephen_DeCanio_Site/DeCanioLeontief07-1.pdf (accessed November 9, 2010). Discusses the basis on which to place value on assets in cost-benefit analysis for climate change studies.

de Quetteville, H. 2008. Global warming forces European ski resorts to offer summer sports. *The Telegraph*, May 26. http://www.telegraph.co.uk/news/world-news/2032879/Global-warming-forces-European-ski-resorts-to-offer-summer-sports.html (accessed April 23, 2011). Provides an overview of what alpine towns in Europe are doing at recreational facilities in response to warmer temperatures.

Dietz, S., D. Anderson, N. Stern, C. Taylor, and D. Zenghelis. 2007. Right for the right reasons: A final rejoinder on the Stern Review. *World Economics*. 8(2): 229–258.

Ebi, K. L., J. Balbus, P. L. Kinney, E. Lipp, D. Mills, M. S. O'Neill, and M. Wilson. 2008. Effects of global change on human health. In: *Analyses of the Effects of Global Change on Human Health and Welfare and Human Systems*. Edited by J. L. Gamble, K. L. Ebi, F. G. Sussman, and T. J. Wilbanks. Synthesis and Assessment Product 4.6. Washington, DC: U.S. Environmental Protection Agency, pp. 39–87. Provides information on the effects climate change will have on pollen production and its effects on society.

Energy Information Administration. 2008. Energy in brief: What are greenhouse gases and how much are emitted by the United States? http://tonto.eia.doe.gov/energy_in_beif/greenhouse_gas.cfm (accessed November 7, 2010). Discusses climate change and its ramifications.

Environmental Protection Agency. 2011. *Inventory of U.S. Greenhouse Gas Emissions and Sinks: 1990–2006*. USEPA 430-R-08-005. Washington, DC: U.S. Environmental Protection Agency. http://www.epa.gov/climatechange/emissions/usinventory report.html (accessed April 1, 2011). Discusses comprehensive inventory of information about GHG emissions.

Fankhauser, S., and R. S. J. Tol. 1996. Climate change costs—recent advancements in the economic assessment. *Energy Policy*. 24(7): 665–673.

Federal Emergency Management Agency. 2008. Community Rating System (CRS) documents. http://training.fema.gov/EMIWeb/CRS/ (accessed March 20, 2011). Discusses FEMA's approach to climate change and insurance risks and mitigation.

Feely, R. A., C. L. Sabine, K. Lee, W. Berelson, J. Kleypas, V. J. Fabry, and F. J. Millero. 2004. Impact of anthropogenic CO_2 on $CaCO_3$ system in the oceans. *Science*. 305(5682): 362–366.

Food and Agricultural Organization. 2008. Climate change will have major impact on fishing industry, says UN agency. *UN News Service*, July 10. http://www.un.org/apps/news/printnewsAr.asp?nid=27330. Discusses how changing temperatures will play havoc with the fishing industry in the coming years.

Grebmeier, J. M., J. E. Overland, S. E. Moore, E. V. Farley, E. C. Carmack, L. W. Cooper, K. E. Frey, J. H. Helle, F. A. McLaughlin, and S. L. McNutt. 2006. A major ecosystem shift in the northern bering sea. *Science*. 311(5766): 1461–1464.

Hamilton, J. M., and R. S. J. Tol. 2004. The impact of climate change on tourism and recreation. In: *Human-Induced Climate Change: An Interdisciplinary Assessment*. Edited by M. Schlesinger, H. S. Kheshgi, J. Smith, F. C. de la Chesnaye, J. M. Reilly, T. Wilson, and C. Kolstad. Cambridge, UK: Cambridge University Press, pp. 147–155.

Hatfield, J., K. Boote, P. Fay, L. Hahn, C. Izaurralde, B. A. Kimball, T. Mader, J. Morgan, D. Ort, W. Polley, A. Thomson, and D. Wolfe. 2008. Agriculture. In: *The Effects of Climate Change on Agriculture, Land Resources, Water Resources and Biodiversity in the United States*. Edited by P. Backlund, O. Ort, W. Polley, A. Thomson, D. Wolfe, M. G. Ryan, S. R. Archer, R. Birdsey, C. Dahm, L. Heath, J. Hicke. D. Hollinger, T. Huxman, G. Okin, R. Oren, J. Randerson, W. Schlesinger, D. Lettenmaier, D. Major, L. Poff, S. Running, L. Hansen, D. Inouye, B. P. Kelly, L. Meyerson, B. Peterson, and R. Shaw. Synthesis and Assessment Product 4.3. Washington, DC: U.S. Department of Agriculture, 21–74.

Helm, D. 2008. Climate-change policy: Why has so little been achieved? *Oxford Review of Economic Policy*. 24(2): 211–238. Provides a review of the Stern Review's mitigation proposals.

Hope, C. W. 2006. The marginal impact of CO_2 from PAGE2002: An integrated assessment model incorporating the IPCC's five reasons for concern. *Integrated Assessment Journal*. 6(1): 19–56.

Intergovernmental Panel on Climate Change. 2001. *Summary for Policymakers, Climate Change 2001: Synthesis Report*. A Contribution of Working Groups I, II, and III to the Third Assessment Report of the Intergovernmental Panel on Climate Change. New York: Cambridge University Press. http://www.ipcc.ch/publications_and_data/publications_and_data_reports.htm (accessed September 9, 2010). Discusses methods for equitably managing the climate change problem for future generations.

Intergovernmental Panel on Climate Change. 2007. *Climate Change 2007: Synthesis Report*. Contribution of Working Groups I, II and III to the Fourth Assessment Report of the Intergovernmental Panel on Climate Change. Edited by R. K. Pachauri and A. Reisinger. Geneva, Switzerland: Intergovernmental Panel on Climate Change. http://www.ipcc.ch.publications_and_data/publications_ipss_fourth_assessment_report_synthesis_report.htm (accessed September 13, 2010). Discusses management issues of climate change.

Janetos, A., L. Hansen, D. Inouye, B. P. Kelly, L. Meyerson, B. Peterson, and R. Shaw. Biodiversity. 2008. In: *The Effects of Climate Change on Agriculture, Land Resources, Water Resources, and Biodiversity in the United States*. Synthesis and Assessment Product 4.3. Washington, DC: U.S. Department of Agriculture, pp. 151–181.

Jowit, J., and P. Wintour. 2008. Cost of tackling global climate change has doubled, warns Stern. *The Guardian*, June 26. Provides an update to Nicolas Stern's Report on climate change from October 2006.

Kafalenos, R. S., K. J. Leonard, D. J. Beagan, V. R. Burkett, B. D. Keim, A. Meyers, D. T. Hunt, R. C. Hyman, M. K. Maynard, B. Fritsche, R. H. Henk, E. J. Seymour, L. E. Olson, J. R. Potter, and M. J. Savonis. 2008. What are the implications of climate change and variability for Gulf Coast transportation? In: *Impacts of Climate Change and Variability on Transportation Systems and Infrastructure: Gulf Coast Study. Phase I*. Edited by M. J. Savonis, V. R. Burkett, and J. R. Potter. Synthesis and Assessment Product 4.7. Washington, DC: U.S. Department of Transportation, pp. 4-1 to 4F-27.

Karl, T. R., J. M. Melillo, and T. C. Peterson (Eds.) 2009. *Global Climate Change Impacts in the United States*. London: Cambridge University Press. Presents a comprehensive report on what types of changes to expect in the United States between now and the end of the century.

Kiely, T. E., D. Donaldson, and A. Grube. 2004. Pesticides industry sales and usage: 2000 and 2001 Market estimates. Washington, DC: U.S. Environmental Protection Agency. http://www.epa.gov/oppbeadl/pestsales/ (accessed January 3, 2011). Discusses costs and usage of pesticides and their economic ramifications under changing climate conditions.

Lea, R. 2006. Just another excuse for higher taxes. *The Telegraph*, October 31. http://www.telegraph.co.uk/comment/personal-view/3633767/Just-another-excuse-for-higher-taxes.html (accessed August 15, 2011).

Maddison, D. J. 2003. The amenity value of the climate: The household production function approach. *Resource and Energy Economics*. 25(2): 155–175.

Mendelsohn, R. O., M. E. Schlesinger, and L. J. Williams. 2000. Comparing impacts across climate models. *Integrated Assessment*. 1(1): 37–48.

Mills, E. 2006. Synergisms between climate change mitigation and adaptation: An insurance perspective. In: *Mitigation and Adaptation Strategies for Global Change*. 12(5): 809–842.

Mills, E. 2009. A global review of insurance industry responses to climate change. *The Geneva Papers*. 34: 323–359. Provides information on insurance programs and responses to climate change risks.

National Research Council. 2008. *Potential Impacts of Climate Change on U.S. Transportation: Transportation Research Board Special Report 290*. Washington, DC: Transportation Research Board. http://onlinepubs.trb.org/onlinepubs/sr/sr290.pdf (accessed December 9, 2010). Discusses the ramifications of climate change on the transportation factor and long-term planning and considerations that must take place to avoid negative feedback.

National Oceanic and Atmospheric Administration. 2008. Midwestern U.S. floods. National Climatic Data Center, July 9. Discusses the flooding that occurred during the spring of 2008 in the Midwestern United States.

Nordhaus, W. D., and J. G. Boyer. 2000. *Warming the World: Economic Models of Global Warming*. Cambridge, MA: MIT Press.

O'Driscoll, P. 2006. Study links extended wildfire seasons to global warming. *USA Today*, July 7. http://www.usatoday.com/weather/climate/2006-07-06-climate-fires_x.htm (accessed August 15, 2011).

Orr, J. C., V. J. Fabry, O. Aumont, L. Bopp, S. C. Doney, R. A. Feely, A. Gnanadesikan, N. Gruber, A. Ishida, F. Joos, R. M. Key, K. Lindsay, E. Maier-Reimer, R. Matear, P. Monfray, A. Mouchet, R. G. Najjar, G. K. Plattner, K. B. Rodgers, C. L. Sabine, J. L. Sarmiento, R. Schlitzer, R. D. Slater, I. J. Totterdell, M. F. Weirig, Y. Yamanaka, and A. Yool. 2005. Anthropogenic ocean acidification over the twenty-first century and its impact on calcifying organisms. *Nature*. 437(7059): 681–686. Discusses the impacts on fisheries in Alaska caused by climate change.

Outdoor Industry Foundation. 2006. *Outdoor Recreation Participation Report*. Boulder, CO: Outdoor Foundation. Presents detailed information on recreation and recreational trends in the United States.

Peet, M. M., and D. W. Wolf. 2000. Crop ecosystem responses to climate change: Vegetable crops. In: *Climate Change and Global Crop Productivity*. Edited by K. R. Reddy and H. F. Hodges. New York: CABI Publishing. Discusses the effect temperature extremes will have on the future of agricultural products.

Peterson, T. C., D. M. Anderson, S. J. Cohen, M. Cortez-Vázquez, R. J. Murnane, C. Parmesan, D. Phillips, R. S. Pulwarty, and J. M. R. Stone. 2008. Why weather and climate extremes matter. In: *Weather and Climate Extremes in a Changing Climate*.

Regions of Focus: North America, Hawaii, Caribbean, and U.S. Pacific Islands. Edited by T. R. Karl, G. A. Meehl, C. D. Miller, S. J. Hassol, A. M. Waple, and W. L. Murray. Synthesis and Assessment Product 3.3. Washington, DC: U.S. Climate Change Science Program, pp. 11–34.

Potter, J. R., V. R. Burkett, and M. J. Savonis. 2008. Executive summary. In: *Impacts of Climate Change and Variability on Transportation Systems and Infrastructure: Gulf Coast Study, Phase I.* Edited by M. J. Savonis, V. R. Burkett, and J. R. Potter. Synthesis and Assessment Produce 4.7. Washington, DC: U.S. Department of Transportation, pp. ES-1 to ES-10. Provides an overview on long-term transportation infrastructure planning.

Quiggin, J. 2007. Stern and his critics on discounting and climate change. risk and sustainable management group/climate change. Working Paper: CO7#1. http://www.uq.edu.au/rsmg/WP/WPC07_1.pdf (accessed August 15, 2011). Presents some criticisms of the Stern Review.

Reddy, S. 2009. The Indian Monsoon and the economy: How a changing climate could turn the waltz into a danse macabre. Natural Resources Defense Council Staff Blog, June 29. http://switchboard.nrdc.org/blogs/sreddy/the_indian_monsoon_and_the_eco.html. Accessed August 15, 2011. Offers information about the effects of climate change on the monsoons

Rehdanz, K., and D. J. Maddison. 2005. Climate and happiness. *Ecological Economics,* 52(1): 111–125.

Ruddock. J. 2008. Letter of Joan Ruddock MP to Andrew Tyrie MP, March. http://www.fnu.zmaw.de/fileadmin/fnu-files/publication/tol/Ruddock.pdf (accessed July 10, 2011). Provides a review of the Stern Review and associated criticisms.

Scott, M. J., and Y. J. Huang. 2007. *Effects of Climate Change on Energy Production and Use in the United States.* Edited by T. J. Wilbanks, V. Bhatt, D. E. Bilello, S. R. Bull, J. Edmann, W. C. Horak, Y. J. Huang, M. D. Levine, M. J. Sale, D. K. Schmaizer, and M. J. Scott. Synthesis and Assessment Production 4.5. Washington, DC: U.S. Climate Change Science Program, pp. 8–44.

Sedjo, R., and B. Sohngen. 2009. The implications on increased use of wood for biofuel production. *Resources for the Future.* Issue Brief #09-04, June. www.rff.org/rff/documents/RFF-IB-09-04.pdf (accessed April 1, 2011). Discusses methods to utilize forest resources to enhance carbon sequestration methods.

Smit, B., and O. Pilifosova. 2001. Adaptation to climate change in the context of sustainable development and equity. In: *Climate Change 2001: Impacts, Adaptation and Vulnerability.* Contribution of Working Group II to the Third Assessment Report of the Intergovernmental Panel on Climate Change. New York: Cambridge University Press. http://www.ipcc.ch/publications_and_caga/publications_and_data_reports.htm (accessed November 12, 2010). Discusses the adaptation abilities of various nations.

Sohngen, B. S. 2009. An analysis of forestry carbon sequestration, as a response to climate change. Frederiksberg, Denmark: Copenhagen Consensus Center. http://fixtheclimate.com/uploads/tx_templavoila/AP_Forestry_Sohngen_v.2.0.pdf

Stern, N. 2007. *The Economics of Climate Change: The Stern Review.* Cambridge, UK: Cambridge University Press.

Sussman, F. G., and J. R. Freed. 2008. *Adapting to Climate Change: A Business Approach.* Arlington, VA: Pew Center on Global Climate Change. http://www.pew climate.org/docUploads/Business-Adaptation.pdf Discusses business adaptation options in light of climate change.

Tol, R. S. J. 2009. The economic effects of climate change. *Journal of Economic Perspectives.* 23(2): 29–51.

Trapp, R. J., N. S. Diffenbaugh, H. E. Brooks, M. E. Baldwin, E. D. Robinson, and J. S. Pal. 2007. Changes in severe thunderstorm environment frequency during the 21st century caused by anthropogenically enhanced global radiative forcing. *Proceedings of the National Academy of Sciences.* 104(50): 19719–19723.

U.S. Global Change Research Program. 2009. *Global Climate Change Impacts in the United States.* New York: Cambridge University Press. http://downloads.globalchange.gov/usimpacts/pdfs/climate-impacts-report.pdf (accessed August 15, 2011).

Wilbanks, T. J., R. Romero Lankao, M. Bao, R. Berkhout, S. Cairncross, J. P. Cero, M. Kasha, R. Wood, and R. Zapata-Marti. 2007. Industry, settlement and society. In: *Climate Change 2007: Impacts, Adaptation, and Vulnerability.* Contribution of Working Group II to the Fourth Assessment Report on the Intergovernmental Panel on Climate Change. Edited by M. L. Parry, O. F. Canziani, J. P. Palutikof, P. J. van der Linden, and C. E. Hanson. New York: Cambridge University Press, 357–390. Discusses the impacts on the insurance industry from climate change.

Ziska, L. H. 2003. Evaluation of the growth response of six invasive species to past, present and future atmospheric carbon dioxide. *Journal of Experimental Botany.* 54(381): 395–404. Explores the human allergic reactions to species whose growth has been enhanced by CO_2 concentrations in the atmosphere.

Suggested Reading

Helm, D., and C. Hepburn. 2010. *The Economics and Politics of Climate Change.* Cary, NC: Oxford University Press.

Hoffman, A. J., and J. G. Woody. 2008. *Climate Change: What's Your Business Strategy? (Memo to the CEO).* Boston, MA: Harvard Business Press.

Rampell, C. 2009. The economic impact of climate change. *Economix,* June 16. http://economix.blogs.nytimes.com/2009/06/16/the-economic-impact-of-climate-change/ (accessed April 13, 2011). Discusses the economic challenges predicted by climate change.

12

The Economics of Mitigation Options

Overview

Throughout the world, countries are adopting policies in an attempt to make progress against climate change. As we have seen so far, positive actions include increasing renewable energy generation and encouraging energy efficiency. The positive effects of these actions are reducing vulnerability to energy price spikes, promoting development of local economies, and improving air quality. This chapter examines cap and trade as a policy tool, how the carbon trading market works in an international arena, the need for global action, and possible economic implications. It also examines the economics of mitigation options and looks specifically at new technology including carbon capture and storage. Specifically, it outlines the process in general, as defined by the Intergovernmental Panel on Climate Change (IPCC), and then examines specialized technology in two unique settings: geological formation sequestration and deep ocean sequestration. In conclusion, this chapter explores the value and necessity of various global adaptation strategies.

Introduction

As the earth's atmosphere continues to warm and more people become aware and educated about climate change—including its effects and ramifications for the future—efforts worldwide are being made to reduce its impacts, find ways to mitigate the situation, and adapt to the present environment, as well as prepare for the future. Every person on this earth will have to, in some way, adapt to the effects of climate change, focusing on ways to help solve the problem, reduce what impacts are possible, find ways to be environmentally responsible, and learn to cope in a positive way with permanent change. On a community, national, and international level there are also adaptation and mitigation options available—some already successfully in operation, others just beginning, and still others on the horizon. The important thing is that we keep progressing and moving ahead, taking the action possible to minimize what impacts we can, prevent what we can, and adapt to the rest. In order to make that possible, our forward actions cannot pause.

Development

Cap and Trade

Cap and trade is defined as "an environmental policy tool that delivers results with a mandatory cap on emissions" (EPA, 2010). The cap is the foundation on which the policy is constructed—it is the permissible carbon emission limit. For an individual country, it is the absolute, nationwide limit on climate change pollution. This measurement is usually set on a scale of billions of tons of CO_2 (or equivalent for other greenhouse gases [GHGs]) released into the atmosphere each year. Once the cap is in place and is being met, then over time, the cap is lowered in order to further cut emissions, the principal objective being to lower it enough over time to avoid the worst consequences of climate change.

The trade portion of the system is a market created by powerful incentives for companies to reduce the pollution they would normally emit. The trade market also works with the individual emitters and provides flexibility as to how they can meet their limits. In order to make all this happen within a country (each country under the Kyoto Protocol has a specific emission reduction level they are working toward), the respective government creates allowances that add up to the total emissions allowed under the cap. Each year, those industries and businesses subject to the cap must turn in allowances equal to their emissions for that year. Examples of industries and businesses that must do this include power plants, manufacturing industries, chemical industries, steel industries, mining companies, processing industries, and any other entity that produces and releases large amounts of CO_2 into the atmosphere. In order for the nation to meet the cap, each of these entities must reduce their emissions. If an entity reduces its emissions enough that it has more allowances than it needs, it can profit by selling the extra allowances. This opportunity gives them the incentive to reduce their emissions below what is mandated by the cap.

If an entity finds it too expensive to reduce its emissions, cap and trade allows it to purchase more allowances from other entities that have reduced their emissions far enough that they have extra allowances. The more a company reduces its emissions, the more money it can either make or save.

Internationally cap and trade works the same as when applied within a single country. Under the Kyoto Protocol, countries required to reduce their emissions are allowed to purchase carbon credits from developing countries or from industrialized countries whose emissions are below the level required. The credits cover emissions of all GHGs, which are expressed as carbon dioxide equivalents (CO_2e).

According to the Environmental Defense Fund (EDF), credits applying to any GHG are a serious limitation to the policy, and they believe credits should only be used for specific types of pollution. The EDF says that CO_2 travels quickly to the upper atmosphere and does not become concentrated in one particular area of the landscape. Emissions such as mercury, however, are usually deposited near where they are emitted, creating hot spots. Because mercury is also a toxin that poses a threat to human health, it should not be included in cap and trade (MacLeod, 2007).

The international trade in carbon credits is intended to promote investment in energy efficiency, renewable energy, and other ways of reducing emissions. In the majority of developed, industrialized countries, GHG-emitting companies have taken on the responsibility of running, regulating, and facilitating the trade of carbon credits in the carbon market. There are two main types of carbon markets: project-based markets and allowance-based markets.

Project-Based Markets

Project-based markets encourage investment in companies or programs that are committed to reducing emissions. These projects are run under the clean development mechanism (CDM) or joint implementation (JI). The CDM is an arrangement under the Kyoto Protocol that allows industrialized countries with a GHG-reduction requirement (called an Annex B party) to invest in projects that reduce emissions in developing countries as an alternative to more expensive emission reductions in their own countries. The critical factor that distinguishes an approved CDM carbon project is that it must prove that its actions have resulted in a reduction of emissions in the developing country that would not have occurred otherwise—a concept called additionality. What the CDM does, in effect, is allow net global GHG emissions to be reduced at a much lower global cost through the financing of emissions reductions projects in developing countries where the costs are much lower than they would be in industrialized countries. The CDM is supervised by the CDM executive board and is overseen by the Conference of the Parties of the United Nations Framework Convention on Climate Change (UNFCCC). According to the UNFCCC, the CDM is a trailblazer. It is the first global environmental investment and credit scheme of its kind, providing a standardized emissions offset plan. An example of a CDM might involve, for example, a rural electrification project using solar panels or the installation of more energy-efficient boilers. The UNFCCC views CDM as a way to stimulate sustainable development and emission reductions, while giving industrialized countries some flexibility in how they meet their emission-reduction or -limitation targets.

The JI in the Kyoto Protocol helps Annex I countries (developed nations) with GHG caps to meet their obligations. Any Annex I country can invest in emission reduction projects (called joint implementation projects) in any other Annex I country as an alternative to reducing emissions domestically. This mechanism allows countries to lower the costs of complying with their respective Kyoto targets by investing in GHG reductions in an Annex I country where reductions are cheaper and then applying the credit for those reductions toward their commitment goal. An example of a JI project could involve replacing a coal-fired power plant with a more efficient combined heat and power plant or a coal-heated building with a geothermal-heated building. JI projects are undertaken in countries that have economies in transition. JI projects differ from CDM projects in that JI projects are done in countries that have an emission-reduction requirement.

Through a JI project, emission reductions are awarded credits called emission reduction units (ERUs), where one ERU represents an emission reduction equaling 1 metric ton of CO_2e. The ERUs come from the host country's pool of assigned emissions credits, known as assigned amount units (AAUs). Each Annex I party has a predetermined amount of AAUs that are calculated on the basis of its 1990 GHG emission levels. By requiring JI credits to come from a host country's pool of AAUs, the Kyoto Protocol ensures that the total amount of emissions credits among Annex I parties does not change for the duration of the Kyoto Protocol's first commitment period. JI offers a flexible, cost-effective means of fulfilling part of their Kyoto commitments, while the host country (receiver) benefits from both foreign investment and technology transfer. A JI project must provide a reduction in emissions by sources or an enhancement of removal by sinks that is additional to what would otherwise have occurred. As far as project-based markets using CDM or JI mechanisms, the principal sellers are in Asia and South America.

Allowance-Based Markets

Allowance-based markets are what enable large companies—such as energy producers—to purchase emission allowances under plans administered by international carbon trading organizations, such as the EU Emissions Trading System. Allowance-based markets enable companies to offset their emissions by purchasing credits from countries that either have no limit placed on their emissions or have kept emissions below the level required. Since 2003, partly spurred on by the EU Emissions Trading System's start in 2005, the carbon-trading business has been growing. Carbon-trading schemes are now opening up worldwide and include the Carbon TradeEx America and the Chicago Climate Exchange, which was established by several large corporations along with the World Resources Institute.

The concept of carbon markets is still fairly new. Today, they only account for roughly 0.5 percent of the annual global GHG emissions. The idea is gaining in popularity and is being recognized as an effective global tool for slowing climate change. Especially encouraging is the fact that carbon trade is now being conducted within the United States, which is not a participant in the Kyoto Protocol (Figure 12.1).

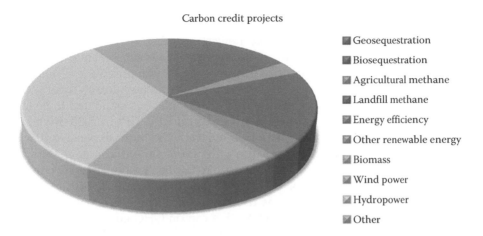

Carbon credit projects

- Geosequestration
- Biosequestration
- Agricultural methane
- Landfill methane
- Energy efficiency
- Other renewable energy
- Biomass
- Wind power
- Hydropower
- Other

FIGURE 12.1
(See color insert.) Trading carbon credits is one way to share the burden of reducing CO_2 emissions globally. Current trading ratios are as follows: geosequestration (5%), biosequestration (11%), agricultural methane (3%), landfill methane (16%), energy efficiency (4%), biomass (3%), wind power (15%), hydropower (32%), other renewable energy (1%), and other (10%).

Carbon credits are sold in 91 metric-ton units. If a business is selling credits but does not have 91 tons, then the carbon-trading company combines more than one available partial unit together to make a salable unit. There is still debate on what is tradable and how concrete an emissions reduction a given practice achieves. To deal with uncertainty, some practices are discounted. On a farm, for example, tradable units considered include the following:

- Capture of methane from a waste lagoon/anaerobic digester
- Practice of no-till to sequester carbon on large acreages
- Reduction of nitrogen application to reduce nitrous-oxide emissions and energy
- Practice of timber stand improvement in woodlands to sequester carbon in trees
- Supply of an energy processor with wood chips, grass for pellets, oil seed for biodiesel, etc., to displace fossil fuels
- Completion of improvements in efficiency, reducing energy use
- Use of wind, solar, or geothermal energy sources to displace fossil fuel use

Although carbon trading is a futures market, the rules of the game are still being developed. Income generated from carbon trading could help pay for adoption of new practices and keep farms or land financially viable.

The Economics of Cap and Trade

According to Nat Keohane, PhD, newest member of the U.S. National Economic Council as of January 2011, serving as advisor on environmental and energy policy to President Obama, aggressive cap and trade is not only affordable, but it is also critical to both the earth and humanity's future. The cost to the economy will be minimal—for example, it is estimated to be less than 1 percent of the U.S. gross domestic product (GDP) in 2030. Keohane also stresses that the longer action is delayed, the more expensive it will be to make emission cuts. In addition, the more time that passes without addressing the issues, the more irreversible damage will be done by climate change. Through the use of economic models, Keohane determined that by continuing with a business as usual approach, the U.S. economy would reach $26 trillion by January 2030. With a cap on GHG emissions, however, the economy will reach the same level only 2–7 months later. Therefore, the impact on the economy would not be significant—"just pennies a day," according to Keohane (Environmental Defense Fund, 2009).

He also stresses that total job loss would be minimal (the manufacturing sector would experience some impact), and the new carbon market would create a multitude of new jobs. Households will be most affected by energy costs, but even there the increase would be modest. Overall costs would be small enough to allow programs to be developed that would take any burden off low-income households.

Dr. Keohane believes that cap and trade is the best means to fight climate change because it not only gives each company the ability to choose how to cut their emissions, but it also gives the economy the most flexibility to reduce pollution in the most cost-effective way. He also says it turns market failure into market success: "Global warming is a classic example of what economists term 'market failure.' GHG emissions have skyrocketed because their hidden costs are not factored into business decisions—factories and power plants pay for fuel but not for the pollution they cause. Putting a dollar value on the pollution fixes that failure and gives industry incentive to pollute less" (Keohane, 2008).

Keohane also says that cap and trade taps American ingenuity and that history clearly shows that Americans can overcome steep challenges. In two short years during World War II, for example, Americans redirected much of the U.S. economy. Manufacturers produced different goods against tight deadlines. Detroit converted car factories to munitions production. Fireworks factories made military explosives. A. C. Gilbert, a maker of model train engines, produced airborne navigational instruments. Therefore, based on past performance, given the right incentives, America can transform the way energy is made as well.

Keohane also cautioned that we must act immediately, or costs and risks will rise. The longer we wait to curb pollution, the steeper the cuts must be to avoid catastrophic climate change. Time is required to develop new technologies and build infrastructure. Plus, developing countries like China and India are waiting for the United States to act before they take action. Therefore there is very little time remaining to cap greenhouse-gas emissions before a large risk of climate catastrophe and heavy economic costs are incurred. If action is taken now, it can be successfully done—affordably.

Economics of Mitigation

According to the International Monetary Fund (IMF), it is possible to fight climate change without having a negative impact on economic growth. Although the IMF reports that "Climate change is a potentially catastrophic global externality and one of the world's greatest collective action problems," in order to curb climate change the IMF suggests a worldwide long-term plan of gradual increases in carbon prices. If this happened, they believe it would bring about the needed shifts in investments and consumption; it would discourage people from buying emission-intensive and energy-inefficient products. For example, if there is a better financial incentive to purchase a fuel-efficient hybrid car over a gas guzzler, that will be the purchase choice. The IMF has also determined that mitigation would not have as drastic an impact on the world economy as some fear (Kato, 2008).

The *Stern Report on the Economics of Climate Change* is one of the best-known reports on the economics of climate change, as discussed in Chapter 11. The *Stern Report* predicts that climate change will have a serious impact on economic growth without mitigation and recommends that 1 percent of the global GDP be invested to mitigate its effects. If this is not done, it could cause a recession equivalent to upwards of 20 percent of the GDP (Stern, 2005).

The insurance industry is also very concerned about the economic implications of climate change. Since 1960, the number of major natural disasters has tripled. Over the past 30 years, the proportion of the global population affected by weather-related disasters has doubled, rising from 2 percent in 1975 to 4 percent in 2001. A 2005 report from the Association of British Insurers concluded that limiting carbon emissions could avoid up to 80 percent of the projected additional annual costs of tropical cyclones by the 2080s. In June 2004, a report issued by the Association of British Insurers stated, "Climate change is not a remote issue for future generations to deal with. It is, in various forms, here already; impacting on insurers' businesses now" (Morris, 2011).

The world's two largest insurance companies—Munich Re and Swiss Re—stated in a study released in 2002 that "The increasing frequency of severe climatic events, coupled with social trends, could cost almost $150 billion each year in the next decade. These costs would, through increased costs related to insurance and disaster relief, burden customers, taxpayers, and industry alike" (McCann and Metts, 2007).

The costs to mitigate climate change depend on several factors. The most important factor is the target level of CO_2. The lower the level (in ppm), the sooner steps must be taken to reach that goal, and the sooner action must be taken and results achieved, the shorter the interval over which the costs must be spread, which makes initial mitigation more expensive. A commonly referenced target level by many countries is 350 ppm (current levels are 392 ppm), but the level is rising an average of 2–3 ppm annually. Nations that signed the Kyoto Protocol are required to lower their emissions to a specific level below their 1990 emissions level.

In terms of potential mitigation costs, the IPCC has estimated annual mitigation costs could range from $78 billion to $1,141 billion, roughly equal to 0.2–3.5 percent of the current world GDP (which is approximately $35 trillion) (Stern, 2005). In 2007, the McKinsey Global Institute used a cost-curve analysis to determine that it was possible to stabilize global GHG concentrations at 450–500 ppm with costs around 0.6–1.4 percent of the global GDP by 2030 (Enkvist et al., 2007). In 2007, the chairman of Lloyd's of London, Lord Peter Levene, stated, "The threat of climate change must be an integral part of every company's risk analysis" (Hussey, 2009).

Former U.S. Vice President Al Gore recently challenged the nation to produce every kilowatt of electricity through wind, solar, and other renewable sources of energy. He believes the cost of switching to clean electricity sources could cost as much as $3 trillion over 30 years in public and private money. Even though this seems like a lot, he also says that the investment will "pay itself back many times over" and that he has "never seen an opportunity for the country like the one that is emerging now" (*ABC News*, 2008).

Another suggestion presented as a way to finance mitigation is to create a new climate change tax. A *New York Times* article from September 16, 2007, outlines how Gilbert Metcalf, an economics professor at Tufts University, describes how revenue from a carbon tax could be used to actually reduce payroll taxes in a way that would leave the distribution of the total tax burden basically unchanged. He proposes a tax of $15 per metric ton of CO_2, together with a rebate of the federal payroll tax on the first $3,660 of earnings for each worker. Proponents of this scheme feel this is a better approach than forcing cars to become more fuel-efficient. Their argument is that when cars get better mileage, the owners will just be tempted to drive them more often, rather than cut back and start using public transportation (Mankiw, 2007).

Another issue concerning the economics of mitigation is the state of the present U.S. economy. Namely, the current recession may very well affect funding for poor nations to fight climate change. African activists have appealed for major polluters—such as the United States—to commit to donating 1 percent of their GDP toward foreign mitigation efforts.

Antonio Hill, a senior policy adviser for the British aid group Oxfam, expressed concern with the situation, especially the fact that wealthy nations were willing to lend developing nations money but were less willing to donate. Hill remarked, "As far as we're concerned this is the moral equivalent of having someone drive a car into your house and offering you a loan to pay for the damages" (Climate Ark, 2008).

Carbon Capture and Storage

Mitigation strategies relating to climate change involve taking positive action to reduce greenhouse gas emissions and also enhance sinks aimed at reducing the extent of climate change. Mitigation is taking action to reduce or eliminate the problem before it becomes a problem, rather than allowing the problem to occur and then adjusting to it. According to the IPCC, in order to stabilize GHG concentrations, greenhouse gases would have to be reduced by at least the percentages listed in Table 12.1.

In a new study conducted by Scott Doney of the Woods Hole Oceanographic Institution, a newly developed computer model indicates that the land and oceans will absorb less carbon in the future if current trends of emissions continue, which could mean significant shifts in the climate system. According to Doney, "Time is of the essence in dealing with greenhouse-gas emissions. We can start now or we can wait 50 years, but in 50 years we will be committed to significant rapid climate change, having missed our best opportunity for remediation." He also stressed that the earth's ability to store carbon in its natural reservoirs is inversely related to the rate at which carbon is added to the atmosphere. In other words, as soon as humans cut greenhouse gas emissions, the easier it will be for the earth to naturally store carbon. He stresses that the study suggests that land and oceans can absorb carbon at a certain rate, but at some point they may not be able to keep up (*ScienceDaily*, 2010).

Presently, about one-third of all human-generated carbon emissions have dissolved in the ocean. How fast the ocean can remove CO_2 from the atmosphere depends on atmospheric CO_2 levels, ocean circulation, and mixing rates. The more CO_2 in the atmosphere, the more there can be in the ocean; faster circulation increases the volume of water exposed to higher CO_2 levels in the air, which increases uptake by the ocean. Climate change, however, will cause ocean temperatures to rise, and warmer water holds less dissolved gas, which means the oceans will not be able to store as much anthropogenic CO_2 as climate change continues. A negative climate change side effect on the oceans is that increasing amounts of CO_2 in the oceans will increase

TABLE 12.1

Greenhouse Gas Reduction Necessary to Stabilize the Atmosphere

Greenhouse Gas	Reduction Required to Stabilize Atmospheric Concentration (%)
Carbon dioxide	60
Methane	15–20
Nitrous oxide	70–80

Source: Metz, B. et al. (eds). *Climate Change 2007: Mitigation of Climate Change.* Cambridge, U.K.: Cambridge University Press, 2007.

their acid content. When CO_2 gas dissolves in ocean water, it combines with water molecules (H_2O) and forms carbonic acid (H_2CO_3). The acid releases hydrogen ions into the water. The more hydrogen ions in a solution, the more acidic it becomes. According to the Woods Hole Oceanographic Institution, hydrogen ions in ocean surface waters are now 25 percent higher than in the preindustrial era, with an additional 75 percent increase projected by 2100.

According to the IPCC, carbon capture and storage (CCS) is a process that involves capturing the CO_2 arising from the combustion of fossil fuels (such as in power generation or refining fossil fuels), transporting it to a storage location, and isolating it long term from the atmosphere (IPCC, 2007). Before CO_2 gas can be sequestered from power plants and other point sources, it must be captured as a relatively pure gas. According to the U.S. Department of Energy, on a mass basis, CO_2 is the nineteenth largest commodity chemical in the United States, and it is routinely separated and captured as a by-product from industrial processes such as synthetic ammonia production and limestone calcinations (Department of Energy, 2011) (Figure 12.2). CCS has the potential to reduce overall mitigation costs, but its widespread application would depend on overall costs, the ability to successfully transfer the technology to developing countries, regulatory issues, environmental issues, and public perception. The capture of CO_2 would need to be applied to large point sources, such as energy facilities or major CO_2-emitting industries to

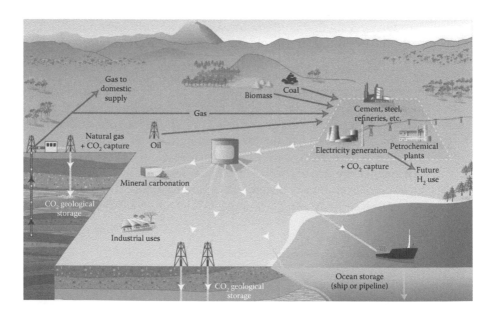

FIGURE 12.2
(See color insert.) The IPCC's schematic diagram of possible CCS systems, showing the sources for which CCS might be relevant, transport of CO_2, and storage options. (Courtesy of Rubin E. et al., IPCC special report: Carbon dioxide capture and storage technical summary, http://www.ipcc.ch/pdf/special-reports/srccs/srccs_technicalsummary.pdf, 2011.)

make it cost effective. Potential storage areas for the CO_2 would be in geological formations (such as oil and gas fields, nonminable coal seams, and deep saline formations), in the ocean (direct release into the ocean water column or onto the deep sea floor), and industrial fixation of CO_2 into inorganic carbonates.

Current technology captures roughly 85–95 percent of the CO_2 processed in a capture plant. A power plant that has a CCS system (with an access to a geological or ocean storage) uses approximately 10–40 percent more energy than a plant of equivalent output without CCS (the extra energy is for the capture and compression of CO_2). The final result with a CCS is that there is a reduction of CO_2 emissions to the atmosphere by 80–90 percent compared with a plant without CCS.

When CO_2 is captured, it must be separated from a gas stream. Techniques to do this have existed for 60 years. Used in the production of town gas by scrubbing the gas stream with a chemical solvent, CO_2 removal is already used in the production of hydrogen from fossil fuels. This practice helps remove CO_2 from contributing to climate change. When the CO_2 is transported to its storage site, it is compressed in order to reduce its volume; when it is compressed it occupies only 0.2 percent of its normal volume. Each year, several million tons of CO_2 are transported by pipeline, ship, and road tanker.

Today, there are several options for storing CO_2. Initially, it was proposed to inject CO_2 into the ocean where it would be carried down into deep water and would stay for hundreds of years. In order for any CCS scheme to be effective, however, it needs to sequester huge amounts of CO_2—comparable to what is currently being submitted into the atmosphere—in the range of gigatons per year. Because of the size requirement, the most feasible storage sites are the earth's natural reservoirs, such as certain geological formations or deep ocean areas.

The technology of injecting CO_2 underground is very similar to what the oil and gas industry uses for the exploration and production of hydrocarbons. It is also similar to the underground injection of waste practiced in the United States. Wells would be drilled into geological formations, and the CO_2 would be injected. This is also the same method used today for enhanced oil recovery. In some areas, it has been proposed to pump CO_2 into the ground for sequestration while simultaneously recovering oil deposits. There are arguments both for and against this strategy; on one hand, recovering oil would offset the cost of sequestration. On the other hand, burning the recovered oil as a fossil fuel adds additional CO_2 to the atmosphere, which offsets some of the positive effects of the sequestration.

Two other strategies involve injecting CO_2 into saline formations or into nonminable coal seams. The world's first CO_2 storage facility, located in a saline formation deep beneath the North Sea, began operation in 1996. Other alternatives have been proposed as well, such as using CO_2 to make chemicals or other products, fixing it in mineral carbonates for storage in a solid form, such as solid CO_2 (dry ice), CO_2 hydrate, or solid carbon. According to the IPCC, another suggested option is to capture CO_2 from flue gas using microalgae to make a product that can be turned into a biofuel (IPCC, 2007). In order to decide where to find feasible sites for carbon sequestration, it is important to know where large carbon sources are geographically distributed in order to assess their potential. This enables managers to estimate the costs of transporting CO_2 to storage sites.

According to the IPCC, more than 60 percent of global CO_2 emissions originate from the power and industry sectors. Geographically, 66 percent of these areas occur in three principal regions worldwide: Asia (30 percent), North America (24 percent), and Western Europe (12 percent). In the future, however, the geographical distribution of emission sources is expected to change. Based on data from the IPCC, by 2050 the bulk of emission sources will be from the developing regions, such as China, South Asia, and Latin America. The power generation, transport, and industry sectors are still expected to be the leading contributors of CO_2.

Global storage options are focusing primarily on geological or deep ocean sequestration. It is expected that CO_2 will be injected and trapped within geological formations at subsurface depths greater than 800 meters where the CO_2 will be supercritical and in a dense liquid-like form in a geological reservoir or injected into deep ocean water with the goal of dispersing it quickly or depositing it at great depths on the ocean floor with the goal of forming CO_2 lakes. Current estimates place both types of sequestration as having ample potential storage space—estimates range from hundreds to tens of thousands of gigatons of CO_2.

Geological Formation Sequestration

Many of the technologies required for large-scale geological storage of CO_2 already exist. Because of extensive oil-industry experience, the technologies for drilling, injection, stimulations, and completions for CO_2 injection wells exist and are being patterned after current CO_2 projects. In fact, the design of a CO_2 injection well is very similar to that of a gas injection well in an oil field or natural gas storage project. Capture and storage of CO_2 in geological formations provides a way to eliminate the emission of CO_2 into the atmosphere by capturing it from large stationary sources, transporting it (usually by pipeline), and injecting it into suitable deep rock formations. Geologic storage of CO_2 has been a natural process within the earth's upper crust for millions of years; there are vast reservoirs of carbon held today in coal, oil, gas, organic-rich shale, and carbonate rocks. In fact, over eons CO_2 has been derived from biological activity, igneous activity, and chemical reactions that have occurred between rocks, and fluids and gases have naturally accumulated in the subsurface layers (Figure 12.3).

The first time CO_2 was purposely injected into a subsurface geological formation was in Texas in the early 1970s as part of an "enhanced oil recovery" effort. Based on the success of this effort, applying the same technology to store anthropogenic CO_2 as a greenhouse gas mitigation option was also proposed around the same time, but not much was done to pursue any

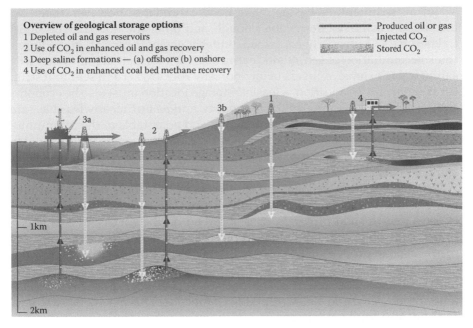

FIGURE 12.3
(See color insert.) CO_2 can be sequestered in deep underground geological formations. (Courtesy of Rubin E. et al., IPCC special report: Carbon dioxide capture and storage technical summary, http://www.ipcc.ch/pdf/special-reports/srccs/srccs_technicalsummary.pdf, 2011.)

actual sequestration. It was not until nearly 20 years later, in the early 1990s, that research groups began to take the idea more seriously. In 1996, Statoil and its partners at the Sleipner Gas Field in the North Sea began the world's first large-scale storage project. Following their lead, by the end of the 1990s, several research programs had been launched in the United States, Europe, Canada, Japan, and Australia. Oil, coal mining, and electricity-generating companies spurred much of the interest in this technology as a mitigation option for waste byproducts in their respective industries.

Since this initial push, environmental scientists—many connected with the IPCC—have become involved in geologic sequestration as a viable option to combat climate change. The significant issues now are whether the technique is: safe, environmentally sustainable, cost effective, and capable of being broadly applied. Geologic storage is feasible in several types of sedimentary basins, such as oil fields, depleted gas fields, deep coal seams, and saline formations. Formations can also be located both on and offshore. Offshore sites are accessed through pipelines from the shore or from offshore platforms. The continental shelf and some adjacent deep-marine sedimentary basins are also potential sites, but the abyssal deep ocean floor areas are not feasible because they are often too thin or impermeable. Caverns and basalt are other possible geological storage areas.

Not all sedimentary basins make good candidates, however. Some are too shallow, some not permeable enough, and others do not have the ability to keep the CO_2 properly contained. Suitable geologic formations require a thick accumulation of sediments, permeable rock formations saturated with saline water, extensive covers of low-porosity rocks to act as a seal, and structural simplicity. In addition, a feasible storage location must also be economically feasible, have enough storage capacity, and be technically feasible, safe, environmentally and socially sustainable, and acceptable to the community. Most of the world's populations are concentrated in regions that are underlain by sedimentary basins. The table lists some of the current geological storage projects around the world (Table 12.2).

The most effective geologic storage sites are those where the CO_2 is immobile because it is trapped permanently under a thick, low-permeability seal, is converted to solid minerals, or is absorbed on the surface of coal micropores or through a combination of physical or chemical trapping mechanisms. If done properly, CO_2 can remain trapped for millions of years. When converting possible local and regional environmental hazards, the biggest danger is that if CO_2 were to seep from storage, human exposure to elevated amounts of CO_2 could cause respiratory problems. This is why these storage facilities are closely monitored.

Announced in November 2007 in *The Salt Lake Tribune*, the U.S. Department of Energy is planning on spending $67 million over the next decade on a technological solution for climate change that will be tested in Utah. CO_2 will be pumped into a 1.6-kilometer rock formation under Carbon County for long-term underground storage. According to Dianne Nielson, energy advisor to Utah Governor Jon Huntsman, Jr., "I don't think such technological methods are as far out as some people say. In Utah, we have a leg up on research and development" (*Salt Lake Tribune*, 2007).

Energy and policy makers in Utah see this opportunity as a way to offset fossil-fuel pollution produced by power plants and energy development. Brian McPherson, an engineering professor employed by Utah Research Science and Technology to develop commercial applications for new technologies from Utah's major state universities, is currently involved in cutting-edge research involving carbon-sequestration engineering.

TABLE 12.2

Current Carbon-Sequestration Projects within Geological Formations

Country	Project Name	Year Begun	Total Storage
Norway	Sleipner	1996	20 Mt
Canada	Weyburn	2000	20 Mt
Algeria	InSalah	2004	17 Mt
United States	Salt Creek	2004	27 Mt
Japan	Minami-Nagoaka	2002	10,000 t
China	Qinshui Basin	2003	150 t

A total of $88 million is being spent on research and testing primarily in the Farnham Dome formation. This is the location of abandoned oil and gas fields in the southeastern corner of Utah near the Navajo Indian Reservation. The project could expand and cover areas that run from New Mexico to Montana, however, because there are additional underground geological basins that have an enormous storage potential—possibly enough to hold up to 100 years' worth of carbon emissions from major sources. In this project, liquefied CO_2 will be injected into muddy layers of rock about 1,524 meters below the desert surface. A layer of impervious shale caps the formation. Farnham Dome already contains a natural pool of CO_2 that formed between 10 and 50 million years ago that was harvested commercially until 1979 for the manufacture of dry ice and soft drinks.

Huntsman, Utah's former governor, led the way during office by actively looking for ways to reduce the state's impacts on climate change. He was one who joined California Governor Arnold Schwarzenegger in signing the Western Climate Initiative, with a goal to reduce greenhouse gas emissions 15 percent region-wide by 2020. According to McPherson, "Carbon sequestration will be a good short-term solution for dealing with climate change as the world shifts to renewable energy sources that are now in their infancy" (*Salt Lake Tribune*, 2007).

This project is also a good illustration of teamwork by professionals from various professional sectors: private, academic, and government. Involved in the project are experts from the University of Utah, Utah Geological Survey, Questar Gas, Rocky Mountain Power, Savoy Energy, Blue Source, Pure Energy Corp., the Navajo Nation, and the New Mexico Institute of Mining and Technology (*Salt Lake Tribune*, 2007).

Currently, there are several CCS projects taking place worldwide, but Brian McPherson's efforts have made significant progress. As McPherson recently reported: "With three active injection sites and a fourth commercial-scale test located in the Farnham Dome formation near Price [Utah] coming online soon, the Southwest Project is as far along or farther than any CCS project in the world" (InnovationUtah, 2008).

Arif Nurmohamed, a director for the British Broadcasting Corporation, is currently filming a special on climate change mitigation and is including the CCS efforts at Farnham Dome. He remarked "If you accept that climate change is a problem and that we are partially to blame through our CO_2 emissions, then the work Brian is doing is very important. If his tests are successful, then carbon sequestration in geologic formations gives us a bridging technology to manage CO_2 emissions while we reduce our need for fossil-fuel based energy" (O'Malley, 2008).

Ocean Sequestration

According to the IPCC, various technologies have been identified to enable and increase ocean CO_2 storage. One suggested option would be to store a relatively pure stream of CO_2 that has been captured and compressed. The CO_2 could be loaded onto a ship and injected directly into the ocean or deposited on the sea floor. CO_2 loaded on ships could be either dispersed from a towed pipe or transported to fixed platforms feeding a CO_2 lake on the sea floor. The CO_2 must be deeper than 3 kilometers, because at this depth, CO_2 is denser than seawater.

Relative to CO_2 accumulation in the atmosphere, direct injection of CO_2 into the ocean could reduce maximum amounts and rates of atmospheric CO_2 increase over the next several centuries. Once released, it would be expected that the CO_2 would dissolve into the surrounding seawater, disperse, and become part of the ocean carbon cycle.

C. Marchetti was the first scientist to propose injecting liquefied CO_2 into waters flowing over the Mediterranean sill into the mid-depth North Atlantic, where the CO_2 would be isolated from the atmosphere for centuries—a concept that relies on the slow exchange of deep ocean waters with the surface to isolate CO_2 from the atmosphere. Marchetti's objective was to transfer CO_2 to deep waters because the degree of isolation from the atmosphere increases with depth in the ocean. Injecting the CO_2 below the thermocline would enable the most efficient storage. In the short-term, fixed or towed pipes would be the most viable method for oceanic CO_2 release, because the technology is already available and proven (Ametistova et al., 2002).

One proposed option would be to send the CO_2 down as "dry ice torpedoes." In this option, CO_2 could be released from a ship as dry ice at the ocean's surface. If CO_2 has been formed into solid blocks with a density of 1.5tm^{-3}, they would sink quickly to the sea floor and could potentially penetrate into the sea-floor sediment. Another method, called "direct flue-gas injection" would involve taking power-plant fire gas and pumping it directly into the deep ocean without any separation of CO_2 from the flue gas. Costs for this are still prohibitive, however.

It would be possible to monitor distributions of injected CO_2 using a combination of shipboard measurement and modeling approaches. Current analytical monitoring techniques for measuring total CO_2 in the ocean are accurate to about ±0.05 percent. According to the IPCC, measurable changes could be seen with the addition of 90 metric tons of CO_2 per 1 km^3. This would mean that 1 metric gigaton of CO_2 could be detected even if it were dispersed over an area of 10^7 km^3 (or 5,000 km × 2,000 km × 1 km), if the dissolved inorganic carbon concentrations in the region were mapped out with high-density surveys before the injection began (Figure 12.4).

In the case of monitoring the injection of CO_2 into the deep ocean via a pipeline, several monitoring techniques could be employed. At the point of entry from the pipeline into the ocean, an inflow plume would be created of high CO_2/low pH water extending from the end of the pipeline. The first monitoring array would consist of sets of chemical, biological, and current sensors and underwater cameras in order to view the end of the pipeline.

Methods of ocean storage

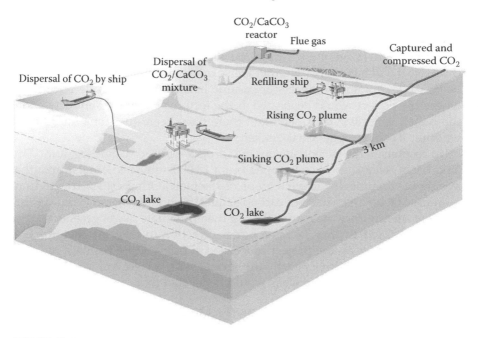

FIGURE 12.4
There are several proposed methods of CO_2 sequestration in the world's oceans. (Courtesy of Rubin E. et al., IPCC special report: Carbon dioxide capture and storage technical summary, http://www.ipcc.ch/pdf/special-reports/srccs/srccs_technicalsummary.pdf, 2011.)

An array of moored sensors would monitor the direction and magnitude of the resulting plume around the pipe. Monitors would also be set along the pipeline to monitor leaks. A shore-based facility would provide power to the sensors and could receive real-time data. In addition, a forward system would monitor the area and could provide data over broad areas very quickly. Moored systems could monitor the CO_2 influx, send the information to surface buoys, and make daily transmissions back to the monitoring facility via satellite.

A major project pumping CO_2 underground in an ocean sequestration storage project is proposed for the ocean region just off the coast of the Pacific Northwest of the United States. In a report from *The Seattle Times* in July 2008, David Goldberg, a geophysicist at Columbia University's Lamont-Doherty Earth Observatory, said, "It's hard to deny the size of the prize" (Doughton, 2008).

The research team involved believes that it is possible to devise a system where CO_2 emissions from power plants could be captured, liquefied, and pumped into porous basalt layers roughly 0.8 kilometers beneath the ocean floor. They currently estimate there is enough storage space in the geologic formation to easily store over a century's worth of CO_2 emissions from the

United States. Taro Takahashi, also a scientist at Lamont-Doherty on the project, says, "In principle, the type of reservoir we propose on the ocean floor is one of the safest—if not the safest—way of storing liquefied CO_2 for a long, long time" (Doughton, 2008).

The proposal has met with some opposition, however. Some complain that the project will be too costly. Others raise questions about whether there will be possible ecological impacts or seismic hazards. The Juan de Fuca Plate, where the project would be located, is the tectonic plate that subducts under the North American continent, and there have been questions raised as to whether it could trigger earthquakes.

In response, the scientists on the project—David Goldberg and Taro Takahashi—say they have mapped out a 78,000 square-kilometer area that avoids the seismically active regions of the plate. Other critics are environmentalists who are critical of carbon sequestration in general, claiming that it is just a temporary measure delaying the immediate use of greener energy alternatives, which they feel is a better direction to take. Angela Slagle, a marine geologist on the project concerned about wildlife habitat, said they would also avoid drilling new hydrothermal vents, where a myriad of unique sea life exists. Instead, the wells would be located more than 161 kilometers from the coastline.

The research team also stresses that undersea basalts can trap and hold CO_2 in several different ways, which in turn provide multiple layers of protection against leaks. Kurt Zenz House, a Harvard researcher who was one of the first to propose undersea carbon storage, says, "Under immense pressure and cold temperature below the seafloor, CO_2 forms a very dense liquid that is much heavier than seawater. In addition, gravity would prevent the liquefied gas from seeping upward, just as it prevents water in a well from flying into the air" (Doughton, 2008).

According to a report in *MSNBC News*, Dr. James E. Hansen (the climate change expert at NASA's Goddard Institutes for Space Studies) says the world has a 10-year window of opportunity to take positive action on climate change to avoid an impending catastrophe. He stresses that it is critical that governments worldwide adopt "an alternative scenario to keep CO_2 emission growth in check and limit the increase in global temperatures to 1°C. I think we have a very brief window of opportunity to deal with climate change... no longer than a decade at the most" (MSNBC News Services, 2006).

In attendance at the Climate Change Research Conference, Hansen stressed that if people continue to ignore the building evidence that climate change is happening and continue their lifestyles in a "business as usual" manner, atmospheric temperatures will rise 2–3°C, and "we will be producing a different planet" (MSNBC News Service, 2006). He also warns that under that warmer world drastic changes would be inevitable, such as the rapid melting of ice sheets (which, he said, would put most of Manhattan under water), prolonged droughts, severe heat waves, powerful hurricanes in new areas, and the likely extinction of 50 percent of the species on earth (MSNBC News Services, 2006).

Adaptation Strategies

In a report issued by NASA in 2009, because of the increasing challenges caused by climate change, several scientists and policy makers in the United States came together to take part in the newly established United States National Assessment on the Potential Consequences of Climate Variability and Change—called the National Assessment for short (USGCRP, 2009).

The National Assessment currently consists of sixteen separate regional projects. Project leaders are charged with assessing their region's most vulnerable aspects—the resources that would be impacted most by climate change. These include resources such as water supply and quality, agricultural productivity, and human health issues. Once potential impacts are identified, strategies are proposed and developed to cope and adapt to climate change impacts should they occur. Michael MacCracken, head of the coordinating national office, says, "The goal of the assessment is to provide the information for communities as well as activities to prepare and adapt to the changes in climate that are starting to emerge" (USGCRP, 2009).

The more successful mitigation strategies applied toward the effects of climate change today, the less human populations will have to adapt in the short and long term. According to the Pew Center on Global Climate Change, however, recognizing that the climate system has a great deal of inertia and is increasing, mitigation efforts alone are now insufficient to protect the earth from some degree of climate change. Even if extreme measures to combat climate change were taken immediately to slow or even stop emissions, the momentum of the earth's climate is such that additional warming is inevitable. Some of the warming that is unstoppable now is due to emissions of greenhouse gases that were released into the atmosphere decades ago. Because of this, humans have no choice but to adapt to the damage that has already been done.

Adaptation is not a simple, straightforward issue for humans or ecosystems. Each system has its own "adaptive capacity." In systems that are well managed (such as in developed countries and regions like the United States, Canada, Western Europe, and Australia), wealth, the availability of technology, responsible decision-making capabilities, human resources, and advanced communication technology help tremendously in successful adaptation to climate change. Societies that are able to anticipate environmental changes and plan accordingly ahead of time are also more likely to succeed.

The ability of natural ecosystems to successfully adapt is another issue, however. Although biological systems are usually able to adapt to environmental changes and inherent genetic changes, the time scales are usually much longer than a few decades or centuries (as is the case with climate change). Even minor changes in climate can be detrimental to natural ecosystems. An example of this is the polar bear in the Arctic. Today, sea ice is melting at a rapid rate, leaving the polar bear with limited areas to breed and hunt. The situation has already become so grave in a short period of time that the polar bear's survival is now in jeopardy. The polar bear is now under consideration as a threatened and endangered species.

Like polar bear habitats, many of the world's ecosystems are stressed by several types of disturbances, such as pollution, fragmentation (isolation of habitat), and invasion of exotic species (this includes weed invasions). These factors, coupled with climate change, are likely to impact ecosystems' natural resiliency and prevent them from being able to adapt over the long term.

As far as human adaptability, some adaptation will involve the gradual evolution of present trends; other adaptations may come as unexpected surprises. Changes will involve sociopolitical, technological, economic, and cultural aspects. Because populations are increasing, more people live in coastal areas, and more people live in floodplains and in drought-prone areas, adaptation measures will be required as climate changes. Fortunately, however, technology has developed to a point that there are better means today to successfully respond to climate change than there were in the past. For example, agricultural practices have evolved to the point where most crop species have been able to be translocated thousands of miles from their regions of origin by resourceful farmers.

A critical key to success is reactive adaptation; how willing will populations be to permanently change behaviors in order to adapt to changing climates and environmental conditions? Topics that populations will have to adapt to will encompass issues such as:

- Rationed water
- Changes in water use habits
- Changes in crop type
- Resource conservation plans
- Mandatory use of renewable energy
- Restricted transportation types

Waiting to act until change has occurred can be more costly than making forward-looking decisions that anticipate climate change, especially with coastal and floodplain development. A "wait and see" approach would be unwise with regard to ecosystem impacts. According to the Pew Center, "Proactive adaptation, unlike reactive adaptation, is forward-looking and takes into account the inherent uncertainties associated with anticipating change. Successful proactive adaptation strategies are flexile; they are designed to be flexible under a wide variety of climate conditions" (Easterling, Hurd, and Smith, 2004).

An extremely important part of adaptation that cannot be overlooked is government influence and public policy. Governments have a strong influence over the magnitude and distribution of climate change impacts and public preparedness. When climate and environmental disasters occur, it is usually government institutions that provide the necessary funding and develop the technologies, management systems, and support programs to minimize the occurrence of a repeat situation. A well-known example of this is the Dust Bowl that occurred in the Midwestern United States in the 1930s. It was through the efforts of the U.S. government that conservation efforts were started in order to properly manage the nation's soil and agriculture to prevent a repeat disaster of that nature.

In view of climate change today and the already unstoppable effects into the future, adaptation and mitigation are necessary (and complementary concepts). Adaptation is a key requirement in order to lessen future damage. It is important to understand, however, that even though society will have to adapt, the losses suffered will be inevitable, and certain geographical areas will experience more extreme losses than others, particularly the developing countries.

According to a report that appeared in *National Public Radio News* in December 2007, Australia is already taking proactive adaptation measures with their drinking water. Recently Perth, Australia, has been faced with drought conditions. Malcolm Turnbull, Australia's minister for the environment and water resources, says, "Over the past 10 years or so, the city has seen a 21 percent decline in rainfall, but the stream flow into dams—the actual amount running into storage—has dropped about 65 percent. We've seen similar declines in stream flow, though not quite so dramatic across southern Australia" (Sullivan, 2007).

Turnbull further refers to Perth as Australia's "canary in the climate change coal mine"—a city scrambling to find other sources of water for a growing population. The city is currently facing extreme economic prosperity and demand from China as it buys Western Australia's natural resources. Because of the population growth in Perth and resource demands, the city's water resources are currently under great demand and pressure. Therefore, Western Australia Water Corp. has turned to the Indian Ocean to help solve the problem.

In order to adapt to the accelerating water demands, the Kwinana Desalination Plant was constructed—the first desalination plant in Australia. The plant is currently producing nearly 151 million liters of drinking water a day, equal to about 20 percent of Perth's daily consumption rate. Simon McKay, the project manager, says the desalination process from seawater to tap water only takes about 30 minutes. He also points out that in the past, desalination plants have been extremely expensive (the Middle East has used them for decades), but new technology has made them cheaper and more efficient.

Unfortunately, they still consume a large amount of energy. For this reason, the operation in Australia runs on wind energy, generated from 48 wind turbines, each as tall as a 15-story building, at the Emu Downs Wind Farm. Kerry Roberts, Emu Down's general manager, says "If you look at the combined output of the wind farm at maximum wind speeds—39–45 kilometers per hour—you're looking at an output of close to 80 megawatts. That's enough power to run Perth's desalination plant 257 kilometers to the south." Gary Crisp of Western Australia Water Corp says, "I predict that desalination will account for at least half of Perth's water in the next 30 years" (Sullivan, 2007).

Although it is understood today that both mitigation and adaptation must occur simultaneously, each country will be faced with different issues to resolve and overcome. Developing nations may face different issues of climate change than developed nations. Varying geographic locations will also experience differing ranges of change. The only way to manage this overwhelming issue is for nations to work together in a global effort to control greenhouse gas emissions, working to understand universal cause-and-effect relationships. Without international cooperation, there is little hope of stopping the problem before it is too late.

Conclusions

Although it is true that many underdeveloped countries are now beginning to cause a climate change problem because they are industrializing and making the same mistakes that currently developed countries once made (which can be corrected with assistance and guidance from developed countries), there is also the issue of the undeveloped countries that are facing the worst effects of climate change—such as sea-level rise—who have not ever significantly contributed to the problem. This suggests an uneven balance of responsibility. If a global climate policy is going to work, many argue that all countries must participate in the solution to some degree, because emerging and developing economies are expected to produce 70 percent of global emissions during the next 50 years. Other experts add that any framework that does not include large and fast-growing economies (China, India, Brazil, and Russia) would be very costly and politically unwise. Others believe that undeveloped countries should not be held accountable. Like most issues, reality—and morality—most likely lies somewhere in between.

This chapter has presented several financial and technological strategies to handle the mitigation of climate change. Whichever methods are used will ultimately depend on the region, available technology, available finances, political policy, and prevailing social paradigm. What is critical is that action be taken immediately to fight climate change in order to lower the negative consequences of sea-level rise, flooding, drought, disease, and other disasters. Perhaps facing the issue realistically is through a combination of some workable means of adaptation, mitigation, and prevention. All three approaches have been discussed in this book, and all are viable, workable components to the ultimate solution. Although changing our personal behavior, mindset, and attitudes is perhaps the most critical element in the mix, we cannot over-look the assistance and good that can come from mitigation efforts, either, and where they are technologically, environmentally, and economically fea-sible, they are also worth looking at. As we have learned, climate change is already in progress, and its effects are not completely eradicable at this point, so adaptation is also a necessity. In this instance, developed countries need to help those that need assistance in a struggling world. The battle needs to be fought—and won—by all.

Discussion

1. Do you believe that every business should consider climate change costs as one of the costs of doing business? If so, what types of costs should they plan for, which costs are preventable, and which can be lowered through planning ahead of time?

2. What type of reaction do you think the American public would have to a climate change tax when much of the country either has not been educated about the subject, shows no interest in making lifestyle changes to mitigate it, and is currently facing personal economic stresses? How do you make the general public aware of the issues?

3. Do you think it's "fair" that developed countries donate 1 percent of their gross domestic product in order to help poor nations fight climate change? Why or why not?

4. Why do you think developed nations are willing to lend developing nations money but are less willing to donate to them?

5. How do you think the rate of responsibility should be placed on developed nations? Those currently developing? Those who will most likely never develop? Should everyone be treated equally or should there be a prorated, sliding scale of responsibility and reparations?

6. Even if seeds have been genetically altered to be more resilient so that they can be grown in broader geographic regions under shifting climates, how are agricultural specialists going to deal with the fact that new geographic areas may not have fertile soils to grow crops in?

7. If you were a resource planner and were trying to anticipate what the world would look like in 20 years and had to develop a resource plan for the area you live in, how would you develop your plan? What factors would be important? How would you prioritize the ele-ments in it? What values would you put on different resources? How would the individual resources interact and impact each other?

References

ABC News. 2008. Gore to issue clean energy challenge. July 17. http://blogs. abcnews.com/politicalradar/2008/07/gore-to-issue-c.html (accessed April 10, 2011). Discusses Al Gore's explanation of the benefits of switching to clean energy.

Ametistova, L., J. Twidell, and J. Briden. 2002. The sequestration switch: Removing industrial CO_2 by direct ocean absorption. *The Science of the Total Environment.* 289(1–3): 213–223. Discusses the initial theory of carbon sequestration in the world's oceans as suggested by C. Marchetti.

Climate Ark. 2008. Climate policies that harm indigeneous peoples: No good. *Climate Ark*, April 3. http://www.climateark.org/blog/2008/04/climate_investment_crucial_ind.asp (accessed April 13, 2011). Discusses the reasons why all countries must participate in climate change mitigation.

Department of Energy. 2011. Carbon capture research. http://fossil.energy.gov/programs/sequestration/capture (accessed April 14, 2011). Discusses the latest research in carbon-sequestration methods.

Doughton, S. 2008. Storing carbon dioxide under NW seafloor proposed. *The Seattle Times*, July 14. http://seattletimes.nwsource.com/html/localnews/2008050976_webcarbonstorage14m.html (accessed April 14, 2011). Discusses the carbon-sequestration project proposed off the coast of the Pacific Northwest that could conceivably store over a century's worth of CO_2.

Easterling, W. E. III, B. H. Hurd, and J. B. Smith. 2004. Coping with global climate change: The role of adaptation in the United States. Pew Center on Global Climate Change. http://www.pewclimate.org/docUploads/Adaptation.pdf (accessed August 16, 2011).

Environmental Defense Fund. 2009. Cost of cutting carbon: Pennies a day. October 2. http://www.edf.org/article.cfm?contentID=5405 (accessed April 11, 2011). Discusses the costs of the cap and trade mitigation option.

Environmental Protection Agency. 2010. Cap and trade. http://www.epa.gov/captrade/ (accessed August 16, 2011).

Enkvist, P.-A., T. Nauclér, and J. Rosander. 2007. A cost curve for greenhouse gas reduction. *The McKinsey Quarterly*. 1: 35–45. Shows cost curve analyses to determine the best method in which to stabilize global GHG concentrations.

Hussey, M. 2009. Cleaning up the corporate act. *Risk Management Professional*, September. http://www.theirm.org/documents/RMP_sep09_Lowres.pdf (accessed April 12, 2011). Discusses how businesses must consider the costs of climate change in their business cost formulas.

InnovationUtah. 2008. Carbon sequestration cliffhanger. USTAR. http://www.innovationutah.com/carbonengineering/documents/BBCDocumentaryonCarbonSequestration.pdf (accessed August 16, 2011).

Intergovernmental Panel on Climate Change. 2007. *WG1: The Physical Science Basis.* Cambridge, UK: Cambridge University Press.

Kato, T. 2008. Implications of climate change for Africa. Fourth Tokyo International Conference on African Development (TICAD IV), May 29. http://www.imf.org/external/np/speeches/2008/052908a.htm (accessed April 11, 2011). Discusses mitigation options that do not negatively impact the economy.

Keohane, N., and P. Goldmark. 2008. What will it cost to protect ourselves from global warming? Environmental Defense Fund. http://www.edf.org/documents/7815_climate_economy.pdf (accessed April 11, 2011). Discusses the economics and opportunities of cap and trade.

MacLeod, M. 2007. How does cap and trade work? Environmental Defense Fund, June 4. http://blogs.edf.org/climate411/2007/06/04/how-does-cap-and-trade-work (accessed April 13, 2011). Discusses the limitations and uses of cap and trade.

McCann, J., and C. Metts. 2007. The effects of global warming. Isbibbio, October 3. http://isbibbio.wikispaces.com/effects_d (accessed April 6, 2011). Discusses the costs of not mitigating CO_2.

Mankiw, N. G. 2007. One answer to global warming: A new tax. *The New York Times*, September 16. http://www.economics.harvard.edu/files/faculty/40_One_Answer_to_Global_Warming.pdf (accessed April 12, 2011). Discusses how one way to finance climate change is through a new tax.

Metz, B., O. R. Davidson, P. R. Bosch, R. Dave, and L. A. Meyer, (eds). 2007. *Climate Change 2007: Mitigation of Climate Change. Contribution of Working Group III to the Fourth Assessment Report of the Intergovernmental Panel on Climate Change*. Cambridge, U.K.: Cambridge University Press. http://www.ipcc.ch/publications_and_data/ar4/wg3/en/contents.html.

Morris, B. 2011. Economic effects of climate change. Climate Change and Global Warming Fact Sheet. http://climate change.110mb.com/effects-economic-climate-change.htm (accessed April 5, 2011). Discusses the global economic ramifications of climate change.

MSNBC News Services. 2006. Warming expert: Only decade left to act in time. MSNBC News, September 14. http://www.msnbc.msn.com/id/14834318/ns/us_news-environment (accessed April 14, 2011). Discusses the changes that must be made now to avoid permanent damage to the environment caused by climate change.

O'Malley, M. 2008. Carbon sequestration cliffhanger. Utah Pulse.com, August 1. http://utahpulse.com/featured_article/carbon-sequestration-cliffhanger (accessed March 22, 2011). Discusses the Farnham Dome CSS project in Utah.

Rubin, E., L. Meyer, H. de Coninck, J. C. Abanades, M. Akai, S. Benson, K. Caldeira, P. Cook, O. Davidson, R. Doctor, J. Dooley, P. Freund, J. Gale, W. Heidug, H. Herzog, D. Keith, M. Mazzotti, B. Metz, B. Osman-Elasha, A. Palmer, R. Pipatti, K. Smekens, M. Soltanieh, K. Thambimuthu, and B. van der Zwaan. 2007. IPCC special report: Carbon dioxide capture and storage, technical summary. http://www.ipcc.ch/pdf/special-reports/srccs/srccs_technical summary.pdf (accessed August 21, 2011).

Salt Lake Tribune. 2007. Utah leading CO_2 sequestration. November 16. Discusses the Department of Energy's plans to test CCS technology in some of Utah's geological formations for long-term carbon storage.

ScienceDaily. 2010. Comprehensive look at human impacts on ocean chemistry. June 21. http://www.sciencedaily.com/releases/2010/06/100617185131.htm (accessed April 10, 2011). Looks at the effects of carbon dioxide on ocean circulation, the carbon cycle, and acidification.

Stern, N. 2005. *The Stern Report: The Economics of Climate Change*. London: Authority of the House of Lords. July 6. http://www.scribd.com/doc/122554/Stern-

Report-The-Economics-of-Climate-Change (accessed April 13, 2011). Famous report that discusses the global economic ramifications of climate change.

Sullivan, M. 2007. Australia turns to desalination amid water shortage. National Public Radio, June 18. http://www.npr.org/templates/story/story.php?storyId=11134967 (accessed April 1, 2011). Discusses Australia's desalination efforts in Perth to provide water to the residents of Australia.

United States Global Change Research Program. 2009. The National Climate Assessment. http://www.globalchange.gov/what-we-do/assessment (accessed April 10, 2011). Discusses climate change impacts and makes evaluations based on climate models.

Suggested Reading

Davoudi, S., J. Crawford, and A. Mehmood (Eds.) 2009. *Planning for Climate Change: Strategies for Mitigation and Adaptation.* London: Earthscan Publications, Ltd.

Hansjürgens, B., and R. Antes. 2010. *Economics and Management of Climate Change: Risks, Mitigation and Adaptation.* New York: Springer.

Jochem, E., J. A. Sathaye, and D. Bouille. 2011. *Society, Behaviour, and Climate Change Mitigation (Advances in Global Change Research).* New York: Springer.

13

Professional Guidance: Climate Research, Modeling, and Data Analysis

Overview

An important aspect of climate change management is the research that goes into the science—the (largely) behind-the-scenes intensive work being spent on every component and aspect of the problem. Climate change is such a complex issue that progress could not be made without the research and testing going on and those researchers presenting their results to the scientific community, so that the policy makers, in turn, become educated about the latest developments and discoveries. The professional guidance provided by climatologists, physical geographers, computer scientists, physicists, geologists, chemists, mathematicians, statisticians, and others who are actively engaged in research provide a critical piece to the big picture. This chapter focuses on those efforts, beginning with an overview of the role of climate-change research and then discussing climate modeling—how it began, its fundamentals, and the challenges that both climatologists and computer programmers face today in its development. The chapter also explores some of the diverse uses of climate models and how they are helping to increase scientific and public knowledge about climate change. It then looks in particular at two examples of how the science of remote sensing is currently being used in climate models to inventory, monitor, and assess natural resources in order to make important decisions concerning the long-term effects of climate change.

Introduction

Knowledge is power. Without knowledge there is no understanding, and without understanding there is no feasible or productive way to solve the problem at hand concerning climate change. Therefore, the role of research is imperative—the more we understand, the better position we are in to make wise and productive decisions toward its solution.

Given the dynamic nature of climate change with the myriad of variables involved, the variable timescales involved between different factors, and the influences and interactions between inputs, mathematical modeling and data analysis are ideal methods of research. If models can be developed that focus on specific variables and can then reliably replicate the past, then their predictions for the future can lend credence to our understanding of a very complex scientific phenomenon. Just as important is the rapid development of computer technology that the world has seen transpire in the past few years. As this technology continues to advance and become more sophisticated, allowing larger amounts of data to be analyzed, the interactions of more variables—and more complicated relationships—can be considered, which also greatly advances our understanding of the issues involved. This combination of an increase in technology's sophistication; increased modeling reliability; better modeling output; better communication channels between scientists and policy makers; and increased interest, education, and involvement with the general population has resulted in some tremendous quality advancements in research and technology occurring in the last few years. This represents an important and very much needed step forward.

Development

Climate Change Research

Climate change research represents an important phase in the big picture of climate change management. Although the research requires a significant economic investment up front, successful breakthroughs can mean critical cost savings in the long run, giving research an important role in the overall solution. Today, climatologists can study climate patterns using sophisticated models of the earth's atmosphere and oceans. As technology has advanced and mathematical models have become more sophisticated through the process of matching observed and modeled patterns, climatologists have been able to tease out the human fingerprints that are associated with the changes, further solidifying the proof that humans are having an impact on the environment, and climate in particular.

One area receiving attention in the research arena is the production of "green carbon." A small business in California is testing an alternative to carbon sequestration (see Chapter 12 for a discussion on carbon-capture and -sequestration technology) that takes waste CO_2 and tailings from mining operations and turns the material into a substance that can be used in a variety of industrial, agricultural, and environmental applications. The resulting substance is called precipitated calcium carbonate (PCC). PCC has been produced in the past via an energy-intensive process, but the green-carbon technology transforms the carbon emissions instead of simply sequestering them. The PCC product can then be used in a variety of products, materials, and industrial processes. One of the biggest markets projected to use PCC is the paper industry as a filler and brightener. The industrial use of PCC is projected to grow to 9 million metric tons by 2010. The company spearheading all of this—Carbon Sciences—plans to take its research and implementation one step further: They plan to apply their green-carbon technology at an ethanol plant, where the entire process will actually reduce the amount of CO_2, making the venture carbon negative (rather than even carbon neutral).

Another important area of climate change research concerns the role and contribution of non-CO_2 greenhouse gases. According to the Goddard Institute for Space Studies (GISS), some computer climate simulations about the future have led to the conclusion that the Kyoto Protocol reductions will have little effect in the twenty-first century and that it may take "30 Kyotos" to reduce climate change to an acceptable level. Because of this, GISS has recommended research on, and the cutback of, non-CO_2 greenhouse gases and *black carbon* (soot) during the next 50 years. Based on GISS's research, non-CO_2 greenhouse gases have had the biggest impact on global warming. Cutting back on them, therefore, will help slow climate change.

The U.S. National Renewable Energy Laboratory is currently involved in biomass research. Biochemical conversion technologies involve three basic steps: converting biomass to sugar or other fermentation feedstock, fermenting the product produced in the first step, and processing the fermentation product to yield fuel-grade ethanol and other fuels, chemicals, heat, and/or electricity. At the National Renewable Energy Laboratory, researchers are trying to improve the efficiency and economics of the biochemical conversion process technologies by concentrating on simplifying the most difficult portions of the process. The current focus is on the pretreatment phase of breaking down hemicellulose to component cellulase enzymes, for breaking cellulose down to its component sugar. Researchers are also focusing on thermochemical conversion technologies that convert biomass to fuels, chemicals, and power using gasification and pyrolysis technologies. Gasification—heating biomass with about one-third of the oxygen necessary for complete combustion—produces a mixture of carbon monoxide and hydrogen, known as syngas. Pyrolysis—heating biomass in the absence of oxygen—produces liquid pyrolysis oil. Both syngas and pyrolysis oil can be used as fuels that are cleaner and more efficient than solid biomass. Both can also be converted into other usable fuels and chemicals.

As research and discoveries continue and cleaner, more efficient energy sources are discovered and implemented, society advances closer to curbing climate change and its resultant harm to the environment.

Climate Modeling

The earth's climate system is too complex for the human brain to grasp. There are so many interrelated forces constantly being influenced by outside factors and constantly shifting, trying to find some balance of equilibrium, that it is simply not possible to write down a list of equations describing how the climate system works and reacts. The earth's climate is not a straightforward process that gets from point A to point B every day in exactly the same way, at the same time, or in the same place. The only consistency about climate is that it is not consistent, and that is because there are so many variables involved, and the patterns of possible interactions are enormous.

One of the key challenges climatologists face today with climate change is that it is important to be able to predict with some sense of confidence how the earth's climate will change from region to region as temperatures rise so that policy makers can make appropriate decisions. Because of the inherent complexity and uncertainty, in order for climatologists to be able to do this they need to rely on climate models. Climate models are systems of differential equations derived from the basic laws of physics, fluid motion, and chemistry formulated to be solved on supercomputers.

The Modeling Challenge: A Brief History

Climatology—a branch of physics—and physics make use of two very powerful tools: experiments and mathematics. Weather and climate are so complex that without computers it would be impossible to mathematically quantify the climate system. Therefore, up until the computer age, there was no way to explain why and how climate behaved as it did. Once the technology developed, it was possible to build and assess quantitative climate models, because climate is based on physical principles.

The first objective of a climate model is to explain—however basically—the world's climates. Early on, the simplest and most widely accepted model of climate change was self-regulation, which means that changes are only temporary deviations from a natural equilibrium. Beginning in the 1950s, an American team began to model the atmosphere as an array of thousands of numbers. To answer the question about carbon, some primitive models were constructed representing the total carbon contained in an ocean layer, in the air, and in vegetation, with elementary equations for the fluxes of carbon between the reservoirs. Regardless of the CO_2 budget, scientists expected that natural feedbacks would operate and automatically readjust the system, restoring the equilibrium. Climatologists also recognized the need for more sophisticated models. They wanted to be able to explain triggers that caused past events, such as ice ages, plate tectonics, and changes in the ocean currents.

In the 1960s, computer modelers made encouraging progress by being able to make fairly accurate short-range predictions of regional weather. Modeling long-term climate change for the entire planet, however, was restricted because of insufficient computer power, ignorance of key processes such as cloud formation, inability to calculate the crucial ocean circulation, and insufficient data on the world's actual climate.

In the 1980s, models had improved enough that Syukuro Manabe, a senior meteorologist at Tokyo University, was able to use them to discover that the earth's atmospheric temperature should rise a few degrees if the CO_2 level in the atmosphere doubled. Through the use of models, by the late 1990s, most experts acknowledged climate change and its effects. One area that scientists were interested in being able to model was that of climate surprises—rapid climate changes.

One of the most well-known models was an energy budget model developed by William Sellers of the University of Arizona in 1969. He computed possible variations from the average state of the atmosphere separately for each latitude zone. Sellers was able to reproduce the present climate and was able to document that it showed extreme sensitivity to small changes. He determined that if incoming energy from the sun decreased by 2 percent (whether because of solar variation or increased dust in the atmosphere), it could trigger another ice age. Based on his results, Sellers suggested that "man's increasing industrial activities may eventually lead to a global climate much warmer than today" (Sellers, 1969).

Because an entire climate cannot be brought inside a laboratory, the only way to carry on an experiment of the entire system is to build a model of the entire system—a proxy. The most unpredictable part of the climate system— and as a result, one of the hardest to model—is the amount of radiation emitted by the sun and the earth. At any given time, water is present in water vapor, the oceans, and locked away in ice. The form and position the water takes changes constantly in response to its interaction between solar and thermal radiation. Clouds (especially low-lying thick clouds) reflect huge amounts of sunlight back into space and keep it from overheating the earth. High-altitude wispy clouds and water vapor absorb greater amounts of outgoing thermal (heat) radiation, which is generated off the earth's surface after it gets warmed by the sun.

In addition to greenhouse gases, clouds and water vapor contribute to keep the earth's average temperature comfortably livable year round. Atmospheric water has a tremendous effect on the earth's climate. For years, researchers have been trying to understand all of the complex interactions: specifically, how clouds and water vapor will act if climate change escalates and the atmosphere gets hotter.

Scientists at National Aeronautics and Space Administration (NASA) have developed several computer models to simulate the interactions between clouds and radiation. The area they are focusing most on is the tropics because that region gets the most sunlight. The results so far have been mixed: Some say in the future low-lying thick clouds will increase, making climate change worse; others say when the earth's surface heats up, cirrus clouds will dissipate and allow more thermal energy to escape to outer space. The reason this is so difficult to model consistently is because clouds are constantly shifting, separating, growing, and shrinking. In addition, the only way to study them is through remote sensing—the utilization of satellite imagery and digital image-processing software.

Today, some of the "simple" models that can be run on desktop computers are comparable to what was once considered state of the art for even the most advanced computers in the 1960s. As a comparison, the computers used by NASA during the Apollo missions occupied an entire room. Today, those same programs can be run on a desktop computer. Computer models of the coupled atmosphere-land surface-ocean-sea ice system are essential scientific tools for understanding and predicting natural and human-caused changes in the earth's climate.

Fundamentals of Climate Modeling

One of the key reasons climate is such a challenge to model is because it is a large-scale phenomena produced by complicated interactions between many small-scale physical systems. According to Gavin A. Schmidt at NASA's GISS, "Climate projections made with sophisticated computer codes have informed the world's policymakers about the potential dangers of anthropogenic interference with earth's climate system. The task climate modelers have set for themselves is to take their knowledge of the local interactions of air masses, water, energy, and momentum, and from that knowledge explain the climate system's large-scale features, variability, and response to external pressures, or 'forcings.' That is a formidable task, and though far from complete, the results so far have been surprisingly successful. Thus, climatologists have some confidence that theirs isn't a foolhardy endeavor" (Schmidt, 2007).

It was not until the 1960s that electronic computers were able to meet the extensive numerical demands of even a simple weather system, such as modeling atmospheric low pressures and storm fronts. Since that time, more components have been added to climate models, making them more robust and complex, such as information characterizing land, oceans, sea ice, atmospheric aerosols, atmospheric chemistry, and the carbon cycle (Figure 13.1).

The Physics of Modeling

The physics involved in climate models can be divided into three categories:

1. Fundamental principles (momentum, properties of mass, and conservation of energy)
2. Physics theory and approximation (transfer of radiation through the atmosphere and equations of fluid motion)
3. Empirically known physics (formulas for known relationships, such as evaporation being a function of wind speed and humidity)

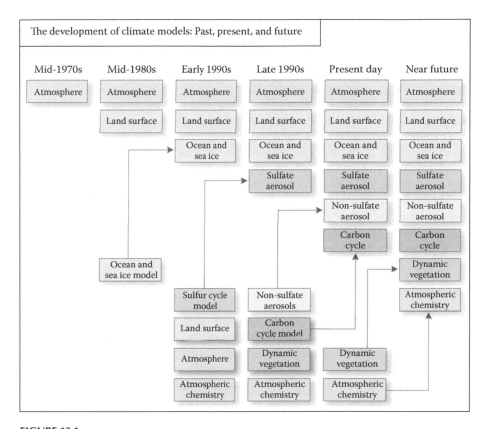

FIGURE 13.1
(See color insert.) The evolution of climate models beginning in the mid-1970s and extending into the near future. (From Casper, J. K., *Global Warming. Climate Management: Solving the Problem*, Facts on File, New York, 2010.)

Each model has its own unique details and requires several expert judgment calls. The most unique characteristic of climate models is that they have emergent qualities. In other words, when combining several interactions within the model, or parameters, the results of the interaction can produce an emergent quality unique to that system that was not previously obvious when looking at each system component separately. For instance, there is no mathematical formula that describes the earth's equatorial intertropical convergence zone of tropical rainfall, which occurs through the interaction of two separate phenomena (the seasonal solar radiation cycle and the properties of convection). As more components are added to a model, it becomes more complex and can have more possible outcomes.

Simplifying the Climate System

All models must simplify complex climate systems. One critical aspect of climate models is the detail in which they can reconstruct the part of the world they are trying to portray. This level of detail is called spatial resolution. If a climate model has a spatial resolution of 250 kilometers, then there are data points draped around the globe like a net with an x/y/z coordinate set spaced on a grid at an interval of 250 kilometers. The z-coordinate—representing the vertical height—can vary, however. The resolution of a typical ocean model, for example, is 125–250 kilometers in the horizontal (x/y) and 200–400 meters in the vertical (z). Equations are generally solved every simulated "half hour" of a model run. Some of the smaller scale, localized processes such as ocean convection or cloud formation have to be generalized in a process called parameterization; otherwise it would be too demanding on the computer system.

There are three major types of processes that need to be dealt with when constructing a climate model:

1. Radiative processes
2. Dynamic processes
3. Surface processes

Radiative processes deal with the transfer of radiation through the climate system, such as absorption and reflection of sunlight: in other words, where the sunlight travels once it is in the system. Dynamic processes deal with both the horizontal and vertical transfer of energy. This can include processes such as convection (the transfer of heat by vertical movements in the atmosphere, influenced by density differences caused by heating from below); diffusion (the spreading outward of energy throughout a system); and advection (the horizontal transport of energy through the atmosphere). Surface processes are those processes that involve the interface between the land, ocean, and ice: the effects of albedo (how reflective a surface is); emissivity (the ability of a surface to emit radiant energy); and surface-atmosphere energy exchanges.

The simplest models have a "zero order" spatial dimension. The climate system is defined by a single global average. Models get more complex as they increase in dimensional complexity, from one-dimensional (1-D) to two-dimensional (2-D) to three-dimensional (3-D) models. Changing the spatial resolution also controls the complexity of the models. In a 1-D model, the number of latitude bands can be limited; in a 2-D model, the number of grid points can be limited by spacing the points farther apart in a coarser grid. How long the model is run and the time intervals it is run on also affect the length and volume of the calculations involved (Figure 13.2).

Modeling the Climate Response

The purpose of a model is to identify the likely response of the climate system to a change in any of the parameters and processes, which control the state of the system. For example, if CO_2 is added into a simulation, the goal of

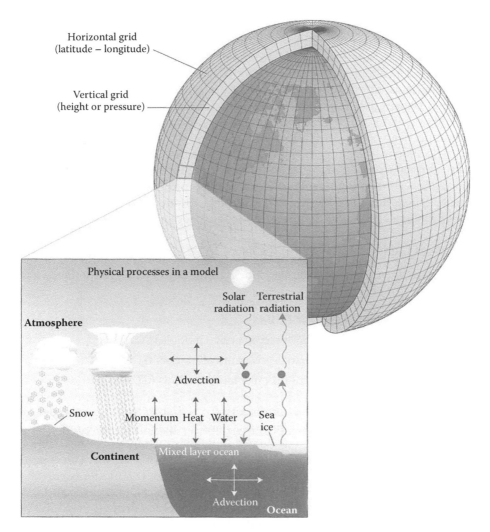

FIGURE 13.2
(See color insert.) A climate model is comprised of a set of x/y/z points placed around the globe at specified intervals in a netlike structure, called its resolution. A small grid with lots of points close together has a high resolution and is more detailed; a large grid with points spread farther apart has a low resolution and less detail. In the model, each point x/y/z intersection has a value associated with it—one value for each variable represented in the model. In this example, each grid point would have a distinct value for solar radiation, terrestrial radiation, heat, water, advection, atmosphere, and so on. (From Casper, J. K., *Global Warming. Climate Management: Solving the Problem*, Facts on File, New York, 2010.)

the model is to see how the climate system will respond to it as the climate system tries to find an equilibrium, or perhaps a model can focus on glacier melt and the results of ocean circulation as a result of the addition of freshwater and its effect on the climate.

Sometimes, complete processes can be omitted from a model if their contribution is negligible to the timescale being looked at. For instance, if a model is looking at a span of time that lasts only a few decades, there is no reason to model deep ocean circulation that can take thousands of years to complete a cycle. Not only would adding this data be useless, but it would slow down the computer processing time and perhaps give erroneous results by trying to make a connection where none exists. This puts an added responsibility on the modeling scientist to have a thorough knowledge of the data and a solid understanding of the interrelationships between the variables.

Types of Climate Models

There are several types of climate models, but they can be grouped into four main categories:

1. Energy-balance models (EBMs)
2. One-dimensional radiative-convective models (RCMs)
3. Two-dimensional statistical-dynamical models (SDMs)
4. Three-dimensional general-circulation models (GCMs)

These four types increase in complexity from first to fourth, to the degree that they simulate particular processes and in their temporal and spatial resolution. The simplest models do not allow for much interaction. The most complicated type—the GCM—allows for the most interaction. The type of model used depends on the purpose of the analysis. If a model is run that requires the study of the interaction between physical, chemical, and biological processes, then a more sophisticated model is normally used. EBMs simulate the two most fundamental processes controlling the state of the climate—the global radiation balance and the latitudinal (equator to pole) energy transfer. Because EBMs are the most simplistic models, they are usually in a 0-D or 1-D format. In the 0-D form, the earth is represented as a single point in space. In 1-D models, the dimension that is added is latitude, meaning that at whichever latitude interval is specified, the values in the model (such as albedo, energy flux, or temperature) would be input at each designated latitude. RCMs can be 1-D or 2-D. Height is the attribute that is characteristic of these models. With the addition of the z-value, RCMs are able to simulate in detail the transfer of energy through the depth of the atmosphere. They can simulate the dynamic transformations that occur as energy is absorbed, scattered, and emitted. They can model and simulate the role and interaction of convection and how energy is transferred through vertical motion in the atmosphere. Also, because of their 2-D capability, RCMs can simulate horizontally averaged energy transfers. These models are helpful when climatologists are interested in understanding the fluxes between terrestrial and solar radiation that are constantly occurring throughout the atmosphere. When heat rates are calculated for different levels in the atmosphere, parameters such as cloud density, albedo, and atmospheric turbidity are taken into account. The model can determine when the lapse rate exceeds its stability and convection (the vertical mixing of the air) takes place—a process called convective adjustment. RCMs are mainly used in studying forcing perturbations, which have their origin within the atmosphere, such as volcanic pollution. SDMs are usually 2-D in form—a horizontal and vertical component. Currently there are many variations of SDMs. These models usually combine the horizontal energy transfer modeled by EBMs with the radiative-convection approach of RCMs.

GCMs are sets of sophisticated computer programs that simulate the circulation patterns of the earth's atmosphere and ocean. The models represent many complex processes concerning land, ocean, and atmospheric dynamics, using both empirical relationships and physical laws. By varying the amounts of greenhouse gases (GHGs) in the model's representation of the atmosphere, future climate can be projected both globally and regionally. GCMs cannot be used reliably, however, for scales smaller than a continent.

In the 1990s, GCMs began modeling the effects of aerosols in the atmosphere, and scientists can now model GCMs for natural particulates (such as from volcanic eruptions) and anthropogenic aerosols from the burning of fossil fuels, sulfates, and organic aerosols through biomass burning. The purpose of GCMs is to describe how major changes in the earth's atmosphere, such as changes in the GHG concentrations, affect climatic patterns including temperature, precipitation, cloud cover, sea ice, snow cover, winds, and atmospheric and ocean currents.

GCMs are not used to predict weather events, and their resolution is too coarse to predict the effects of local geographic features, such as specific mountains, that may influence climate. They have proven very useful, however, for examining long-term climatic trends, patterns, and responses to significant change. They are still notably complex when compared with the actual climate system though. The Met Office Hadley Centre, the foremost climate change research center in Britain, currently uses the climate models illustrated in Table 13.1.

Testing a Model: Modeling Trouble Spots

Models are tested at two different levels—at a small scale (Did the wind patterns go in the right direction?), which includes the individual parameters; and at a large scale (Did the atmosphere warm up?), where the predicted emergent features can be assessed. The best way to test a climate model is to hindcast it—testing the model to see whether it can forecast changes in climate that have already occurred. This is accomplished by plugging in

TABLE 13.1

Met Office Hadley Centre Model Configurations

Atmosphere	3-D atmosphere model (AGCM)	AGCM plus "slab" ocean	Atmospheric chemistry	Coupled atmosphere-ocean model (AOGCM) = AGCM + OGCM	Regional climate model (RCM)
Land surface					
Ocean	3-D ocean model (OGCM)		Carbon cycle		

AGCM: Atmosphere general circulation model. AGCMs consist of three-dimensional representations of the atmosphere coupled to the land surface and cryosphere. They are useful for studying atmospheric processes, the variability of climate, and climate's response to changes in sea-surface temperature.

AGCMs plus "slab" ocean: This model predicts changes in sea-surface temperatures and sea ice by treating the ocean as though it were a layer of water of constant depth (usually 50 meters), heat transports within the ocean being specified and remaining constant while climate changes.

OGCMs: The ocean general circulation model is the ocean counterpart of an AGCM, a three-dimensional representation of the ocean and sea ice.

Carbon-cycle models: The terrestrial carbon cycle is modeled within the land-surface scheme of the AGCM, and the marine carbon cycle within the OGCM.

Atmospheric chemistry models: Three-dimensional global atmospheric chemistry models look at the destruction of ozone and methane in the lower atmosphere.

AOGCMs: Coupled atmosphere-ocean general circulation models are the most complex models, consisting of an AGCM and an OGCM. Some models also include the biosphere, carbon cycle, and atmospheric chemistry.

RCMs: Regional climate models are those with resolutions of about 50 kilometers, designed to be used in smaller regional areas.

previously measured parameters, such as ocean temperature and solar variability from past years, and running it in the virtual atmosphere of the climate model. The model is run forward through the past and into the present to predict changes in other atmospheric parameters—such as clouds and radiation balance. Ideally, the model should come up with the same values for clouds and radiation balance that are known to exist.

The 1991 eruption of Mount Pinatubo in the Philippines provided a good laboratory for model testing. Not only was the subsequent global cooling of 0.5°C accurately forecast soon after the eruption, but the radiative, water vapor, and dynamic feedbacks included in the models were quantitatively verified. This is as close to a controlled lab experience as climate change can expect to get. According to NASA, there are currently over a dozen facilities worldwide that are developing climate models. Over the past 20 years, the models have progressively become more sophisticated. Although errors overall between them appear to be unbiased, there are characteristics between the models that are similar, such as patterns of tropical precipitation.

Confidence and Validation

Although climate models should help clarify complex natural processes, the confidence placed in them should always be questioned. All climate models, by their very nature, represent a simplification of actual complicated processes. One thing that makes climate models so complex and difficult is that they often represent processes that occurred over timescales so long ago that it is impossible to test model results against real-world observations. Also, model performance can be tested through the simulations of shorter timescale processes, but short-term performance may not necessarily reflect long-term accuracy. Because of the possibility of error, climate models must be used with caution, and the user must realize that a certain amount of uncertainty is present in the model. Margins of uncertainty must be attached to any model projection.

Validation of climate models (testing against real-world data) provides the only objective test of model performance. As an example, with prior GCMs, some validation exercises in the past have detected a number of deficiencies in various simulations, such as:

- Modeled stratospheric temperatures tended to be too low.
- Modeled mid-latitude westerlies tended to be too strong, and easterlies tended to be too weak.
- Modeled subpolar low-pressure systems in the winter tended to be too deep and displaced too far to the east.
- Day-to-day variability tended to be lower than in the real world.

Finding these discrepancies in models and correcting them are part of the process that enables the creation of stronger models. The process is iterative; no model is its strongest after the first run.

Modeling Uncertainties and Challenges

Because modeling is still in its infancy, it has numerous challenges. This section details the unknowns of modeling, including solar variability, the presence of aerosols, the characteristics of clouds, nature's unpredictability, error amplification, and other uncertainties.

Solar Variability

Solar variability is important in modeling the earth's climate. The total energy output of the sun varies over time, causing warming and cooling cycles of the earth's atmosphere. NASA satellites have confirmed that the sun's energy output varies in sync with the 11-year sunspot cycle of magnetic changes in the sun. Satellite data exist back to the 1970s, giving climatologists only about 30 years of continuous data. They go farther back,

however, and look at climate variations over century-long intervals by analyzing the association of brightness changes with surface magnetic changes
because records of the sun's magnetism are available for several past centuries. Climatologists have records of lengths of sunspot cycles that are useful
proxies as indicators of changes in the sun's brightness. Comparisons can be
calculated between sunspot cycle length and surface temperatures. Records
have been constructed back to 1750.

The sun's magnetic record can also be converted to estimate brightness
changes and input into a climate simulation. According to scientists at the
George C. Marshall Institute, using the sun's magnetic records has shown
that brightness changes have had a significant impact on climate change.
Periods of a brighter sun could contribute to warming of the earth's atmosphere (Baliunas, 1996).

Aerosols

Pollutants such as sulfur dioxide make model predictions difficult. Aerosols
form a haze that absorbs or reflects sunlight and causes a cooling effect,
which offsets some of the predicted greenhouse warming. Aerosols can also
change the properties and behavior of clouds. The theoretical effect of aerosols in modeling has been to cool the climate in both the present and the
future, but so far, climatologists have had a difficult time getting models
of aerosols to be consistent. Furthermore, as pollution issues are dealt with
and aerosol content in the atmosphere diminishes, scientists need a solid
understanding of their effect on climate change in order to be able to model
changes associated with their reduction.

Clouds

Because clouds are a smaller-scale phenomena (they are generally smaller
than the model's resolution) and transient—they come and go rather
quickly—they are one of the most difficult properties to account for in climate models. One thing scientists are struggling with is how clouds will
change in the future, specifically, how their composition, structure, and
extent will change as the earth's surface continues to get hotter. Cloud behavior is extremely difficult to predict because there are so many variables that
constantly change over time and space, such as surface temperature, air temperature, wind currents, varying amounts of water vapor, and abundance of
aerosol particles.

According to NASA, all meteorological models inevitably fail at some
point because of the sheer complexity of the earth's system. To support this,
chaos theory shows that weather will never be predictable with any significant accuracy for longer than 2 weeks, even with a nearly perfect model and
nearly perfect input data. Today, climate models are still in their early stages
of development—similar to the status of weather prediction 30 years ago.

Clouds have a very important role to play in climate models, so climatologists are trying to understand their dynamic nature, enabling better accommodation for them in models. According to NASA, clouds are the critical arbiters of the earth's energy budget. Clouds cover 60 percent of the planet at any given time; they play a major role in how much sunlight reaches the earth's surface, how much is reflected back into space, how and where warmth is spread around the globe, and how much heat escapes from the surface and atmosphere back into space. This makes clouds a key component of the earth's climate system, and until scientists understand cloud physics better, they will not be able to construct accurate global climate models.

Scientists at NASA have discovered that some clouds cool the surface by reflecting sunlight, and other types warm the surface by allowing sunlight to pass through and then trap the heat radiated by the surface. This proves there is a physical feedback loop between sea-surface temperature and cloud formation—each influences the other. Concerning climate change, a key question for climatologists and modelers is how tropical clouds will change if tropical sea-surface temperatures warm significantly. One research team came up with a hypothesis that the earth has a built-in mechanism for changing the structure and distribution of certain types of clouds in the tropics to release more radiant energy into outer space as the surface warms.

One concept that has been proposed is called the iris hypothesis. NASA uses remote-sensing satellites to obtain global measurements of the amount of sunlight reflected on the earth and the amount of heat emitted up through the top of the atmosphere to calculate the bottom line on the earth's energy budget. By doing this, scientists can determine which components of the earth's system are most responsible for climate change. In the early 1980s, Richard Lindzen, a theoretician and professor of meteorology at the Massachusetts Institute of Technology (MIT), was interested in modeling how climate responds to changes in water vapor and cloud cover (Lindzen, 1992). He began looking closely at the presence of water vapor as a greenhouse gas and the effect it was having on global warming. The warmer the atmosphere becomes, the more water vapor it can hold. As the atmosphere absorbs CO_2 and the temperature rises, the additional heat allows the atmosphere to absorb even more water vapor. The water vapor further enhances the earth's greenhouse effect in a progressive cycle. NASA scientists estimate that doubling the levels of CO_2 in the atmosphere is comparable to a 13 percent increase in water vapor. In the tropics, clouds moisturize the air around them, and clouds are a major source of moisture.

Lindzen and his researchers focused on cloud cover using the Japanese Geostationary Meteorological Satellite-5 (GMS-5; Japanese name Himawari-5) to collect their measurements. The area they focused on was the area bordered by the Indonesian archipelago, the center of the Pacific Ocean, Japan, and Australia, because the area contains the world's largest and warmest body of water called the Indo-Pacific Warm Pool. What Lindzen wanted to determine was what type and extent of clouds are correlated to what ranges in sea-surface temperature. Lindzen said, "We wanted to see if the amount of cirrus associated with a given unit of cumulus varied systematically with changes in sea-surface temperature. The answer we found was, yes, the amount of cirrus associated with a given unit of cumulus goes down significantly with increases in sea surface temperature in a cloudy region" (Herring, 2002).

What they discovered was that the earth has a natural adaptive infrared iris—a built-in check and balance mechanism that may be able to counteract climate change to some extent. The iris hypothesis suggests that, similar to the way the iris in a human eye contracts to allow less light to pass through the pupil under bright light, an area covered by high cirrus clouds contracts to allow more heat to escape into outer space when the environment gets too warm.

Although Lindzen is still trying to figure out exactly how the process works, his hypothesis is that the amount of cirrus precipitated out from cumulus depends upon what percentage of the water vapor that is rising in a deep convective cloud condenses and falls as rain drops. Most of the water vapor condenses, but not all of it rains out. Some of the moisture rises in updrafts and forms thin, high cirrus clouds. Lindzen feels his discovery is important because if the amount of CO_2 is doubled in the atmosphere but there is no feedback within the system, then there is only one degree of warming. However, climate models predict a much greater climate change because of the positive feedback of water vapor. What needs to be added to the model is the negative feedback (the infrared iris), which can be anywhere from a fraction of a degree to one degree—the same order of magnitude as the warming. Not all scientists agree with Lindzen's model, and other scientists have not been able to reproduce it. It has garnered some attention, however. As more data are collected and more models are run, if repeatable results are obtained, then his theories may be pursued further.

Nature's Inherent Unruly Tendencies

According to Dr. Orrin H. Pilkey, a coastal geologist and emeritus professor at Duke University, and Dr. Linda Pilkey-Jarvis, a geologist at Washington State Department of Geology, depending too much on computer models may not be completely reliable because "nature is too complex and depends on too many processes that are poorly understood or little monitored—whether the process is the feedback effects of cloud cover on global warming or the movement of grains of sand on a beach" (Dean, 2007).

One thing they criticize about mathematical models is that there are too many fixed mathematical values applied to phenomena that change often. Another modeling weakness is that formulas may include coefficients (also called fudge factors according to Dr. Pilkey) to ensure that they come out right. In addition, sometimes modelers fail to verify that a project performed as predicted, considering nature's possible unruly outcomes. On the other hand, Dr. Pilkey also cautions against moving too far in the other direction, especially when modeling climate change. According to him, "Experts' justifiable caution about model uncertainties can encourage them to ignore accumulating evidence from the real world" (Dean, 2007). The Pilkeys also stress that "It is important to remember that model sensitivity assesses the parameter's importance in the model, not necessarily in nature. If a model itself is a poor representation of reality, then determining the sensitivity of an individual parameter in the model is a meaningless pursuit" (Dean, 2007).

What they suggest, perhaps alongside, if not in replacement of models, is adaptive management. With this approach, policy makers can start with a model of how an ecosystem works but make constant observations in the field, altering their policies as conditions change. The problem with this approach is that because of management, funding, and policy issues, these requirements are often hard—if not impossible—to achieve. When models are used, they do have some basic recommendations for how to better use them: studies should pay more attention to nature to accumulate information on how living things and their environments interact; modelers should state explicitly what assumptions they have made; modelers should seek to discern general trends instead of giving a model more analytical power than it probably has; and models should be complemented with observations from the field. According to Dr. Pilkey, "If we wish to stay within the bounds of reality we must look to a more qualitative future—a future where there will be no certain answers to many of the important questions we have about the future of human interactions with the earth" (Dean, 2007).

Error Amplification

If a compass heading is set even a half degree off, the farther the boat travels, the farther off course it becomes, the error growing in magnitude the longer the boat progresses. In large, complex models, such as GCMs, if there is an initial input error—however tiny—in the physics of climate data, as the model runs, it can accumulate, adding up through the millions of numerical operations to give an impossible final result. This can render a model completely useless if the error is not initially caught and fixed. One approach in fine-tuning large climate models is to construct simpler models of the interactions between biological systems and gases. By improving the interactions of the individual components within the system, potential errors can be culled out and corrected before being added to a large model where even a small measurement can eventually become amplified into a major error.

Modeling Uncertainties and Drawbacks

One of the biggest drawbacks climate modelers face today is that direct, observational data is extremely limited. Global temperatures have only been collected and monitored for about 100 years. Many climate modelers believe climate modeling is still in its infancy and with many hurdles to overcome, not only in the mathematics of modeling itself and computer development, but also in understanding climate processes themselves. In some areas, uncertainties have actually grown. Some of that uncertainty is reflected in the comments of three climate modelers: Gerald North of Texas A&M University says, "The uncertainties are large." Peter Stone of MIT says, "The major climate prediction uncertainties have not been reduced at all." The cloud physicist Robert Charlson, professor emeritus at the University of Washington, Seattle, says, "To make it sound like we understand climate is not right" (Kerr, 2001).

Stone takes it further when referring to the "politically charged atmosphere" of climate change today and the fact that the inherent uncertainties in modeling are being focused on and used as fuel to dismiss them because possibly they are making climate change appear worse than it is. He comments, "We can't fully evaluate the risks we face. A lot of people won't want to do anything. I think that's unfortunate. Greenhouse warming is a threat that should be taken seriously. Possible harm could be addressed with flexible steps that evolve as knowledge evolves. By all accounts, knowledge will be evolving for decades to come" (Kerr, 2001).

Climate modeling has three basic challenges to improve accuracy:

1. Detecting consistently rising temperatures
2. Attributing that warming to rising greenhouse gases
3. Projecting warming into the future

Michael Mann, a climatologist at the University of Virginia, said the first challenge has already been resolved by the Intergovernmental Panel on Climate Change (IPCC) in their 2007 report. He credits part of their increased confidence to more sophisticated and effective statistical techniques for analyzing sparse observations. Concerning the rising GHG challenge, David Gutzler of the University of New Mexico says, "Attributing the warming to greenhouse gases is much harder. To pin the warming on increasing levels of greenhouse gases requires distinguishing greenhouse warming from the natural ups and downs of global temperature" (Kerr, 2007).

The IPCC's 1995 report said data "suggested" a human influence toward the rising GHGs. In their recent report, however, their attribution statement was much stronger: "[M]ost of the observed warming over the last 50 years is likely (66–90 percent) to have been due to the increase in greenhouse gas concentrations" (IPCC, 1995). The climate modeler Jerry D. Mahlman, the recently retired director of the National Oceanic and Atmospheric Administration's Geophysical Fluid Dynamics Laboratory in Princeton, New Jersey, comments on the IPCC's 2007 report, "I'm quite comfortable with the confidence being expressed. The report states that confidence in the models has increased. Some of the model climate processes, such as ocean heat transport, are more realistic; some of the models no longer have the fudge factors that artificially steadied background climate; and some aspects of model simulations, such as El Niño, are more realistically rendered. The improved models are also being driven by more realistic climate forces. A sun subtly varying in brightness and volcanoes spewing sun-shielding debris into the stratosphere are now included whenever models simulate the climate of the past century" (Casper, 2010). According to Mahlman, other modeling uncertainties that still need to be improved include the role of atmospheric aerosols, lack of enough data, cloud behavior, anthropogenic effects, global cooling, future pollution control, and future social behavior.

Jeffrey Kiehl of the National Center for Atmospheric Research in Boulder, Colorado, says, "A number of uncertainties are still with us, but no matter what model you look at, all are producing significant warming beyond anything we've seen for 1,000 years. It's a projection that needs to be taken seriously" (Kerr, 2001).

Other Unknowns

Other current modeling challenges include the carbon cycle, future economics, past and future temperatures, cooling effects, abrupt weather events, and future thermohaline circulation. The direct effect of CO_2 on global warming is presently accounted for in current models, but what needs better clarification is to what extent CO_2 influences global temperatures because of its secondary influences. For example, models still need to determine how much of the anthropogenic CO_2 actually makes it into the atmosphere. Scientists know that not all human-attributed CO_2 emissions end up in the atmosphere; some are absorbed by the earth's natural carbon cycle and end up in the oceans and terrestrial biosphere (plants and soils) instead. Because the earth's carbon cycle is extremely complicated, scientists still need to better understand how the carbon sources and sinks work in the cycle in order to enable climate models to better represent that attribute. Another problem is trying to predict future CO_2 emissions because they will be influenced by worldwide growth patterns. The role of developing countries and their fossil fuel use will become critical, as will the rate at which countries switch to renewable energy sources. The enforcement of pollution controls and the rate of deforestation will have effects that are difficult to predict.

Temperature is also a difficult variable to determine. Future global temperature is difficult to predict because the atmosphere is so sensitive to the concentration of aerosols and CO_2. Because of this sensitivity, even small input errors can accumulate into misleading modeling results. The cooling effects from particles in the atmosphere, such as aerosols, sulfur emissions, and volcanic eruptions, can have significant local or regional impacts on temperature. In models this can affect albedo and reflection values. To help manage for this, global cooling parameters may need to be added to the model. Abrupt weather events are not currently predictable because present-day models' spatial resolutions are too coarse. As an example, in some climate models, New Zealand is only represented by ten data points—not nearly enough resolution to study small-scale spatial events like changes in air currents.

Modeling of the thermohaline circulation (ocean conveyor belt) faces uncertainties because of the complexities controlling deepwater formation, the interrelationship between large-scale atmospheric forcing with warming and evaporation at low latitudes, and cooling and increased precipitation at high latitudes. Uncertainty also lies in trying to model the addition of fresh-water from the Arctic to the tropical Atlantic. Rates and direction of flow and convection are extremely difficult to predict at this point. According to NASA, other challenges are extreme events such as hurricanes and heat waves, the turbulent behavior of the near-surface atmosphere, and the effects of ocean eddies. Concerning climate models overall, NASA scientist Gavin Schmidt says, "Climate models are unmatched in their ability to quantify otherwise qualitative hypotheses and generate new ideas that can be tested against observations. The models are far from perfect, but they have success-fully captured fundamental aspects of air, ocean, and sea-ice circulations and their variability. They are, therefore, useful tools for estimating the con-sequences of humankind's ongoing and audacious planetary experiment" (Schmidt, 2007).

Two Case Studies

This section provides two examples of how the science of remote sensing is used in two separate studies analyzing climate change variables and their effect on the landscape, vegetation, and the economic sectors involv-ing a biomass inventory and invasive species assessment, and a range-land grazing and ecosystem health analysis over time. Both projects are being conducted by the Bureau of Land Management (Department of the Interior), an agency that is responsible for the administration of 107 million hectares of public lands in the United States, located primarily in twelve Western States (Utah, Colorado, Nevada, New Mexico, Arizona, California, Wyoming, Oregon, Washington, Montana, Alaska, and Idaho). Because studies are costly, a large number of them are either conducted directly by governments, completed by contractors but funded by the federal gov-ernment, or conducted by research units associated with universities or professional organizations and often supplemented with funding from government entities or public and/or private donations. This type of work is critical for the advancement of our understanding of climate change. Ongoing modeling and research is often the source of breakthroughs in new knowledge and solutions to problems, making the long hours spent in research well worth it.

Because each project involves remote sensing, a brief explanation of the science, technology, and applications is provided first. Remote sensing is the small- or large-scale acquisition of information about an object or phenomenon by the use of either recording or real-time sensing device(s) that are wireless, or not in physical or intimate contact with the object (generally collected by aircraft, satellite, or unmanned remote vehicles). A variety of devices are used for gathering information on a given object or area. The technology offers an economical, efficient way to consistently collect data and monitor vast landscapes over long periods of time. It is also the most economical method of collecting data for random sites that are not easily accessible in the field, making it a viable tool, able to assist with a wide variety of spatial applications. Successive photos of an area taken from a series of years in the past can be used for monitoring and change detection. Imagery can also be used to inventory, classify, and map features such as vegetation, soils, landforms, drainages, transportation systems, health of the landscape, erosion, fire damage, and so forth. Imagery in digital form can be acquired in various resolutions, providing different levels of detail, such as pixels (single-cell units comprising the image) ranging in size from representing a few centimeters on the ground to kilometers. The imagery can portray different portions of the electromagnetic spectrum, as well (see Chapter 2). Typically standard color (as our eyes perceive objects), infrared, or thermal imagery is utilized, but imagery can also be accessed in radar wavelengths. Its versatility allows for many diversified applications; the exact specifications chosen are dependent on the needs of the particular project. Once the appropriate imagery is obtained, specialized computer software can be utilized to display, analyze, and classify the data enabling it to be interpreted and decisions to be made.

Remote sensing imagery can then be used as one of the data layers within a geographic information systems (GIS) environment so that information gained from the imagery can then be used in conjunction with other data—such as land ownership, transportation networks, wildlife data, natural resource data, rights-of-way information, mineral and mining data, and many other data types—in order to make further decisions and additional models once specific spatial relationships are identified and their significance determined (Figure 13.3) (Casper, 2010).

Case Study 1: Forest Inventory and Species Invasion

Pinyon-pine and juniper woodlands are widespread on the Colorado Plateau between about 1,524 and 2,134 meters in elevation. The Colorado pinyon pine (*Pinus edulis*) is the most common pine species in this woodland type, and Utah juniper (*Juniperus osteosperma*) is the most common juniper. Annual precipitation is typically from 25 to 38 centimeters in pinyon-juniper woodlands, and tree species in these communities have evolved both drought and cold resistance. Pinyons dominate at higher elevations and form more

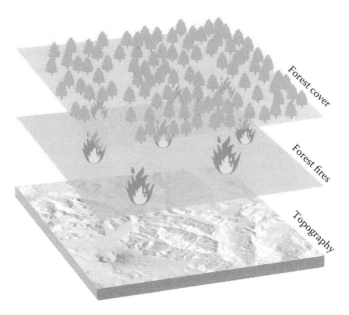

FIGURE 13.3

A GIS is able to take multiple layers of data from different sources and different scales, project them to a common map projection, and overlay them onto each other for spatial reference. It is then possible to utilize powerful analytical programs within the GIS system to analyze relationships and interactions among and between the various data enabling a wide range of decisions to be made. (From Casper, J. K., *Global Warming. Climate Management: Solving the Problem*, Facts on File, New York, 2010.)

closed-canopied stands that exhibit forest-like dynamics and species composition, often growing with oaks and alderleaf, mountain mahogany, and limited grasses. Juniper tends to grow at lower elevations and in more arid areas. In some areas, big sagebrush has become dominant because of overgrazing. The long history of livestock grazing in many pinyon-juniper woodlands on the Colorado Plateau has both diminished and altered the herbaceous vegetation, leading to widespread desertification. Woodland communities have expanded considerably over the past century in many parts of the Colorado Plateau. Tree densities have increased, and pinyon-juniper have expanded upslope into ponderosa pine forests and downslope into grass and shrub communities (Figure 13.4).

The study took place in an area called South Beaver, near Cedar City, Utah (the southwestern portion of Utah, which is the southwestern portion of the United States). Located in the physiographic region of the Colorado Plateau, the area is typically arid, with hot, dry summers, cold winters, and periods of drought often an issue. The project was part of a forest inventory, with the primary goal to map the vegetation units of the dominant species, which were pinyon and juniper. In this area, they are often invasive species to other species and have a tendency to encroach and overtake other vegetative

FIGURE 13.4
Pinyon and juniper inhabit a mountainside in the South Beaver study area near Cedar City, Utah. The goal of the project was to map its distribution, density, and proximity to roads, and determine areas of encroachment and migration over time. (Courtesy of Andrew Dubrasky, Bureau of Land Management, Cedar City Field Office, Utah.)

environments. Pinyon and juniper also provide a source of biomass that is suitable for the biofuel industry. There were two questions being asked in this study that pertained to climate change:

1. What is the current distribution of pinyon and juniper (where has it migrated)?
2. What are the current densities as it relates to a source of future potential biofuel?

Natural color digital aerial photographs were put together like a mosaic using Imagine® image-processing software, and from that several hundred field sites were selected from which to gather field data (predominant vegetation type). While the field data was being collected, the imagery was entered into eCognition®, another image-processing software, along with other digital data, such as elevation, slope, aspect, a vegetation index, and a brightness index, and a segmentation was run, which looked at the spectral data of the image and divided it into polygons based upon the data input from each of the individual data layers (Figure 13.5). Two approaches to classification

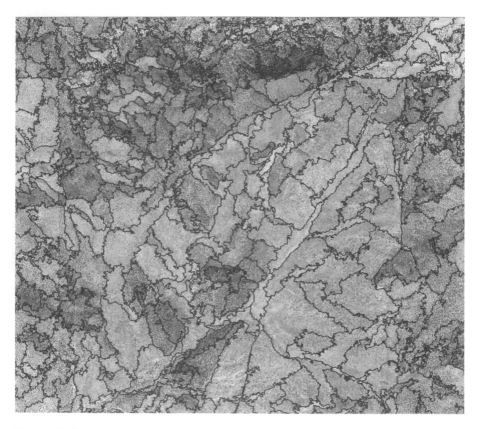

FIGURE 13.5
An example of a map segmentation generated by eCognition® image-processing software. The computer looks at all the input variables (such as imagery, elevation, slope, aspect, vegetative index, soil type, etc.) and determines the "best fit" for the polygon lines based on the raw values and designated weights of importance of each variable for each pixel (cell) in the image. It is a sampling of these polygons that are then field checked and identified, and the resultant data are used in the software's supervised classification program to create a final classification map on which decisions are based.

were used. The first method involved analyzing the spectral signatures of different points on the image and fitting spectral curves to the resulting spectral values of the points, shown on the top portion of the next figure. The second method utilized the field data gathered for the preselected points in a traditional supervised classification, as shown in the bottom portion of the illustration (Figure 13.6).

The final results not only measured the vegetation distribution of pinyon/ juniper, but of all species, and also provided density classifications of the pinyon/juniper in classes of 25, 50, 75, and 100 percent, enabling land managers to answer both questions of the study. In order to harvest the pinyon/ juniper for biofuels, proximity analyses to existing navigable roads were conducted and a cost/benefit analysis was calculated to determine which

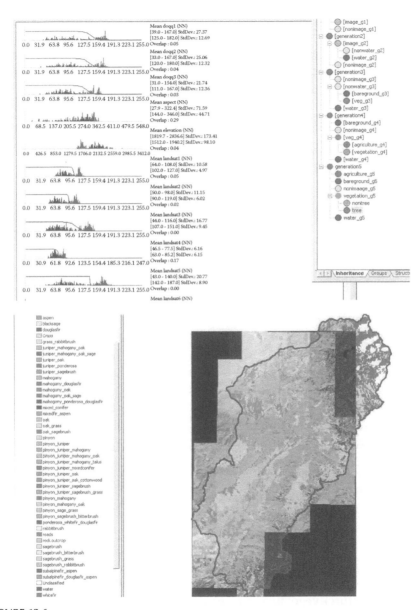

FIGURE 13.6
(See color insert.) The top portion illustrates how the different classes are determined during a classification relying on spectral signatures. In this more complicated approach, classes are determined by finding natural breaks in spectral signatures and fitting curves to the resultant spectral data. The bottom approach is another option used that is less complicated, called supervised classification, in which field data points of known information are identified on the digital map and attributed. Once those have been coded, the image-processing software will compare the spectral signatures of all the other pixels to the known ones and group them according to those that are most similar. Both approaches were used in the project.

areas are the most economically feasible. In terms of species migration and invasion, their current distribution was compared with older aerial imagery to determine a baseline distribution (for example, from the 1970s) and now can be monitored as species migrate northward as temperatures climb and conditions become drier.

Case Study 2: Monitoring Rangelands and Drought

Remote sensing technology is often used in rangeland monitoring activities for a wide variety of activities, such as:

- Inventory and classification of rangeland vegetation
- Determination of carrying capacity of rangeland plant communities
- Determination of forage and browse utilization
- Determination of range readiness
- Monitoring of erosion
- Census gathering of wildlife and evaluation of rangeland for wild-life-habitat values
- Measurement of the improvement of range sites
- Implementation of intensive grazing-management systems

In this ongoing study, a combination of multiscale imagery is currently being used in a nested approach. Large regions are being monitored with satellite imagery and aerial photography. Broad-scale issues that are monitored include:

- Rainfall patterns (indicative of climate change)
- Drought conditions
- Major vegetation types
- Land cover/land use

Large-scale imagery (greater detail imagery) is also used and allows the assessment of erosion (indicative of overgrazing and compromised vegetative root structure), vegetative species, and canopy cover. Ground monitoring using remote sensing techniques provides unbiased, economical sampling. Remote sensing is utilized to determine whether the current designated carrying capacity is adequate or whether the level of grazing is too low or high. It also provides support on both short- and long-term bases. Therefore, the four-step process of this project is as follows (Figure 13.7):

1. Imagery and data collection
2. Building a multiscale image base

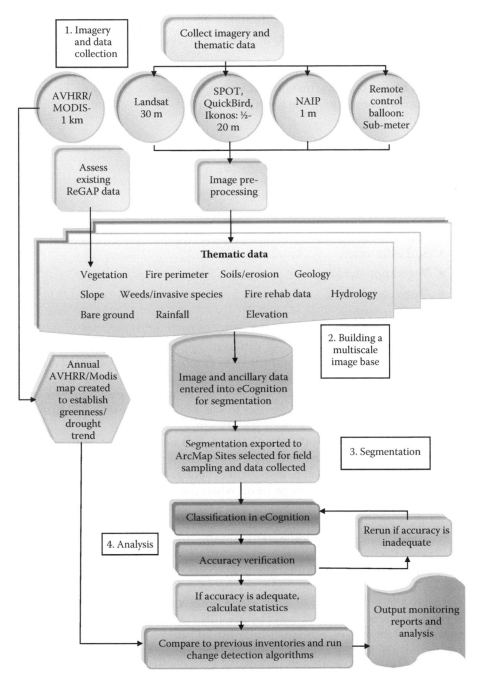

FIGURE 13.7
(See color insert.) Summary of the range-monitoring process. It is imperative to have a working model in place before beginning a project of this nature to avoid costly or repetitive mistakes.

3. Image segmentation using digital image-processing software

4. Image analysis, classification, and interpretation

Regarding imagery and data collection, remote sensing imagery is being used to create a baseline inventory of what the present ground and resource conditions currently are. The identified elements of consideration consisted of the following:

- Species cover
- Species composition
- Phenology by seasons
- Base cover
- Top canopy cover
- Soil infiltration
- Vegetation density
- Open soil surface
- Erosion/rilling

When remotely sensed data at different spatial, spectral, and temporal resolutions are used in the appropriate combination of spectral bands, it can bring out the features most pertinent for detecting change. Over time, current datasets for vegetation and land cover will be generated and compared against older datasets in order to see where changes have occurred—quantified by the image-processing software.

Any future imagery and data collected will be compared with that collected previously in order to establish a documentable record of the landscape's health (Figure 13.8). Subsequent sets of ground and remotely sensed data serve as a measure of range trend when compared to the initial base line sample created from the original imagery and ground data designating time 0 and then provide evidence to range managers of the direction the trend is moving—either upward, downward, or remaining stable—enabling them to plan accordingly.

Regarding building a multiscale image base, combining information from both high- and low-altitude sensors provides a cost-effective data-based rangeland monitoring strategy to assist range managers. Because remotely sensed imagery can "see" wavelengths in the electromagnetic spectrum that cannot be seen by humans (the infrared portion of the electromagnetic spectrum; see Chapter 2, this volume), when a change in vegetation cover resulting from some disturbance occurs, that change is reflected in the spectral signature. Both digital satellite imagery and aerial photography can be used to locate changes. Finding where the spectral signature differs from the expected or normal spectral signature of the site helps identify

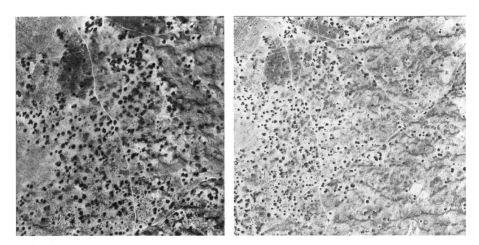

FIGURE 13.8
Change detection between different vintages of imagery can be an extremely powerful tool during the interpretation process and in understanding the scope of changes that have occurred on the landscape over time. In this example, the image on the left was taken in 1993, the one on the right in 2009. Notice the differences between the two images, such as the increase in number of roads and the changes in native vegetation. A range manager would then have to ask what this means in terms of ecosystem health , as well as proper management of herds.

areas of overgrazing, weed invasion, or other damage. It is important to take into account the time of year the imagery is collected. Season is correlated with the vegetation's phenology. When the perennial range grasses are green they give a much different spectral signature than when they are dry and mature, so keeping the acquisition date of the imagery consistent is important.

This particular study utilizes imagery from Landsat (30-meter resolution) and NAIP (1-meter resolution) and imagery acquired from an unmanned aerial vehicle (UAV) (submeter, highly detailed resolution, e.g., 6-centimeter). Time of acquisition is desirable at green-up (late spring/early summer) where possible on an annual basis. With the UAV—also referred to as a remote-controlled airplane—it is possible to obtain video, aerial photographic, digital photography, multispectral imagery, hyperspectral radiometric, LIDAR, and radar data. Although still somewhat experimental, UAVs can play an important role in measuring and monitoring vegetation health and structure of the vegetation/soil complex in rangelands. This particular study is attempting to open the door for further use of UAVs in other natural resource projects.

Regarding segmentation, classification, and analysis, once the data has been collected and the imagery preprocessed, it can be merged with ancillary, thematic data (vegetation, burn perimeters, soil data, geologic formations, slope, weeds/invasive species, ecosystem rehab data, rainfall data, elevation, etc.) and segmented (have polygons drawn on it for image classification purposes). Once segmented with polygons, a land-cover classification map is prepared. Several iterations are often run before the final version is reached for a particular area. Resulting products from remote sensing analysis include:

- Vegetation maps
- Greenness/dryness index maps (seasonal: March, June, September, December)
- Change detection maps and analyses

These products then assist land managers in making more informed decisions concerning whether to allow lands to be grazed, and if so, which land-management techniques to follow in order to maintain the health of the environment. It also serves as a monitoring tool for climate change, such as monitoring drought and desertification trends or the encroachment of invasive species and making responsible land management decisions.

Conclusions

As models become more developed, more questions will be able to be answered, which will enable managers to make more informed decisions and create better-balanced policies. One thing that does need to be kept in mind, however, is that a model is just a mathematical representation, created to replicate reality. Replicating full reality concerning climate change will probably never be possible simply because of the sheer number of variables involved and the complicated nature of natural systems in general. This is important to remember when interpreting the results of models: They are to be used only as a source of guidance that can prove very helpful in gaining an understanding of natural phenomenon. As with most scientific endeavors, the more knowledge that is gained, the more questions that are also uncovered, and the more the researcher realizes they need to ask. As the process is refined, more knowledge is gained, and the process repeats itself once again. Also of note are the breakthroughs and discoveries that have been made because of this technology and the models created from it. The scientific community has moved faster and further in a short period of time than ever before, so this is an exciting time to be involved in this field, on the cutting edge, so to speak, of fresh and critical discoveries that are affecting the future of humanity. Each research project started and completed puts humanity closer to understanding this critical, yet complex, problem called climate change.

Discussion

1. With technology advancing as rapidly as it is, where would you like to see climate modeling go in the near future? Why? What would be the timeliest problems for it to solve now?
2. What natural component do you think is the most problematic to capture when designing and building a model in terms of accurately capturing reality? Why? How could the process be improved?
3. What examples of error amplification can you think of and what effects would it have on interpretation and the results of the analysis?
4. Do you see working models as a way to convince the public of the realities of climate change? Why or why not?
5. If you were in charge of a modeling and research organization, what type of models would you like to see developed with regard to climate change?

References

Baliunas, S. 1996. Uncertainties in climate modeling: Solar variability and other factors. George C. Marshall Institute. http://www.marshall.org/article.php?id=12 (accessed August 16, 2011).

Casper, J. K. 2010. *Global Warming. Climate Management: Solving the Problem.* New York: Facts on File, Inc. Provides a well-rounded outlook on the management issues associated with solving the climate change issue.

Dean, C. 2007. The problems in modeling nature, with its unruly natural tendencies. *New York Times*, February 20. http://www.nytimes.com/2007/02/20/science/20book.html (accessed August 16, 2011).

Herring, D. 2002. Does the earth have an iris analog? NASA Earth Observatory. http://earthobservatory.nasa.gov/Features/Iris/ (accessed August 16, 2011).

Intergovernmental Panel on Climate Change. 1995. Climate change 1995: The science of climate change. In *Contribution of Working Group I to the Second Assessment Report of the Intergovernmental Panel on Climate Change.* Edited by J. T. Houghton, L. G. Meira Filho, B. A. Callander, N. Harris, A. Kattenberg, and K. Maskell. Cambridge, UK: Cambridge University Press.

Kerr, R. A. 2001. Rising global temperature, rising uncertainty. *Science.* 292(5515): 192–194. Discusses modeling GHGs and stands behind Obama's efforts to stop their continuation, spread, and destruction.

Lindzen, R. S. 1992. Global warming: The origin and nature of the alleged scientific consensus. *Regulation,* 15(2). http://www.cato.org/pubs/regulation/regv15n2/reg15n2g.html

Schmidt, G. A. 2007. The physics of climate modeling. NASA Science Briefs, January. http://www.giss.nasa.gov/research/briefs/schmidt_04 (accessed February 16, 2011). Discusses modeling complications with small-scale physical systems.

Sellers, W. D. 1969. A global climatic model based on the energy balance of the earth–atmosphere system. *Journal of Applied Meteorology.* 8: 392–400.

Suggested Reading

Hicke, J. A., and J. A. Logan. 2009. Mapping whitebark pine mortality caused by a mountain pine beetle outbreak with high spatial resolution satellite imagery. *International Journal of Remote Sensing.* 30: 4427–4441.

McGuffie, K. 2005. *A Climate Modelling Primer.* Hoboken, NJ: John Wiley & Sons.

Neelin, D. J. 2011. *Climate Change and Climate Modeling.* Cambridge, UK: Cambridge University Press.

Trenberth, K. E. 2010. *Climate System Modeling.* Cambridge, UK: Cambridge University Press.

14

The Clock Is Ticking: Changing for the Benefit of Future Generations

Overview

This chapter looks at the future and how several leading scientists expect the world to change under the influence of increased climate change. It then looks at the predicted winners and losers in the future as the earth continues to heat up. In conclusion, it looks at what new technology is on the horizon to help manage for a better tomorrow.

Introduction

Almost daily there is a news report reflecting a climate change issue: drought, floods, wildfires, hurricanes, heat waves, melting glaciers, polar bears starving, and multiple species of wildlife being threatened with extinction. Although it is not possible to forecast exactly when, where, and how severe warming's impacts will be, there is enough evidence available today to understand that many of the impacts from climate change will be severe and will result in disasters with enormous economic and human costs.

Each day corrective action is delayed puts life on earth at greater risk. What is important to realize is the climate system's inertia. Because it responds slowly, positive action taken today will not be realized for decades to come. In addition, the longer the delay, the greater the risks become and the more difficult it will be to respond effectively. Even worse, if the delay becomes too long, it may never be possible to stabilize the climate at a safe level for life to exist as it presently does. Tipping points become a serious issue—when the system tips or shifts into an entirely new state, such as the major collapse of ice sheets causing rapid sea-level rise or massive thawing of permafrost releasing huge amounts of stored methane into the atmosphere.

Unfortunately, climate change has progressed enough that no amount of cutting back on greenhouse gas (GHG) emissions will allow some ecosystems to return to the way they once were. If emissions are cut back now on an aggressive basis, scientists at National Aeronautics and Space Administration (NASA) believe it is still possible to avoid the worst consequences of climate change. Unlike the targeted 5 percent outlined in the Kyoto Protocol, the European Union has said that it will actually require a reduction of 60–80 percent to prevent dangerous climate change (Hare, 2008).

On the positive side, scientists do understand what the world is up against and are trying to educate the public to make the right choices, and the public does seem to be responding (although slowly) to the green movement. Solving the problem will take the concerted effort of everyone. There will have to be change in the future design of buildings, transportation, energy systems, leadership, innovation, and investments from governments and businesses. Both public and individual commitments are critical in order to achieve success.

Development

A Look toward the Future

Based on several emission scenarios run by the Intergovernmental Panel on Climate Change (IPCC) and NASA, global temperature is projected to increase by approximately 0.2°C per decade for the next two decades. Even if GHGs were kept steady at 2000 levels, because of the inertia of the oceans— the long time it takes them to store and release heat—there is already a suggested warming in the pipeline of 0.1°C per decade.

If GHG emissions continue at the current rate, or become even greater, climate models suggest that changes in the global climate system during this century will be even larger than those that were observed during the twentieth century. Another critical factor, as mentioned previously, is the warmer it gets, the less CO_2 the land and ocean are physically able to store. This means that any increasing concentrations in CO_2 will remain in the atmosphere. At this point, the IPCC projects that from now to 2090, the global average surface air warming will most likely range from 1.1 to 6.4°C. The ranges are attributed to the differences in the models and energy-use scenarios used.

Global average sea level is projected to rise by 18 to 59 centimeters by 2099. Scientists caution, however, that models do not include uncertainties about some climate mechanisms because there is still a lack of knowledge. For example, one of the key uncertainties is ice flow from Greenland and Antarctica. There are still mechanisms that control the flow and dynamics of the ice that scientists are trying to understand. If the speed of future ice flow increases, it will affect future scenarios that may not be accounted for at this point.

The geographical patterns in climate changes are expected to remain similar to those observed over the past several decades. The areas expected to be affected with warming the most are the high northern latitudes (polar region) and over the earth's landmasses. The least amount of warming is expected over the Southern Ocean and parts of the north Atlantic Ocean.

There are several other predicted changes that will also occur by 2099:

- Snow cover and sea ice will continue to shrink, endangering polar bears and other arctic animals, and permafrost will melt, releasing methane.
- As CO_2 increases in the atmosphere, the oceans will become more acidic.
- There will be increasing frequent heat waves, hot extremes, and heavy precipitation events.
- More intense and frequent hurricanes are likely to occur.
- There will be movement of extratropical storm tracts toward the poles, with changes in wind, precipitation, and temperature patterns.
- A greater amount of precipitation in high latitudes and less rain in most subtropical land regions is expected.
- A slowing of the Atlantic Ocean circulation (the major transporter of global heat) will likely occur.

IPCC scientists believe that warming and sea-level rise will continue for centuries even if GHG emissions were to become stabilized because of the long timescales associated with climate processes and feedbacks. One significant uncertainty is that climate change is expected to affect the earth's carbon cycle, but exactly how and by how much is not known at this point. Even if GHG emissions stopped and were stabilized by 2100, the earth's atmosphere would still warm by approximately 0.5°C by 2200. The thermal expansion of the oceans alone would cause an increase of 30 to 80 centimeters of global sea-level rise by 2030. The ocean would continue to warm for many centuries after that.

Greenland's ice sheet is projected to keep melting and also cause sea levels to rise after 2100. NASA scientists say that if it were to continue to melt for thousands of years until it completely melted, it would cause global sea levels to rise about 7 meters. It is not well understood yet what exactly will happen to the Antarctic ice sheets. Their vulnerability could increase through dynamic processes related to ice flow—these details are not included in current models but have been observed in the field. Future models will need to include field observations as they occur and become better understood. Their melting could also add to the global sea level. Current global models so far suggest that the Antarctic ice sheet will stay too cold through-out this century for large-scale surface melting. Some models project that the ice sheet could even increase in mass because of increased snowfall events. Scientists at NASA's Goddard Institutes for Space Studies believe that past, present, and future emissions of CO_2 will contribute to warming and sea-level rise for more than the next thousand years because of the length of time the greenhouse gases remain in the atmosphere (Hobish, 2010) (Figure 14.1).

In a *LiveScience* time line published on April 19, 2007, changes were pre-dicted for the environment if climate change continued (Thompson and Than, 2007) (Table 14.1). With predictions like these, it is not hard to see why scientists and environmentalists alike are lobbying for proactive decisions to be made now. Prolonging action any further is only going to make future adaptation more difficult.

Winners and Losers

In the United States, the U.S. Global Change Research Program (USGCRP) operates the U.S. National Assessment of the Potential Consequences of Climate Variability and Change (National Assessment). It breaks the United States into regional geographic sections (e.g., Pacific Northwest,

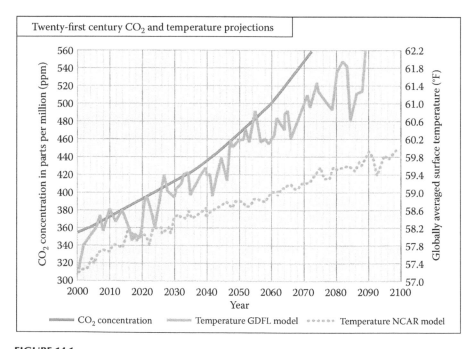

FIGURE 14.1

CO_2 is projected to continue to rise throughout this century, and temperatures, according to two different climate models, will continue to climb under the influence of climate change.

TABLE 14.1

Future Changes in the Earth's Environment under the Effects of Continued Climate Change

2020
• Flash floods will increase across Europe (IPCC).
• Less rainfall could reduce agricultural yields up to 50 percent in some areas of the world (IPCC).
• World population will reach 7.6 billion people (U.S. Census Bureau).
2030
• Up to 18 percent of the world's coral reefs will probably die because of climate change and other environmental stresses (IPCC).
• World population will reach 8.3 billion people. (U.S. Census Bureau).
• Warming temperatures will cause temperate glaciers on equatorial mountains in Africa to disappear (IPCC).
2040
• The Arctic Sea could be ice-free in the summer, and winter ice depth may shrink drastically (IPCC).
2050
• Small alpine glaciers will very likely disappear completely, and large glaciers will shrink by 30–70 percent (IPCC).
• As biodiversity hot spots are more threatened, one-fourth of the world's plant and vertebrate animal species could face extinction (IPCC).
2070
• As glaciers disappear and areas affected by drought increase, electricity production for the world's existing hydropower stations will decrease. The hardest hit will be Europe, where hydropower potential is expected to decline on average by 6 percent; around the Mediterranean, the decrease could be as much as 50 percent (IPCC).
• Warmer and drier conditions will lead to more frequent droughts, more wildfires, and more frequent heat waves, especially in Mediterranean regions (IPCC).
2080
• Some parts of the world will be flooded. Up to 20 percent of the world's population lives in river basins that will be hit with increasing flood hazards. Up to 100 million people could experience coastal flooding each year. The most vulnerable are the densely populated low-lying areas (IPCC).
• Between 1.1 and 3.2 billion people will experience water shortages, and up to 600 million will go hungry (IPCC).
2100
• Atmospheric CO_2 levels will be higher than they have been for the last 650,000 years (IPCC).
• Ocean pH levels will very likely decrease by as much as 0.5 pH units—the lowest it has been in the last 20 million years (IPCC).
• Thawing permafrost will make earth's land area a new source of carbon emissions—it will emit more CO_2 into the atmosphere than it absorbs (IPCC).
• New climate zones will appear on up to 39 percent of earth's land surface (IPCC).
• One-fourth of all plant and land animal species could become extinct (IPCC).

Southeast, West, Midwest, Great Plains, Alaska, and so forth) and generates reports for each region detailing climate impacts and practical methods of adaptation.

The reports are backed by both scientists and policy makers in the hope that constructive progress will be made toward scientific understanding and social action. The way the program works, the National Assessment currently consists of sixteen ongoing regional projects. For each of the regional studies, teams of scientists, resource planners, and other involved parties meet to assess the region's most critical vulnerabilities in areas such as agricultural productivity, coastal areas, water resources, forests, and human health. In addition to looking at potential impacts, the teams also work together to identify possible strategies that can be used to adapt and respond to climate change. The overall goal of the project is to help those in the United States prepare for future climate change. According to Michael MacCracken, who heads the national office, "The goal of the assessment is to provide the information for communities as well as activities to prepare and adapt to the changes in climate that are starting to emerge" (USGCRP, 2009).

A *New York Times* article from April 2, 2007, outlined which countries will be hit the hardest as climate change progresses. There will be what they refer to as winners and losers (Bronzan and Carter, 2007). It is known that the industrialized nations are the largest producers of GHGs. In general, it is the industrialized countries that are also the best equipped to deal with the effects of climate change and to mitigate by financing adaptive measures. Unfortunately, it is the poorer nations lying closer to the tropics— even though they have not contributed to the GHG emission problem as significantly—that will be dealt the majority of the worst side effects, such as drought, crop failure, heat waves, flooding, and sea-level rise (Figure 14.2).

The *Times* article mentioned several geographic areas worldwide that are already in the process of adapting to climate change. In Shishmaref, Alaska, for example, on a low-lying island, the town is in the process of relocating because the island is already being eroded by changes in sea level. The estimated costs to relocate are estimated at $180 million. The shoreline has receded 0.9–1.5 meters per year and is especially vulnerable when tidal high water is combined with the intense wave action of the Chukchi Sea during storms. The community is relocating to an area on the mainland that is accessible to the sea and will provide the community with the subsistence lifestyle they are used to, allowing them to hold on to their tribal culture.

The U.S. corn belt is genetically modifying crop varieties that are designed for drought and pest resistance so that farmers will be able to sustain their yields in the hotter, drier years to come. London is currently making improvements to their flood-protection infrastructure on the Thames River to guard against flooding events as the climate warms. On Sylt Island in Germany, a pilot project is under way to build more resilient dikes out of rocks that are precoated with flexible polyurethane. This keeps the dikes from being weathered by the North Sea by both absorbing the force of the breaking waves and slowing down the water masses.

In Andermatt, Switzerland, one ski resort has had to construct a ramp each ski season in order to gain access to a steadily receding glacier. The ramp has now been protected with a reflective cover to keep it from melting. Venice, Italy, which is extremely prone to flooding from sea-level rise, is constructing

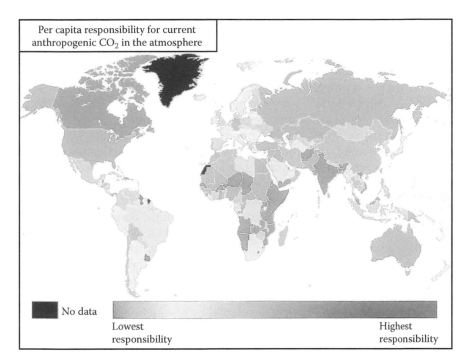

Per capita responsibility for current anthropogenic CO_2 in the atmosphere

No data

Lowest responsibility

Highest responsibility

FIGURE 14.2
(See color insert.) This map shows the per capita responsibility for GHGs worldwide. In many cases, the largest offenders are often the most wealthy, industrialized nations that are likely to encounter the least in losses overall because they have many of the resources necessary to mitigate the negative effects. Unfortunately, the countries that are likely to encounter the greatest losses are the undeveloped countries and those located close to sea level that have not contributed significantly to the climate change problem. (From Casper, J. K., *Global Warming. Climate Management: Solving the Problem*, Facts on File, New York, 2010.)

floodgates to protect the city's infrastructure during extremely high tides. In northern China in a very dry region, a project is under construction to divert water hundreds of miles from the Yangtze River in the south. In Perth, Australia, they have finished construction on a desalination plant to serve as a backup source of water to offset shrinking natural supplies as a result of prolonged drought conditions. As climate change continues, locations will have to continue to adapt to changing conditions as they arise.

New Technologies

In an April 2008 *Scientific American* article, Jeffrey D. Sachs, head of the Earth Institute at Columbia University, said that, "Even with a cutback in wasteful energy spending, our current technologies cannot support both a decline in carbon dioxide emissions and an expanding global economy. If we try to restrain emissions without a fundamentally new set of technologies, we will end up stifling economic growth, including the development prospects for billions of people" (Sachs, 2008). What Sachs said is needed is a huge investment of resources and effort into new technologies that are low carbon, and this will not happen with the kind of effort toward research that has occurred so far. It will require the serious, dedicated involvement of determined government leadership and resources; a program so intense and focused, he referred to it as a "Manhattan-like Project."

As researchers learn more about climate change and gain a better understanding of the complex interactions of the climate system, this knowledge coupled with technology should lead to even better solutions. Over the past 30 years, computing power has increased by a factor of one million. Models today are becoming much more complex and realistic. As a better understanding is reached of the nature of feedbacks from the carbon cycle and their constraints on the climate response, models are becoming much more sophisticated. New "petascale" computer models depicting detailed climate dynamics are now building the foundation for the next generation of complex climate models. New advanced computing abilities will help climatologists better understand the links between weather and climate.

This new technology is being developed by researchers at the University of Miami Rosenstiel School of Marine and Atmospheric Science, the National Center for Atmospheric Research in Boulder, Colorado, the Center for Ocean-Land-Atmospheric Studies in Calverton, Maryland, and the University of California at Berkeley. They are using a $1.4 million award from the National Science Foundation to generate the new models. The scientists at these institutions say that the development of powerful supercomputers capable of analyzing decades of data in the blink of an eye marks a technological milestone capable of bringing comprehensive changes to science. The speed of supercomputing is measured in how many calculations can be performed in a given second. Petascale computers can make 1,000,000,000,000,000 calculations per second, an enormous amount of calculations even when comparing it with an advanced supercomputer. The "peta's" capabilities represent a breakthrough and a golden opportunity for climatologists to advance earth system science and help to improve the quality of life on the planet. Jay Fein, National Science Foundation program director, says, "The limiting factor to more reliable climate predictions at higher resolution is not scientific ideas, but the computational capacity to implement those ideas. This project is an important step forward in providing the most useful scientifically based climate change information to society for adapting to climate change" (National Science Foundation, 2008).

One thing researchers have learned recently through modeling is that climate cannot be predicted independently of weather. They have discovered that weather has a profound impact on climate. Now that they have discovered this, they expect to be able to greatly improve weather and climate predictions and climate change projections. In addition, with the increase in computing capabilities, one of the team members—Ben Kirtman, a meteorologist at the Rosenstiel School of Marine and Atmospheric Science—has developed a new weather- and climate-modeling strategy that he calls "interactive ensembles," which is designed to isolate the interactions between weather and climate (Kirtman and Shukla, 2002).

The interactive ensembles for weather and climate modeling are currently being applied to one of the United States' main climate change models—the National Center for Atmospheric Research's Community Climate System Model, the current operational model used by the National Oceanic and Atmospheric Administration's climate forecast system. The Community Climate System Model is also a model used by hundreds of researchers and is one of the climate models that was used in the Nobel Prize–winning IPCC assessments.

The research currently being done serves as a pilot program for the implementation of even more complicated computational systems, which, today, still remain a scientific and engineering challenge. According to Kirtman, "This marks the first time that we will have the computational resources available to address these scientific challenges in a comprehensive manner. The information from this project will serve as a cornerstone for petascale computing in our field and help to advance the study of the interactions between weather and climate phenomena on a global scale" (Kirtman and Shukla, 2002). Models will continue to play an important role in the future, and, as more of the interactions between the atmosphere, geosphere, biosphere, and hydrosphere are understood along with the carbon cycle, greater insight will become available as to how more efficiently to combat climate change (Figure 14.3).

FIGURE 14.3
Continual warming temperatures will eventually affect every person on the earth. (From Wasatch Images, Salt Lake City, Utah.)

The Final Choice

The defining moment will be when people realize that it is a personal decision that must be made by every individual on earth, and the decisions will be as individual and varied as people themselves. Each choice will be a compilation of personal values, beliefs, character, and goals for the future. Each individual's choice will count; each will have equal weight in this war against time.

In the end, each person will have to study the issues and make up their own mind. Each person will have to assess how their actions may affect the lives of their children, grandchildren, and future generations.

> A plant takes from the soil only what it needs.
> In the same way, we too should only take from the
> Earth what we need to flourish.
>
> *—Chiara Lubich to young people*

The solution to climate change is tied to each individual on this earth. Ultimately, the solution boils down to one question: How much are you willing to sacrifice to do your part?

Conclusions

We can be told over and over everything someone else knows about climate change. We can be told repeatedly why switching from fossil fuels to renewable clean fuels is a wise choice. We may know in our minds that it is true. Our colleagues may tell us over and over. We may tell the same piece of information to others repeatedly, but when it comes down to it, one thing matters—what are YOU going to do? This is one time you cannot hide behind someone else's good works and claim immunity. The actions of every person count—EVERY person, and you are responsible for your own actions. The way you interact socially with the issue counts. The way you approach the issue politically and support those in leadership positions that will take the right approaches to the future, counts. The way you deal with the situation economically counts. Each person holds a key to the future—not only your future, but your children's and grandchildren's future. Right now is more important than any other time to learn the issues, learn the science behind them, and learn how to solve the problems, because they will not go away unless you do. This is one problem that will take the collective efforts of everyone's understanding on an individual basis while working together on a group basis—and knowing when we succeed, we will be appreciated for a long time to come from those not able to say thank you yet.

Discussion

1. How would you propose to get others committed toward taking positive steps to solve the climate change problem?
2. What do you think keeps people from making personal life choices that could greatly help the environment?

References

Bronzan, J., and S. Carter. 2007. Winners and losers in a changing climate. *New York Times*, April 2. http://www.nytimes.com/2007/04/02/us/20070402_climate_graphic.html (accessed August 11, 2011).

Casper, J. K. 2010. *Global Warming. Climate Management: Solving the Problem*. New York: Facts on File, Inc. Provides a well-rounded outlook on the management issues associated with solving the climate change issue.

Hare, B. 2008. The EU, the IPCC, and 2°C. Speech given at Potsdam Institute for Climate Impact Research, Potsdam, Germany, Aug. 10.

Hobish, M. K. 2010. Evidence for global warming: Degradation of earth's atmosphere; temperature rise; glacial melting and sealevel rise; ocean acidity; ozone holes; vegetation response. In: *Remote Sensing Tutorial*, ed., N. Short. NASA, Earth Systems Science. http://rst.gsfc.nasa.gov/Sect16/Sect16_2.html.

Kirtman, B. P., and J. Shukla. 2002. Interactive coupled ensemble: A new coupling strategy for CGCMs. *Geophysical Research Letters*. 29(10): 1367. http://www.seas.harvard.edu/climate/seminars/pdfs/kirtman_shukla_2002.pdf (accessed April 28, 2011).

National Science Foundation. 2008. Climate computer modeling heats up: New "petascale" computer models lead to better understanding of weather-climate links. Press Release 08-153, September 4. http://www.nsf.gov/news/news_summ.jsp?cntn_id=112166 (accessed August 11, 2011).

Sachs, J. D. 2008. Keys to climate protection. *Scientific American*, March 18. http://www.scientificamerican.com/article.cfm?id=technological-keys-to-climate-protection-extended (accessed August 11, 2011).

Thompson, A., and K. Than. 2007. The frightening future of earth. *LiveScience*, April 19.

U.S. Global Change Research Program. 2009. The National Climate Assessment. *USGCRP News*. http://www.globalchange.gov/what-we-do/assessment (accessed August 11, 2011).

Suggested Reading

Crate, S. A., and M. Nuttall. 2009. *Anthropology and Climate Change: From Encounters to Actions*. Walnut Creek, CA: Left Coast Press.

Lomborg, B. 2010. *Smart Solutions to Climate Change: Comparing Costs and Benefits*. Cambridge, UK: Cambridge University Press.

Pielke, R., Jr. 2010. *The Climate Fix: What Scientists and Politicians Won't Tell You About Global Warming*. New York: Basic Books.

Appendix

United Nations Framework Convention on Climate Change (UNFCCC) Member Nations

Afghanistan
Albania
Algeria
Angola
Antigua and Barbuda
Argentina
Armenia
Australia
Austria
Azerbaijan
Bahamas
Bahrain
Bangladesh
Barbados
Belarus
Belgium
Belize
Benin
Bhutan
Bolivia
Bosnia and Herzegovina
Botswana
Brazil
Brunei
Bulgaria
Burkina Faso
Burundi
Cambodia
Cameroon
Canada
Cape Verde
Central African Republic

Chad
Chile
China
Colombia
Comoros
Cook Islands
Costa Rica
Côte d'Ivoire
Croatia
Cuba
Cyprus
Czech Republic
Democratic Republic of the Congo
Denmark
Djibouti
Dominica
Dominican Republic
Ecuador
Egypt
El Salvador
Equatorial Guinea
Eritrea
Estonia
Ethiopia
European Union
Federated States of Micronesia
Fiji
Finland
France
Gabon
Gambia
Georgia
Germany
Ghana
Greece
Grenada
Guatemala
Guinea
Guinea-Bissau
Guyana
Haiti
Honduras
Hungary
Iceland
India

Indonesia
Iran
Ireland
Israel
Italy
Jamaica
Japan
Jordan
Kazakhstan
Kenya
Kiribati
Kuwait
Kyrgyzstan
Laos
Latvia
Lebanon
Lesotho
Liberia
Libya
Liechtenstein
Lithuania
Luxembourg
Madagascar
Malawi
Malaysia
Maldives
Mali
Malta
Marshall Islands
Mauritania
Mauritius
Mexico
Moldova
Monaco
Mongolia
Montenegro
Morocco
Mozambique
Myanmar
Namibia
Nauru
Nepal
Netherlands
New Zealand
Nicaragua

Niger
Nigeria
Niue
North Korea
Norway
Oman
Pakistan
Palau
Panama
Papua New Guinea
Paraguay
Peru
Philippines
Poland
Portugal
Qatar
Republic of Macedonia
Republic of the Congo
Romania
Russia
Rwanda
Saint Kitts and Nevis
Saint Lucia
Saint Vincent and the Grenadines
Samoa
San Marino
Sao Tome and Principe
Saudi Arabia
Senegal
Serbia
Seychelles
Sierra Leone
Singapore
Slovakia
Slovenia
Solomon Islands
South Africa
South Korea
Spain
Sri Lanka
Sudan
Suriname
Swaziland
Sweden
Switzerland

Syria
Tajikistan
Tanzania
Thailand
Timor-Leste
Togo
Tonga
Trinidad and Tobago
Tunisia
Turkey
Turkmenistan
Tuvalu
Uganda
Ukraine
United Arab Emirates
United Kingdom
United States
Uruguay
Uzbekistan
Vanuatu
Venezuela
Vietnam
Yemen
Zambia
Zimbabwe

Glossary

Adaptation: An adjustment in natural or human systems to a new or changing environment. Adaptation to climate change refers to adjustment in natural or human systems in response to actual or expected climatic changes.

Aerosols: Tiny bits of liquid or solid matter suspended in air. They come from natural sources such as erupting volcanoes and from waste gases emitted from automobiles, factories, and power plants. By reflecting sunlight, aerosols cool the climate and offset some of the warming caused by greenhouse gases.

Albedo: The relative reflectivity of a surface. A surface with high albedo reflects most of the light that shines on it and absorbs very little energy; a surface with a low albedo absorbs most of the light energy that shines on it and reflects very little.

Anthropogenic: Made by people or resulting from human activities. This term is usually used in the context of emissions that are produced as a result of human activities.

Atmosphere: The thin layer of gases that surrounds the earth and allows living organisms to breathe. It reaches 400 miles (644 km) above the surface, but 80 percent is concentrated in the troposphere—the lower 7 miles (11 km) above the earth's surface.

Baseline: The emission of greenhouse gases that would occur without the contemplated policy intervention or project activity.

Biodiversity: Different plant and animal species.

Biomass: Plant material that can be used for fuel.

Bleaching (coral): The loss of algae from corals that causes the corals to turn white. This is one of the results of global warming and signifies a die-off of unhealthy coral.

Carbon cycle: A complex series of processes through which all of the carbon atoms in existence rotate. It is the biogeochemical cycle by which carbon is exchanged among the biosphere (living things), pedosphere (soil component), geosphere (earth processes), hydrosphere (water processes), and atmosphere of the earth. One of the most important cycles of all the natural earth processes, it allows for carbon to be recycled and reused.

Carbon dioxide (CO_2): A colorless, odorless gas that forms when carbon atoms combine with oxygen atoms. Carbon dioxide is a tiny, but vital, part of the atmosphere. The heat-absorbing ability of carbon dioxide is what makes life possible on earth.

Carbon dioxide equivalent (CO_2e): The universal unit of measurement used to indicate the global warming potential of each of the six greenhouse gases. Carbon dioxide—a naturally occurring gas that is a by-product of burning fossil fuels and biomass, land-use changes, and other industrial processes—is the reference gas against which the other greenhouse gases are measured.

Carbon sink: An area where large quantities of carbon are built up in the wood of trees, calcium carbonate rocks, animal species, or the ocean, or any other place where carbon is stored. These places act as reservoirs, keeping carbon out of the atmosphere.

Certified emission reductions (CERs): A unit of greenhouse gas emission reductions issued pursuant to the Clean Development Mechanism of the Kyoto Protocol and measured in metric tons of carbon dioxide equivalent.

Chlorofluorocarbons (CFCs): Gases that were once widely used as coolants in refrigerators and air conditioners, as foaming agents for insulation and food packaging, and as cleaning agents in certain industries. They are long-lasting compounds that absorb heat energy more effectively than carbon dioxide. When they enter the upper atmosphere, they destroy ozone (which protects life on earth from harmful ultraviolet radiation). An international treaty calls for all production of CFCs to stop by the year 2010.

Climate: The usual pattern of weather that is averaged over a long period of time.

Climate feedback: An interaction mechanism between processes in the climate system, when the result of an initial process triggers changes in a second process that in turn influences the initial one. A positive feedback intensifies the original process, and a negative feedback reduces it.

Climate forcings: Factors that affect the earth's climate. These factors are called "forcings" because they drive or "force" the climate system to change. The forcings that were probably the most important during the last millennium were: changes in the output of energy from the sun, volcanic eruptions, and changes in the concentration of greenhouse gases in the atmosphere. Positive forcings warm the earth, and negative forcings cool it.

Climate model: A quantitative way of representing the interactions of the atmosphere, oceans, land surface, and ice. Models can range from relatively simple to extremely complicated.

Climatologist: A scientist who studies the climate.

Concentration: The amount of a component in a given area or volume. In global warming, it is a measurement of how much of a particular gas is in the atmosphere compared to all of the gases in the atmosphere.

Condensation: The process that changes a gas into a liquid.

Conference of Parties (COP): The meeting of parties to the United Nations Framework Convention on Climate Change.

Coriolis Force: A representative artifact of the earth's rotation. In physics, it is an apparent deflection of moving objects when they are viewed from a rotating reference frame. In a reference frame with clockwise rotation, the deflection is to the left of the motion of the object; in one with anti-clockwise rotation, the deflection is to the right.

Deforestation: The large-scale cutting of trees from a forested area, often leaving large areas bare and susceptible to erosion.

Ecosystem: A community of interacting organisms and their physical environment.

Electromagnetic radiation (EMR): A form of energy exhibiting wave-like behavior as it travels through space. EMR has both electric and magnetic field components, which oscillate in phases perpendicular to each other and perpendicular to the direction of energy propagation. Electromagnetic radiation is classified according to the frequency of its wave. In order of increasing frequency and decreasing wavelength, these are radio waves, microwaves, infrared radiation, visible light, ultraviolet radiation, x-rays, and gamma rays.

Electromagnetic spectrum: The range of all possible frequencies of electromagnetic radiation emitted from the sun. The "electromagnetic spectrum" of an object is the characteristic distribution of electromagnetic radiation emitted or absorbed by that particular object. The electromagnetic spectrum extends from low frequencies used for radios to gamma radiation at the short-wavelength end. The spectrum covers wavelengths from thousands of kilometers down to a fraction of the size of an atom.

El Niño: A cyclic weather event in which the waters of the eastern Pacific Ocean off the coast of South America become much warmer than normal and disturb weather patterns across the region. Its full name is El Niño Southern Oscillation (ENSO). Every few years, the temperature of the western Pacific rises several degrees above that of waters to the east. The warmer water moves eastward, causing shifts in ocean currents, jet-stream winds, and weather in both the Northern and Southern Hemispheres.

Emission reductions (ERs): The measurable reduction of release of greenhouse gases into the atmosphere from a specified activity or over a specified area and in a specified period of time.

Emissions: The release of a substance (usually a gas when referring to the subject of climate change) into the atmosphere.

Evaporation: The process by which a liquid, such as water, is changed to a gas.

Feedback: A change caused by a process that, in turn, may influence that process. Some changes caused by global warming may hasten the process of warming (positive feedback); some may slow warming (negative feedback).

Fossil fuel: An energy source made from coal, oil, or natural gas. The burning of fossil fuels is one of the chief causes of global warming.

Glacier: A mass of ice formed by the buildup of snow over hundreds and thousands of years.

Global warming: An increase in the temperature of the earth's atmosphere, caused by the buildup of greenhouse gases. This is also referred to as the "enhanced greenhouse effect" caused by humans.

Global warming potential (GWP): The cumulative radiative forcing effects of a gas over a specified time resulting from the emission of a unit mass of gas relative to a reference gas (usually CO_2).

Greenhouse effect: The natural trapping of heat energy by gases present in the atmosphere, such as carbon dioxide, methane, and water vapor. The trapped heat is then emitted as heat back to the earth.

Greenhouse gas: A gas that traps heat in the atmosphere and keeps the earth warm enough to allow life to exist.

Gulf Stream: A warm current that flows from the Gulf of Mexico across the Atlantic Ocean to northern Europe. It is largely responsible for Europe's milder climate.

Industrial Revolution: The period during which industry developed rapidly as a result of advances in technology. This took place in Britain during the late eighteenth and early nineteenth centuries.

Infrared: Invisible heat radiation that is emitted by the sun and by virtually every warm substance or object on earth.

Intergovernmental Panel on Climate Change (IPCC): An organization consisting of 2,500 scientists that assesses information in the scientific and technical literature related to the issue of climate change. The IPCC was established jointly by the United Nations Environment Programme and the World Meteorological Organization in 1988.

Joint implementation (JI): Mechanism provided by Article 6 of the Kyoto Protocol, whereby a country included in Annex I of the UNFCCC and the Kyoto Protocol may acquire emission reduction units when it helps to finance projects that reduce net emissions in another industrialized country (including countries with economies in transition).

Kyoto Protocol: Commits industrialized country signatories to reduce their greenhouse gas (or "carbon") emissions by an average of 5.2 percent compared with 1990 emissions, in the period 2008–2012. The Kyoto Protocol was adopted at the Third Conference of the Parties to the United Nations Convention on Climate Change held in Kyoto, Japan, in December 1997.

Land use: The management practice of a certain land cover type. Land use may be such things as forest, arable land, grassland, urban land, and wilderness.

Land-use change: An alteration of the management practice on a certain land cover type. Land-use changes may influence climate systems because they impact evapotranspiration and sources and sinks of greenhouse gases. An example of land-use change is removing a forest to build a city.

Methane: A colorless, odorless, flammable gas that is the major ingredient of natural gas. Methane is produced wherever decay occurs and little or no oxygen is present.

Monsoon: Heavy rains that occur at the same time each year.

Nitrogen: A gas that takes up 80 percent of the volume of the earth's atmosphere. Nitrogen is also an element in substances such as fertilizer.

Nitrous oxide: A heat-absorbing gas in the earth's atmosphere. Nitrous oxide is emitted from nitrogen-based fertilizers.

Nuclear power: The electricity produced by a process that begins with the splitting apart of uranium atoms, yielding great amounts of heat energy.

Ozone: A molecule that consists of three oxygen atoms. Ozone is present in small amounts in the earth's atmosphere at 14–19 miles (23–31 km) above the earth's surface. A layer of ozone makes life possible by shielding the earth's surface from the most harmful ultraviolet rays. In the lower atmosphere, ozone emitted from auto exhausts and factories is an air pollutant.

Parts per million (ppm): The number of parts of a chemical found in one million parts of a particular gas, liquid, or solid.

Permafrost: Permanently frozen ground in the Arctic. As global warming increases, this ground is melting.

Photosynthesis: The process by which plants make food using light energy, carbon dioxide, and water.

Protocol: The terms of a treaty that have been agreed to and signed by all parties.

Proxy data: Data that paleoclimatologists gather from natural recorders of climate variability, e.g., tree rings, ice cores, fossil pollen, ocean sediments, coral, and historical data. By analyzing records taken from these and other proxy sources, scientists can extend our understanding of climate far beyond the 140-year instrumental record.

Radiation: The particles or waves of energy.

Reforestation: A process that increases the capacity of the land to sequester carbon by replanting forest biomass in areas where forests have been previously harvested.

Renewable: Able to be replaced or regrown, such as trees, or a source of energy that never runs out, such as solar energy, wind energy, or geothermal energy.

Resources: The raw materials from the earth that are used by humans to make useful things.

Satellite: Any small object that orbits a larger one. Artificial satellites carry instruments for scientific study and communication. Imagery taken from satellites is used to monitor aspects of global warming such as glacier retreat, ice-cap melting, desertification, erosion, hurricane damage, and flooding. Sea-surface temperatures and measurements are also obtained from man-made satellites in orbit around the earth.

Sequestration: Capture of carbon dioxide in a manner that prevents it from being released into the atmosphere for a specified period of time.

Simulation: A computer model of a process that is based on actual facts. The model attempts to mimic, or replicate, actual physical processes.

Sustainability/sustainable living: Applying policies, principles or management strategies to secure the continuity of economic, social, governmental and environmental aspects of human society, and the environment. Although it preserves biodiversity and ecosystems, sustainable living enables people and economies to meet their needs and express their greatest potential in the present, while planning and acting to maintain these ideals in the long term.

Sustainable development: Meeting the needs of the present without compromising the ability of future generations to meet their own needs.

Temperate: An area that has a mild climate and different seasons.

Thermal: Related to heat.

Thermohaline circulation: The part of the large-scale ocean circulation that is driven by global density gradients created by surface heat and freshwater fluxes.

Tropical: A region that is hot and often wet (humid). These areas are located around the earth's equator.

Tundra: A vast treeless plain in the Arctic with a marshy surface covering a permafrost layer.

Ultraviolet radiation: A portion of the sun's electromagnetic spectrum, consisting of very short wavelengths and high energy. The atmosphere's ozone layer protects life on earth from the damaging effects from UV radiation.

Urban heat-island effect: The term *heat island* describes built-up areas that are hotter than nearby rural areas. The annual mean air temperature of a city with 1 million people or more can be 1–3°C warmer than its surroundings. In the evening, the difference can be as high as 12°C. Heat islands can affect communities by increasing summertime peak energy demand, air-conditioning costs, air-pollution and greenhouse gas emissions, heat-related illness and mortality, and water quality.

Weather: The conditions of the atmosphere at a particular time and place. Weather includes such measurements as temperature, precipitation, air pressure, and wind speed and direction.

Index

T - #0326 - 071024 - C16 - 234/156/21 - PB - 9780367381967 - Gloss Lamination